T0258883

Media Networks

Architectures, Applications, and Standards

OTHER TELECOMMUNICATIONS BOOKS FROM AUERBACH

Bio-Inspired Computing and Networking
Edited by Yang Xiao
ISBN 978-1-4200-8032-2

Communication and Networking in Smart Grids
Edited by Yang Xiao
ISBN 978-1-4398-7873-6

Delay Tolerant Networks: Protocols and Applications
Edited by Athanasios Vasilakos, Yan Zhang, and Thrasyvoulos Spyropoulos
ISBN 978-1-4398-1108-5

Designing Green Networks and Network Operations: Saving Run-the-Engine Costs
Daniel Minoli
ISBN 978-1-4398-1638-7

Emerging Wireless Networks: Concepts, Techniques and Applications
Edited by Christian Makaya and Samuel Pierre
ISBN 978-1-4398-2135-0

Game Theory for Wireless Communications and Networking
Edited by Yan Zhang and Mohsen Guizani
ISBN 978-1-4398-0889-4

Green Mobile Devices and Networks: Energy Optimization and Scavenging Techniques
Edited by Hrishikesh Venkataraman and Gabriel-Miro Muntean
ISBN 978-1-4398-5989-6

Information and Communication Technologies in Healthcare
Edited by Stephan Jones and Frank M. Groom
ISBN 978-1-4398-5413-6

Integrated Inductors and Transformers: Characterization, Design and Modeling for RF and MM-Wave Applications
Egidio Ragonese, Angelo Scuderi, Tonio Biondi, and Giuseppe Palmisano
ISBN 978-1-4200-8844-1

IP Telephony Interconnection Reference: Challenges, Models, and Engineering
Mohamed Boucadair, Isabel Borges, Pedro Miguel Neves, and Olafur Pall Einarsson
ISBN 978-1-4398-5178-4

Media Networks: Architectures, Applications, and Standards
Edited by Hassnaa Moustafa and Sherali Zeadally
ISBN 978-1-4398-7728-9

Mobile Opportunistic Networks: Architectures, Protocols and Applications
Edited by Mieso K. Denko
ISBN 978-1-4200-8812-0

Mobile Web 2.0: Developing and Delivering Services to Mobile Devices
Edited by Syed A. Ahson and Mohammad Ilyas
ISBN 978-1-4398-0082-9

Multimedia Communications and Networking
Mario Marques da Silva
ISBN 978-1-4398-7484-4

Music Emotion Recognition
Yi-Hsuan Yang and Homer H. Chen
ISBN 978-1-4398-5046-6

Near Field Communications Handbook
Edited by Syed A. Ahson and Mohammad Ilyas
ISBN 978-1-4200-8814-4

Physical Principles of Wireless Communications, Second Edition
Victor L. Granatstein
ISBN 978-1-4398-7897-2

Security of Mobile Communications
Noureddine Boudriga
ISBN 978-0-8493-7941-3

Security of Self-Organizing Networks: MANET, WSN, WMN, VANET
Edited by Al-Sakib Khan Pathan
ISBN 978-1-4398-1919-7

Service Delivery Platforms: Developing and Deploying Converged Multimedia Services
Edited by Syed A. Ahson and Mohammad Ilyas
ISBN 978-1-4398-0089-8

TV Content Analysis: Techniques and Applications
Edited by Yannis Kompatsiaris, Bernard Merialdo, and Shiguo Lian
ISBN 978-1-4398-5560-7

TV White Space Spectrum Technologies: Regulations, Standards and Applications
Edited by Rashid Abdelhaleem Saeed and Stephen J. Shellhammer
ISBN 978-1-4398-4879-1

AUERBACH PUBLICATIONS
www.auerbach-publications.com
To Order Call: 1-800-272-7737 • Fax: 1-800-374-3401
E-mail: orders@crcpress.com

Media Networks

Architectures, Applications, and Standards

Edited by
Hassnaa Moustafa and Sherali Zeadally

CRC Press
Taylor & Francis Group
Boca Raton London New York

CRC Press is an imprint of the
Taylor & Francis Group, an **Informa** business

CRC Press
Taylor & Francis Group
6000 Broken Sound Parkway NW, Suite 300
Boca Raton, FL 33487-2742

© 2012 by Taylor & Francis Group, LLC
CRC Press is an imprint of Taylor & Francis Group, an Informa business

No claim to original U.S. Government works

Printed in the United States of America on acid-free paper
Version Date: 20120322

International Standard Book Number: 978-1-4398-7728-9 (Hardback)

Visit the Taylor & Francis Web site at
http://www.taylorandfrancis.com

and the CRC Press Web site at
http://www.crcpress.com

Contents

Preface ix

Editors xi

Contributors xiii

Introduction xvii

PART I   DIGITAL TV

1   Digital TV 3
     HASSNAA MOUSTAFA, FARHAN SIDDIQUI,
     AND SHERALI ZEADALLY

2   Open-IPTV Services and Architectures 29
     EMAD ABD-ELRAHMAN AND HOSSAM AFIFI

3   Mobile TV 57
     SHERALI ZEADALLY, HASSNAA MOUSTAFA, NICOLAS BIHANNIC,
     AND FARHAN SIDDIQUI

4   Connected TV: The Next Revolution? 77
     ERWAN NÉDELLEC

5   3DTV Technology and Standardization 91
     GILLES TENIOU

6   Digital TV Architecture Standardization 101
     OLIVIER LE GRAND

PART II   MEDIA CONTENT DELIVERY AND QUALITY
            OF EXPERIENCE

7   Collaboration between Networks and Applications in the
     Future Internet 121
     SELIM ELLOUZE, BERTRAND MATHIEU, TOUFIK AHMED,
     AND NICO SCHWAN

v

8   Information-Centric Networking: Current Research
    Activities and Challenges 141
    BERTRAND MATHIEU, PATRICK TRUONG, JEAN-FRANÇOIS PELTIER,
    WEI YOU, AND GWENDAL SIMON

9   Toward Information-Centric Networking: Research,
    Standardization, Business, and Migration Challenges 163
    WEI KOONG CHAI, MICHAEL GEORGIADES, AND SPIROS SPIROU

10 Content Delivery Network for Efficient Delivery of Internet Traffic 187
    GILLES BERTRAND AND EMILE STÉPHAN

11 Content Delivery Networks: Market Overview and
    Technology Innovations 209
    BERTRAND WEBER

12 Quality of Experience in Future Media Networks: Consumption
    Trends, Demand for New Metrics and Evaluation Methods 223
    ADAM FLIZIKOWSKI, MAREK DĄBROWSKI, MATEUSZ MAJEWSKI,
    AND KRZYSZTOF SAMP

13 QoE-Based Routing for Content Distribution Network
    Architecture 255
    HAI ANH TRAN, ABDELHAMID MELLOUK, AND SAID HOCEINI

14 QoE of 3D Media Delivery Systems 275
    VARUNA DE SILVA, GOKCE NUR, ERHAN EKMEKCIOGLU,
    AND AHMET M. KONDOZ

PART III   USER-CENTRICITY AND IMMERSIVE TECHNOLOGIES

15 Perceived QoE for User-Centric Multimedia Services.......................295
    NIKOLAOS ZOTOS, JOSE OSCAR FAJARDO, HARILAOS KOUMARAS,
    LEMONIA BOULA, FIDEL LIBERAL, IANIRE TABOADA, AND MONICA
    GORRICHO

16 Immersive 3D Media 347
    ERHAN EKMEKCIOGLU, VARUNA DE SILVA, GOKCE NUR,
    AND AHMET M. KONDOZ

17 IPTV Services Personalization 375
    HASSNAA MOUSTAFA, NICOLAS BIHANNIC, AND SONGBO SONG

18 Context-Awareness for IPTV Services Personalization 401
RADIM ZEMEK, SONGBO SONG, AND HASSNAA MOUSTAFA

19 Metadata Creation and Exploitation for Future Media Networks..........423
THOMAS LABBÉ

20 Semantically Linked Media for Interactive User-Centric Services 445
VIOLETA DAMJANOVIC, THOMAS KURZ, GEORG GÜNTNER,
SEBASTIAN SCHAFFERT, AND LYNDON NIXON

21 Telepresence: Immersive Experience and Interoperability.................471
BRUNO CHATRAS

22 E-Health: User Interaction with Domestic Rehabilitation Tools.......489
LYNNE BAILLIE

23 Societal Challenges for Networked Media 503
PIERRE-YVES DANET

Index 511

Preface

Recent advances in networking technologies, personal, entertainment and home equipment, and multimedia services have dramatically changed users' consumption models for multimedia and audiovisual services creating a new category of users known as prosumers (producers–consumers). Consequently, media networks is an emerging subject that currently attracts the attention of research and industrial communities due to the expected large number of services and applications in the short-term accompanied by a strong change in users' consumption model and style.

Media Networks: Architectures, Applications, and Standards studies media networks with special attention devoted to video and audiovisual services and aims to be a comprehensive and essential reference on media networks and audiovisual domain, to fill a gap in the market on media networks, and to serve as a useful reference for researchers, engineers, students, and educators. Industrial audiences are also expected to be up-to-date with the current standardization activities in this domain, the deployment architectures, network technologies, technical challenges, users' experience, and killer applications.

This book helps in learning the media network, which is an emerging type of network, and in acquiring a deep knowledge of this network and its technical and deployment challenges through covering media networks basics and principles, a broad range of architectures, protocols, standards, advanced audiovisual and multimedia services, and future directions.

The book is divided into three parts: Part I focuses on digital TV in Chapters 1 through 6; Part II covers media content delivery and quality of experience (QoE) in Chapters 7 through 14; and Part III gives special attention to user-centricity and immersive technologies that take into account advanced services personalization, immersive technologies architectures and applications, e-health, and societal challenges in Chapters 15 through 23.

This book has the following salient features:

- Provides a comprehensive wide-scale reference on media networks and audiovisual domain

- Covers basics, techniques, advanced topics, standard specifications, and future directions
- Contains illustrative figures enabling easy reading

We owe our deepest gratitude to all the chapter authors for their valuable contribution to this book and their great efforts. All of them were extremely professional and cooperative. We express our thanks to Auerbach Publications (Taylor & Francis Group) and especially Richard O'Hanley for soliciting the ideas in this book and working with us for its publication, and Jennifer Ahringer for her huge efforts in the production process. Last but not least, a special thank you to our families and friends for their constant encouragement, patience, and understanding throughout this project.

The book serves as a comprehensive and essential reference on media networks and is intended as a textbook to teach this emerging type of network, or to help readers acquire a deep knowledge of these networks and the technical and deployment challenges that must be overcome to enable us to carry out and continue research in this area.

We welcome and appreciate your feedback and hope you enjoy reading the book.

Hassnaa Moustafa
Sherali Zeadally

Editors

Hassnaa Moustafa has been a senior research engineer at France Telecom R&D (Orange Labs), Issy Les Moulineaux (France) since January 2005. She obtained her tenure in computer science (HDR) in June 2010 from the University of Paris XI, her PhD in computer and networks from Telecom ParisTech in December 2004 and her master's in distributed systems in September 2001 from the University of Paris XI. Her research interests include mobile networks, basically *ad hoc* networks and vehicular networks. Routing, security, authentication and access control are the main areas of her research interests in these types of networks. Moreover, she is interested in NGN, IPTV, services' convergence and personalization.

Dr. Moustafa is a regular member of the IETF and a member of the IEEE and IEEE ComSoc. She manages a number of research projects at France Telecom and has many publications in a number of international conferences including *ICC, Globecom, PIMRC, VTC, Mobicom*, among others, and in a wide number of international journals. Dr. Moustafa edited books and coauthored a wide range of book chapters with CRC Press. She has also served as chairperson and co-chairperson at several international conferences and workshops and has coedited several journal special issues. In addition, she served as a TPC member for a wide number of international conferences and workshops including *IEEE ICC, IEEE Globecom*, and several *ACM* conferences, and served as a peer reviewer for several international journals including *IEEE Transactions on Wireless Communications, IEEE Transactions on Parallel and Distributed Systems, IEEE Transactions on Mobile Computing, IEEE Transactions on Broadcasting, Wiley Wireless Communications and Mobile Computing Journal, Springer Telecommunication Systems Journal, Springer Annals of Telecommunications, Elsevier Computer Networks*, and *IEEE ComSoc*.

Sherali Zeadally received a bachelor's degree in computer science from the University of Cambridge, England, and a doctorate in computer science from the University of Buckingham, England, in 1996. He is currently an associate professor in the Department of Computer Science and Information Technology at the University of the District of Columbia, Washington, DC, and serves on the editorial boards of over 18 peer-reviewed international journals. He has served as a guest editor for over 20 special issues of various peer-reviewed scholarly journals, and as a technical program committee member for more than 180 refereed conferences/symposia/workshops. He is a fellow of the British Computer Society and a fellow of the Institution of Engineering Technology, UK.

Contributors

Emad Abd-Elrahman
TELECOM & Management SudParis
Evry, France

Hossam Afifi
TELECOM & Management SudParis
Evry, France

Toufik Ahmed
Laboratoire Bordelais de Recherche en
    Informatique
University of Bordeaux
Bordeaux, France

Lynne Baillie
Multimodal Interaction Design
    Research Group
Glasgow Caledonian University
Scotland, United Kingdom

Gilles Bertrand
France Telecom
Orange Labs
Paris, France

Nicolas Bihannic
France Telecom
Orange Labs
Lannion, France

Lemonia Boula
Media Networks Laboratory
NCSR Demokritos
Athens, Greece

Wei Koong Chai
Department of Electronic and
    Electrical Engineering
University College London
London, United Kingdom

Bruno Chatras
France Telecom
Orange Labs
Paris, France

Marek Dąbrowski
Telecom Poland
Warsaw, Poland

Violeta Damjanovic
Salzburg Research
Salzburg, Austria

Pierre-Yves Danet
France Telecom
Orange Labs
Lannion, France

Erhan Ekmekcioglu
Multimedia Communications Systems
  Research Group
University of Surrey
Surrey, United Kingdom

Selim Ellouze
France Telecom
Orange Labs
Lannion, France

Jose Oscar Fajardo
Department of Electronics and
  Telecommunications
University of the Basque
  Country
Leioa, Spain

Adam Flizikowski
ITTI Ltd.
Poznań, Poland

Michael Georgiades
PrimeTel PLC
Cyprus

Monica Gorricho
Ericsson España S.A.U.
Madrid R&D Center, Technology
  and Innovation
Madrid, Spain

Olivier Le Grand
France Telecom
Orange Labs
Lannion, France

Georg Güntner
Salzburg Research
Salzburg, Austria

Said Hoceini
Department of Network and Telecom
  and LiSSi Laboratory
University Paris-Est Creteil—IUT
  Creteil/Vitry
Vitry sur Seine, France

Ahmet M. Kondoz
Multimedia Communications
  Lab
University of Surrey
Surrey, United Kingdom

Harilaos Koumaras
Institute of Informatics and
  Telecommunications
NCSR Demokritos
Athens, Greece

Thomas Kurz
Salzburg Research
Salzburg, Austria

Thomas Labbé
France Telecom
Orange Labs
Lannion, France

Fidel Liberal
University of the Basque
  Country
ETSI de Bilbao
Bilbao, Spain

Mateusz Majewski
ITTI Ltd.
Poznań, Poland

Bertrand Mathieu
France Telecom
Orange Labs
Lannion, France

Abdelhamid Mellouk
Department of Network and Telecom
 and LiSSi Laboratory
University Paris-Est Creteil—IUT
 Creteil/Vitry
Vitry sur Seine, France

Hassnaa Moustafa
France Telecom
Orange Labs
Paris, France

Erwan Nédellec
France Telecom
Orange Labs
Lannion, France

Lyndon Nixon
STI International
Vienna, Austria

Gokce Nur
Multimedia Communications
 Lab
University of Surrey
Surrey, United Kingdom

Jean-François Peltier
France Telecom
Orange Labs
Lannion, France

Krzysztof Samp
ITTI Ltd.
Poznań, Poland

Sebastian Schaffert
Salzburg Research
Salzburg, Austria

Nico Schwan
Bell Labs
Stuttgart, Germany

Farhan Siddiqui
Computer Information System Faculty
Walden University
Minneapolis, Minnesota

Gwendal Simon
Telecom Bretagne
Bretagne, France

Varuna De Silva
Multimedia Communications
 Research Group
University of Surrey
Surrey, United Kingdom

Songbo Song
France Telecom
Orange Labs
Paris, France

Spiros Spirou
Intracom Telecom
Athens, Greece

Emile Stéphan
France Telecom
Orange Labs
Lannion, France

Ianire Taboada
Media Networks Laboratory
NCSR Demokritos
Athens, Greece

Gilles Teniou
France Telecom
Orange Labs
Lannion, France

Hai Anh Tran
Department of Network and Telecom
and LiSSi Laboratory
University Paris-Est Creteil—IUT
Creteil/Vitry
Vitry sur Seine, France

Patrick Truong
France Telecom
Orange Labs
Lannion, France

Bertrand Weber
France Telecom
San Francisco, California

Wei You
France Telecom
Orange Labs
Lannion, France

Sherali Zeadally
Department of Computer Science and
Information Technology
University of the District of Columbia
Washington, District of Columbia

Radim Zemek
France Telecom
Tokyo, Japan

Nikolaos Zotos
NCSR Demokritos
Athens, Greece

Introduction

Recent advances in networking technologies, personal, entertainment, home equipments, and multimedia services have dramatically changed users' consumption models for multimedia and audiovisual services creating a new category of users known as prosumers (producers–consumers). Users are now empowered with low-cost technologies that enable them to create and contribute a whole range of different media types and applications and making them available through various ways such as UGC (User Generated Contents) on YouTube or Dailymotion. On the other hand, the users' consumption model for video and audiovisual services has changed; consumption of video services during mobility and through different kinds of portable devices (such as tablet devices) is a drastic increase. 3D content is also of great interest, and users' content sharing is becoming increasingly popular. Entertainment and social services are experiencing an unprecedented growth while interactivity and personalization of services are also achieving a great success. All these dramatic changes in technologies, user behaviors, and expectations are making network operators, equipment manufacturers and service and content providers work hard to cope with this evolutional trend in the media domain and to change their traditional business models to adapt to this evolution and handle the emergence of multiple actors. In addition, over-the-top (OTT) players are continuously introducing new actors who provide new multimedia and audiovisual services through numerous innovative applications that are accessible through the Internet (including peer-to-peer applications).

Consequently, media networks are an emerging area that is currently attracting the attention of research and industrial communities because of the expected rapid growth of a large number of services and applications accompanied by strong changes in users' consumption model, habits, and lifestyle.

This book, *Media Networks: Architectures, Applications, and Standards*, explores media networks focusing on video and audiovisual services. Our goal is to present a comprehensive and essential reference on media networks and the audiovisual domain, to fill a gap in the market on media networks, and to serve as a useful reference for researchers, engineers, students, and educators. Industrial audiences are also expected to be up-to-date with the current standardization activities on this domain, the deployment architectures, network technologies, technical challenges, users' experience, and killer applications.

The book will help newcomers to the area of media networks, which are an emerging type of network, to acquire a deep knowledge of these networks and their technical and deployment challenges. The book covers a range of topics related to media networks including fundamental definitions, basics and principles, a broad range of architectures, protocols, standards, advanced audiovisual and multimedia services, and future directions.

The book is broadly divided into three parts: Part I focuses on Digital TV in Chapters 1 through 6. Then, Part II covers media content delivery and Quality of Experience (QoE) in Chapters 7 through 14. Finally, Part III gives special attention to User-Centricity and Immersive Technologies that take into account Advanced Services Personalization, Immersive Technologies Architectures and Applications, E-Health, and societal challenges in Chapters 15 through 23.

In Part I of the book, Chapter 1 presents Digital TV technologies along with their deployment architectures as well as the role of the major contributors (such as network operators, service providers, content providers, and manufacturers) and the future trends. Chapter 2 gives an overview on Open IPTV general concept, services, architectures, content delivery, market aspects, and business models. Chapter 3 presents an overview of Mobile TV along with recent Mobile TV standards that have been recently deployed, identifying some of the technological and deployment challenges that need to be addressed to achieve cost-effective service distribution together with business models for Mobile TV. Chapter 4 is about connected TV. It presents an overview of connected TV technology, various past efforts made in the past in the connected devices area and gives an overview on some of the related standardization efforts. Chapter 5 presents an overview on 3D video. This chapter identifies some of the requirements that must be met to enable 3DTV services over an end-to-end delivery chain. It describes related standardization efforts and identifies subsequent challenges to introduce new 3D video services with enhanced 3D quality of experience. Chapter 6 describes standardization activities on digital TV including IPTV, Mobile TV, and Content Delivery Networks (CDNs).

In Part II of the book, Chapter 7 introduces the Future Internet discussing Future Internet Media applications, demonstrating the new vision of collaboration between Internet services and underlying networks. Chapters 8 and 9 cover Information-Centric Networks (ICN) focusing on ICN solution approaches highlighting the major challenges that need to be considered to enable ICN-related applications (such as Web, VoIP), technical aspects (such as Naming and Addressing, Routing, Security, Resources Management, and Content Caching) and related standardization efforts. Chapters 10 and 11 focus on CDN introducing the recent changes in the Internet eco-systems that have led to an increased interest in CDNs. They present CDNs architectures that are being deployed while considering the different technical challenges and standardization efforts related to CDN issues, and also present an overview on CDN actors and market trends in this area. User satisfaction and experience are becoming increasingly important factors for many consumer-related businesses and media networks and services are no exception. To evaluate customer satisfaction and user experience, QoE is being used as a metric. Chapters 12 through 14 focus on QoE in Future Media Networks. These chapters describe QoE-related topics which include different transport protocols, compression technologies, users' consumption trends, QoE evaluation methods, QoE introduction in CDN architectures, and QoE considerations in 3D media delivery systems.

In Part III of the book, Chapter 15 focuses on perceived QoE in user-centric multimedia applications. Chapter 16 identifies the deficiencies in current delivery mechanisms for immersive 3D media and presents solutions based on content-aware processing, coding, and adaptation techniques. Chapters 17 and 18 focus on services personalization presenting an overview on IPTV services personalization and the context-awareness principle used in services personalization. Chapters 19 and 20 consider metadata exploitation and semantically linked media for user-centric services in Future Media networks, respectively. Chapter 21 gives an overview on Telepresence systems. This chapter presents features differentiating them from conventional videoconferencing systems and describes technical constraints and requirements of Telepresence systems in addition to providing a review of ongoing standardization efforts. Chapter 22 discusses the current trends in Media Networks aimed at improving the effectiveness of healthcare particularly the technological aspects that aim to encourage and engage users in their own rehabilitation in community and home settings. Finally, Chapter 23 concludes the book by describing some of the societal challenges for Networked Media.

DIGITAL TV

I

Chapter 1

Digital TV

Hassnaa Moustafa, Farhan Siddiqui,
and Sherali Zeadally

Contents

1.1   Introduction..4
1.2   History and Current Status...4
1.3   Digital TV Market..6
      1.3.1   Digital TV Deployment and Investment Trends Worldwide 6
      1.3.2   IPTV Market...7
1.4   IPTV..10
1.5   IPTV Architecture Components and Different Standards 11
      1.5.1   High-Level Functional Architecture of IPTV.............11
      1.5.2   Initiatives to Standardize IPTV 12
      1.5.3   IPTV Operation 14
1.6   IPTV Services 16
      1.6.1   Basic Channel Service......................................16
      1.6.2   Enhanced Selective Services.............................16
      1.6.3   Interactive Data Services...................................18
1.7   IPTV Deployment Challenges and Success Factors 19
1.8   Future of Digital TV..23
1.9   Conclusion 24
Acronyms..25
References 26

1.1 Introduction

Digital TV market evolution is presenting new entertainment services and business opportunities for network operators, service providers, and contents providers. *Consumers* are increasingly demanding instant access to digital content through various terminals. Another emerging trend is that *communication devices* are becoming entertainment devices, while the continuous proliferation of fixed and mobile broadband is providing faster and more reliable access to digital contents. Furthermore, *Content providers* are looking for new channels of distribution and revenue streams. The advances in digital TV technology enable a new model for service provisioning, moving from traditional broadcaster-centric TV services to a new interactive and user-centric model.

This chapter provides a detailed review of digital TV technologies along with the evolution of analogue TV to digital TV. We describe the history of Internet Protocol Television (IPTV) together with the real motivations for this technology. We also review the various architectural components of IPTV and present related IPTV standardization activities. An additional important contribution of this chapter is the focus on the different digital TV competitors along with deployment challenges that need to be overcome to fully enable digital TV technology anywhere, anytime, from any device.

1.2 History and Current Status

Today's industry is migrating from conventional analog TV to a new era of digital TV technology through upgrading of the current wired/wireless networks and the deployment of advanced digital platforms. In this context, IPTV has emerged as a technology delivering a stream of video content over a network that uses the Internet Protocol (IP). IPTV is mainly seen as traditional TV delivered over the IP using a broadband network, which is most commonly high-speed Digital Subscriber Line (xDSL). The delivery chain of IPTV is similar to that of cable, satellite or terrestrial transmission, and under the control of an operator. The history of IPTV dates back to the 1990s when several works and trials were carried out for the digital television over Asymmetric Digital Subscriber Line (ADSL). The deployment of IPTV technology began in the late 1990s with the implementation of the broadband ADSL. During this period many industrial companies started to develop IPTV solutions such as Set-Top-Box (STB) terminals, service platforms, and Video on Demand (VoD) servers. IPTV is currently seen as a part of the triple-play and quadruple-play bundles that are typically offered from network operators worldwide.

IPTV provides the traditional linear TV programs in the form of Broadcast (BC) services together with VoD, and Personal Video Recording (PVR) services. Moreover, some value-added services include Electronic Program Guide (EPG), widgets, traffic, and weather services. IPTV can be generally defined as a system that enables the delivery of real-time television programs, movies, and other types of interactive video content and multimedia services (audio, text graphics, data) over a managed IP-based network to provide the required level of Quality of Service (QoS), Quality of Experience (QoE), and interactivity [1]. The type of service providers involved in deploying IPTV services range from large telephone companies and private network operators in different parts of the world to cable and satellite TV carriers.

An emerging parallel approach for digital TV is Internet TV (sometimes called Internet Video or unmanaged IPTV). Unlike IPTV, Internet TV is any video delivered over the public Internet to PCs and some dedicated boxes while relying on the same core base of IP technologies. The debate on which is the winner among IPTV and Internet TV is long and no predictions could be done so far. The Internet TV approach is mainly different from the classical IPTV approach in that it leverages the public Internet to deliver video for end-users without using secure dedicated managed private networks as those used by IPTV. Consequently, Internet TV could not assure a guarantee for the TV QoE because some of the IP packets may be delayed or lost and the video streams are hence transmitted to the user in a best effort fashion with a resolution that might be low. In contrast, the Internet TV has no geographical limitations where TV service can be accessed from any part of the world as long as one has Internet access. Internet TV also consists of some real-time offerings (such as sports and political events) as well as a large proportion of User Generated Content (UGC), like YouTube (UGC is almost called Internet Video). Delivery of this content is mostly under the control of content producers, aggregators, and Over The Top (OTT) operators.

Advances in digital TV technologies enable a new model for service provisioning, moving from traditional broadcaster-centric TV services to a new user-centric TV model. The change in the users' behaviors from active to passive, the content digitization allowing easier distribution, and the support of broadband-enabled technologies have changed our TV experience. Digital TV allows TV services to evolve into true converged services, blending aspects of communications, social media, interactivity, and search and discovery in new ways. These efforts address the growing consumer desire for greater personalization and customization of TV experiences and are promising in allowing low-cost services for end users and a revenue system for broadcasters based on personalized advertising methods, as well as new business opportunities for network operators, service providers, and many more actors. Indeed, the digital TV market evolution is enabling new entertainment services: (i) consumers are increasingly demanding instant access to digital content through various terminals; (ii) communication devices are becoming entertainment devices; (iii) the increasing proliferation of fixed and mobile broadband networks are providing faster and more reliable access to digital content, (iv) content providers are looking for new channels of distribution and revenue streams. The participation of several actors along the digital TV value chain has an important potential in its market evolution, allowing content/service providers and network operators to create new business opportunities and to promote smart services, while enhancing users' interactions with the digital TV system and the QoE which in turn increases users' satisfaction and the acceptability of services.

1.3 Digital TV Market

1.3.1 Digital TV Deployment and Investment Trends Worldwide

The Internet has fundamentally changed our habitual ways of interacting with content and each other. The user is changing his/her media consumption from passive to interactive mode and gets used to personalized, sociably usable, and interactive content. In this context, the digital media industry is considered as a basic element in the entertainment sector. Digital TV is being used to deliver entertainment services where Set-Top-Boxes (STBs) are seen as the key to digital TV service distribution around the world. Indeed, the digital TV market has been a great success throughout the world. In the next few years, digital sales of STBs are expected to reach almost 150 million worldwide. In the United States, IPTV subscribers in Q1/2009 reached 583,000 combined for U.S. Verizon and AT&T compared with 114,000 subscribers for the two largest cable operators, Comcast and Time Warner during the same period. The United States, together with Japan, Canada, China, South Korea, and Germany are considered the leaders in terms of High-Definition Television (HDTV). However, the penetration of HDTV in the Digital TV market is lower than 10% for TV homes worldwide and the majority of these HDTV subscribers are located in the United States. This is slowly changing with the prices coming down somewhat and an increasing focus from broadcasters on developing High Definition (HD) content. In Europe, France presents one of the Europe's key markets for telecom convergence and has Europe's largest IPTV sector, adopting triple play bundles. The United Kingdom has established itself as a leader in the European digital media marketplace, where digital TV is expected to grow on average 12% every year [2]. During 2007, the triple play model in Europe expanded dramatically with the increasing number of incumbents and second-tier operators who developed services to attract new customers and reduce loss of customers. In addition, the quad-play offers have also emerged in 2007, mainly in the United Kingdom [2]. The entrance of the telecommunication companies into the television business has created new and significant competition for the traditional pay television platforms. The installed base of homes with network entertainment centers worldwide (2009–2011) is illustrated in Table 1.1.

According to estimates from Bernstein Research, video platforms from AT&T and Verizon are set for steady growth. Fiber Optic Service (FiOS) TV from Verizon is projected to hit the 2 million mark by the end of 2011 and reach as many as 4.7 million customers in 2012, according to the Wall Street Bernstein firm. U-verse TV from AT&T is predicted to hit the 801,000 customer mark by the end of the

Table 1.1   Homes with Network Entertainment Worldwide

Year	Homes (million)
2009	115
2010	140
2011	180

Source: Data from BuddeComm Estimates, 2008.

year, and U-verse TV could hit an estimated 5.5 million subscribers in 2012, according to Bernstein's predictions (Figure 1.1).

1.3.2 IPTV Market

With the advent of the digital TV industry and the wide deployment of broadband networks, interactive TV (iTV) is attracting a lot of interest. Many TV programs now have an interactive element to them. The Multimedia Home Platform allows the progress of Television-commerce (T-commerce–electronic commerce using digital television) around the world. iTV and interactive content are also being incorporated into mobility, with further progress expected throughout the next two years [2]. In this context, IPTV is considered as a revolutionary delivery system that can incorporate interactivity and other value-added applications running over an existing broadband IP platform. Telecommunication companies view IPTV as a tremendous opportunity for future revenue growth in terms of telco investment. In fact, many IPTV market forecasters predict that in the coming years global IPTV subscribers will generate substantial increases in revenues. IPTV is projected to be a

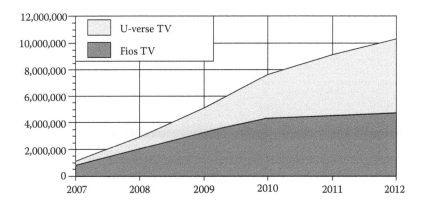

Figure 1.1   Telecos platform growth projections. (Data from Bernstein Research (4/2008) The Bridge 2008.)

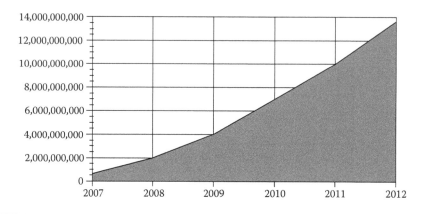

Figure 1.2   US IPTV revenues. (Data from Strategy Analytics (3/2008) The Bridge 2008.)

fast-growing sector over the next 4 years. The US IPTV market is expected to grow to about 15.5 million by 2013 (Figure 1.2). While IPTV holds less than 5% share of the total television households in 2009, the percentage is expected to approach 13% in 2013. The Multimedia Research Group (MRG) [3], a market research firm that publishes market analyses of new technologies, is forecasting that the number of global IPTV subscribers will grow from 26.7 million in 2009 to 81 million in 2013, a compound annual growth rate of 32% (as illustrated in Figure 1.3) [4].

According to forecasts from the Telecommunication Management Group (TMG), IPTV has succeeded in countries with relatively low pay-TV penetration but with a high broadband usage, which is the case among some countries in Europe, home to the largest number of IPTV subscribers worldwide. In terms of service

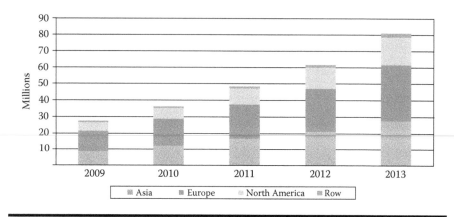

Figure 1.3   Global IPTV subscriber forecast. (Data from Global IPTV subscribers, Copyright ÓMRG, Inc. 2009.)

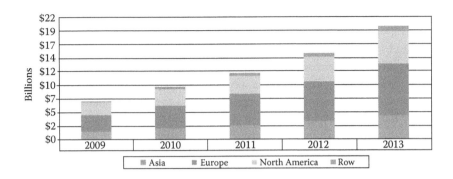

Figure 1.4 Global IPTV service revenue forecast. (Data from Global IPTV service revenue. Copyright Ó MRG, Inc. 2009.)

revenue, the global IPTV market was $6.7 billion in 2009 and is expected to grow to $19.9 billion in 2013, a compound annual growth rate of 31% (as illustrated in Figure 1.4). By 2013, Europe and North America will generate a larger share of global revenue due to very low Average Revenue Per User (ARPUs) in China and India, the fastest growing (and ultimately, the biggest markets) in Asia [5].

Service providers are also expected to increase their expenditures on IPTV equipment over the next 5 years, with spending reaching as high as $8.9 billion in 2013 (Figure 1.5). The IPTV world forum has highlighted the operators' need to focus on content and on making this content more accessible to consumers.

Indeed, operators have been increasing the amount of content they offer to their subscribers; most IPTV services now have portfolios of dozens of TV channels, supplemented by, on average, 5000 VoD assets. However, in many cases, this investment in content has failed to result in a corresponding increase in revenue because consumers are unable to find the programs they want due to poor user interfaces. To address this issue, Belgacom (which runs one of the most successful IPTV services in Western

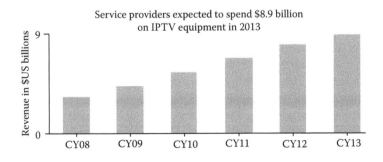

Figure 1.5 Service providers investments in IPTV. (© Infonetics Research, IPTV and Switched Digital Video Equipment, Services and Subscribers. Quarterly Market Size, Share and Forecasts, March 2009.)

Europe, in terms of penetration of its broadband subscriber base) created its own dedicated TV channels for its football content, as well as three themed zones (Adrenaline, Family, and Premiere) for its movie VoD content: each zone offers easy access to four recommended titles, which are refreshed regularly. The growing importance of making content more accessible to consumers was also underlined by the emerging content recommendation engine market [6], using algorithms that analyze data collected from a range of sources, including user profiles and ratings, patterns of content consumption, and recommendations from social networks. These engines can be used not only to suggest relevant viewing choices to the subscriber, but also to target advertising dynamically. Consequently, this will enable IPTV operators to increase their revenues for valuable content both through increased subscription of VoD services and through advertising revenue. Content monetization and customization are still in the early stages of development and, in order for them to mature, various issues still need to be addressed—such as the fact that the TV continues to be a communal, rather than an individual device, and the various concerns about privacy and targeted advertising.

1.4 IPTV

IPTV has been driven several drivers and has become a practical reality that is both commercially and technologically successful. These drivers include advances in digital technologies and consumer electronic devices, broadband evolution, Web services enrichment, social networking popularity, increased entertainment demands, and various business and commercial needs.

The digitization of television allowed for starting a new TV era where IPTV will play a crucial role. Indeed, most satellite, terrestrial, and cable TV providers have started to switch their delivery platforms from analog to digital and most video production studios are now using digital technologies to record and store content with improvements in compression techniques for digital video content. As a result, there is no longer a need to support legacy analog technologies and the adoption of IP-based video content is gaining popularity. In addition, the pervasiveness of the Internet has brought the need for high speed, always on Internet, access to the home. This need is being satisfied through broadband access technologies such as DSL, cable, fiber, and fixed wireless networks. The adoption of broadband Internet access by many households in turn has become a very powerful motivation for consumers to start subscribing to IPTV services. It is worth noting that people's homes and lifestyles are evolving, adopting a range of new technologies that help making life easier and keeping consumers entertained. Digital entertainment devices such as gaming consoles, multiroom audio systems, digital STBs, and flat screen televisions are now quite common, and the migration of Standard Definition (SD) Television to HDTV has increased consumers demand on entertainment and multimedia services, all allowing for continuous evolution of home networks and creating the notion of the *Home Cinema*. Inside the home, technologies enable the "whole home" ecosystem to communicate, and emerging Over The Top (OTT) networking technologies expand the reach of the "home" everywhere in the community and the world, with or without reliance on traditional core network infrastructure [7]. Besides, the dramatic reduction in costs of PCs, the number of households that own multiple PCs and smart devices is rapidly increasing creating a rich environment for home multimedia network services. Moreover, the mobile user experience and the vast offers of services for smart phones (including unlimited Internet access) increases the demand of mobile users for TV and video streaming services on mobile.

Web services are also observing an explosive growth with low-cost application developments and low cost of disk space both local and "in cloud." This triggers the IPTV evolution with advanced services as Personal Video Recording (PVR) capabilities on one hand, and on the other, the Internet TV is in continuous evolution allowing not only video delivery over Internet but also User Generated Content (UGC).

Finally, some business and commercial drivers also play an important role in the evolution of IPTV technology. The increased competition is forcing many telecommunication companies to start the process of offering IPTV services to their subscribers and to extend its limit to include broadband Internet connections and digital telephony forming what is known as the *triple-play* bundle (that currently observes an increasing penetration rate through billions of subscribers worldwide) as well as the *quadruple-play* bundle through adding wireless Internet connectivity. Besides, satellite and terrestrial companies are allowed to provide their customer bases with IP-based pay TV services as a means of opening new market opportunities and contributing to this new business challenge.

1.5 IPTV Architecture Components and Different Standards

The IPTV chain is mainly composed of four domains: (i) the *Consumer Domain* interfacing with the end user, (ii) the *Network Provider Domain* allowing the connection between the consumer domain and the service provider domain, (iii) the *Service Provider Domain* which is responsible for providing consumers with the services, and (iv) the *Content Provider Domain* which owns the content or is licensed to sell contents. In practice, the service provider could act as a content provider.

1.5.1 High-Level Functional Architecture of IPTV

The IPTV functional architecture (Figure 1.6) identifies five different functions in an IPTV system. These functions include Content Provisioning, IPTV control, Content Delivery, End system, and IPTV System Management/Security [8].

- *Content provision:* This function prepares IPTV contents. After the contents are ingested from content sources into the Content Provision, they are converted

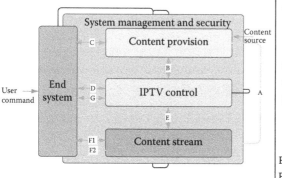

Figure 1.6 IPTV functional architecture. (Adapted from ITU Telecommunications Standardization Sector: Focus Group on IPTV Draft, *Classifications of IPTV Service and its Meaning,* **FG IPTV-ID-0026, 2006.)**

into the designated format and encrypted for right management by the Content Provision Functions. After this process, the contents are transferred to the Content Delivery functions.

■ *IPTV control:* This function is responsible for IPTV service preparation control and other functions such as packing contents into services, generating content distribution policy, and publishing these deliverable services. IPTV Control is also responsible for issuing the content license to users according to their subscriptions.

■ *Content delivery:* This function delivers contents which are packed in IPTV services to the end system. To support services such as VoD and Personal Video Recoding (PVR), and to achieve transport efficiency, the contents are stored/cached within the Content Delivery. When the end user requests a portion of the content, the IPTV Control functions request the Content Delivery functions to provide the content data.

■ *End system:* The end system is responsible for collecting control commands from users, and interacting with IPTV Control functions to get service information, content license and keys for decryption. The end system also provides the capability for content acquisition, content decrypting, and decoding.

■ *IPTV system management and security:* System Management and Security functions are responsible for the overall system status monitoring, configuration, and security aspects.

1.5.2 Initiatives to Standardize IPTV

With the advancements in IPTV technologies, the delivery of interactive and personalized multimedia services including IPTV services over Next Generation

Networks (NGNs) becomes the objective of several standardization bodies such as ITU-T (International Telecommunication Union-Telecommunication) [9], ETSI/ TISPAN (Telecoms & Internet converged Services & Protocols for Advanced Networks) [10], Open IPTV Forum [11], Alliance for Telecommunications Industry Standard (ATIS) [12], and 3GPP [13].

Two main architectures have been approved for IPTV delivery, namely by the ITU-T and the ETSI/TISPAN. The *ITU-T* formed a Focus Group on IPTV (FG IPTV) with the mission of coordinating and developing the global IPTV standard architecture based on a client–server architecture with the addition of a service delivery platform while considering the following key areas: Digital Rights Management (DRM), QoS/QoE metrics, Metadata, and interoperability and test. The *ETSI/TISPAN* emerged with the mission to develop specifications for next generation wireless and fixed networking infrastructures, defining IPTV as a Next Generation Network (NGN) service and utilizes the IP Multimedia Subsystem (IMS). This architecture can connect to legacy networks via gateways that form part of the IMS or any other Session Initiation Protocol (SIP)-based infrastructure. On the other hand, the *ATIS* has launched a subgroup called IPTV Interoperability Forum (IIF) to mainly produce an overall reference architecture for deploying IPTV services while focusing on four major areas, infrastructure equipment, content security, interoperability testing, and QoS.

Besides, the *Open IPTV Forum* has the objective to work with other standardization bodies to define end-to-end specifications for delivering IPTV services across a variety of different networking architectures. Mobility service entirely based on the IP Multimedia Subsystem (IMS) is considered within the Open IPTV Forum and is a set of specifications from the 3GPP for delivering IP multimedia to mobile users.

As for the digital TV services transmission, the Digital Video Broadcast (DVB) [14] is specifying technologies allowing interoperable STBs within the DVB-IP Group [15] to provide a sort of Plug and Play STBs receiving IPTV service over IP. The DVB has also defined the Multimedia Home Platform (MHP) [16] which is a Java-based middleware for IPTV which is interoperable with some propriety IPTV middleware. With the publication of the first set of DVB standards for IPTV, services can be launched that benefit from the advantages that come with open standards. The industry has particularly welcomed the standardization of the Broadband Content Guide, similar to the Electronic Programming Guide (EPG) used in "traditional" digital TV, and the Service Discovery and Selection mechanism. Due to the standardized information availability, the Service Discovery and Selection (SD&S) mechanisms allow a STB to efficiently recognize the multicast and unicast offerings of IPTV service operators on a broadband network. Many DVB member companies have participated in the working groups and are now integrating DVB-IPTV into their product lines. DVB's interactive middleware specifications, Multimedia Home Platform (MHP) and Globally Executable MHP (GEM) also include IPTV profiles. GEM-IPTV has been deployed in more than 900,000 Set-Top Boxes in South Korea.

In improving the (de)compression and coding of content, the Moving Pictures Experts Group (MPEG) [17] is developing several specifications that are relevant to IPTV, such as the MPEG-E standard [18] comprising of various Application Programming Interfaces (APIs). In parallel, the MPEG-7 [18] is representing specific low-level features of the content, such as visual (e.g., texture, camera motion) or audio (e.g., melody) features and also metadata structures for describing and annotating content. MPEG-21 [18], also called Multimedia Framework, defines an open framework to enhance the management of digital media resources exchange and consumption, aiming to achieve functions such as digital content creation, distribution, user privacy protection, terminals, and network resource extraction.

1.5.3 IPTV Operation

An IPTV service requires a video head-end. This is the point in the network at which linear (e.g., broadcast TV) and on-demand content is captured and formatted for distribution over the IP network. Typically, the head-end ingests national feeds of linear programming via satellite either directly from the broadcaster or programmer or via an aggregator. Some programming may also be ingested via a terrestrial fiber-based network. *Video encoders* and *VoD servers* are the major sources of video content for IPTV services. The video head-end is composed of the following components [19]:

- *Video encoder:* A video encoder can encode real-time video analog signals from a content provider or a live event location to a digital format based with given video compression technology such as MPEG2/4. The encoder also deals with on-demand content stored or re-distributed at different video on-demand servers after the encoding and other processing, such as digital rights and encryption.
- *Live video broadcast server:* A live video broadcast server is in charge of reformatting and encapsulating video streams in case video streams with different formats from a video encoder or pre-encoded video file are received. The server also interfaces the core network and transmits the video signal over the core network toward the access network.
- *Video on demand server:* A VoD server hosts on-demand content with streaming engines and is equipped with a large storage capacity.

After encoding, each channel is encapsulated into IP packets and is transmitted over the network. There are two major parts of the transport network in general— *core* and *access* networks.

- *Core networks:* These connect the access networks to customer premises and can be simply a single national distribution network running Gigabit Ethernet or IP/Multi Protocol Label Switching (MPLS) plus various regional distribution networks running carrier grade Ethernet. Managed content is usually centralized and processed within the national distribution network before being delivered to different access networks. However, a wider range of choices for the unmanaged content by other content providers can be made, and the unmanaged content is fed into the national distribution network to the customers through the Internet.

■ *Access networks:* These serve as a critical part of the transport network and are used to reach each individual customer at his or her home through the STB. Sometimes referred to as "the last mile," the broadband connection between the service provider and the household can be accomplished using a variety of technologies. The technologies available today are mainly xDSL and Coaxial Hybrid Fiber Cable (HFC) or fiber techniques such as Fiber-To-The-Node (FTTN), to extend the reach to customer communities before xDSL or cable wiring. Telecom service providers often make use of the DSL technology to serve individual households. Fiber technology such as Passive Optical Networking (PON) is also sometimes used to connect homes. IPTV networks will use variants of Asymmetrical DSL (ADSL) and Very-High-Speed DSL (VDSL) to provide the required bandwidth to run an IPTV service to the household. The service provider puts a device (such as a DSL modem) at the customer premises to deliver an Ethernet connection to the home network [20].

Given that the bandwidth of the access networks usually is very limited, to cater to all of the customers for simultaneous access of the TV channels, multicasting has been widely adopted to enable a scalable delivery of video data for IPTV. Instead of unicasting multiple flows of live content across the whole transport network, a goal of multicasting is to conserve bandwidth and minimize unnecessary packet duplication. A single transmission of each unique video data is shared among a group of customers who demand the same live content (e.g., thousands of viewers tuning in to a sporting event). Data are replicated only at appropriate branching locations, such as a regional edge router when it is necessary to fork another sub-stream to reach another group of customers or an individual customer. The grouping of encoded video streams, representing the channel line, is transported over the service provider's IP network. Each of these networks is unique to the service provider and usually includes equipment from multiple vendors. These networks can be a mix of well-engineered existing IP networks and purposefully built IP networks for video transport. At the network edge, the IP network connects to the access network.

■ The home network distributes the IPTV service throughout the home. The end point in the home network, to which user devices are connected, is the Set-Top Box (STB). A STB is usually installed with middleware client software to obtain the program guide data, decode MPEG2, MPEG4 video data, and display on the screen. Alternatively, a Web browser can obtain the program guide data from a central server. A STB can also be integrated with a DSL or cable modem or even with an IEEE 802.11 switch for home Internet access networking [20].

1.6 IPTV Services

IPTV services, as described in ITU's IPTV document [21], can be classified into three categories such as basic channel service, enhanced selective service, and interactive data service [21]. The basic channel services are composed of Audio and Video (A/V) channels, Audio channels, and A/V with data channels. These services are broadcasted similar to traditional TV channel services. The enhanced selective services encompass the Near VoD broadcasting, Real VoD, Electronic Program Guide (EPG), Personal Video Recorder (PVR), Business to Business (B2B), Customer to Customer (C2C), Multiangle service, and so on. Enhanced selective services are transmitted through the basic channel service for the customer's convenience and provide a wide choice of multimedia content. Finally, interactive data service is made up of T-information, T-commerce, T-communication, T-entertainment, and T-learning. More specific service types are described in Table 1.2 [21].

1.6.1 Basic Channel Service

- *Audio and video:* the audio and video channel service provides a picture quality similar to Digital Versatile Disk (DVD) or higher quality display in many resolutions. AV signals are transmitted without uplink feedback.
- *Audio only:* the Audio-only channel delivers only audio signals from one point to another. An audio waveform submitted to the channel input results in a similar (not necessarily identical) waveform at the channel output. Audio signals are transmitted without uplink feedback.
- *Audio, video, and data:* the audio, video, and data channel services are A/V channel services combined with interactive data for the related or supplementary information of program using a bi-directional link. Users can enjoy the downlink A/V channel streaming and accesses more detailed information via the uplink channel simultaneously.

1.6.2 Enhanced Selective Services

- *Near VoD broadcasting:* Near VoD service is the consumer video service used by multichannel broadcasters using high-bandwidth distribution mechanisms. Multiple copies of a program are broadcasted at short time intervals (typically 10–20 min) providing convenience for viewers, who can watch the program without tuning in at a scheduled point in time.
- *Real VoD:* Real VoD services allow users to select and watch video content over a network as part of an interactive and enhanced selective service. Real

Table 1.2 IPTV Service Classifications

Classifications	Services
Basic Channel Service	Audio and video Audio only Audio, video, and data
Enhanced Selective Service	Near VoD (Video on Demand) broadcasting Real VoD MoD (Music on Demand) including Audio book. EPG (Electronic Program Guide) PVR (Personal Video Recorder) B2B hosting (Business-to-business hosting) C2C hosting (Customer-to-customer hosting) Multiangle service
Interactive Data Service	T-information (news, weather, traffic, advertisement, etc.) T-commerce (security, banking, shopping, auction, and ordered delivery) T-communication (mail, messaging, SMS, channel chatting, VoIP, Web, video conference, and video phone) T-entertainment (photo album, game, karaoke, and blog) T-learning (education for children, elementary, middle and high school student, languages, and estate)

VoD services are either "stream" content, allowing viewing while the video is being downloaded, or "download" it in which the program is brought in its entirety to a set-top box or similar device before viewing starts.

■ *Music on demand (MoD):* MoD including audio book services that allow users to select and appreciate music or audio contents over an IP network as an interactive enhanced selective service much like VoD.

■ *EPG:* An Electronic Program Guide is an on-screen guide to scheduled service programs, contents and additional descriptive information, allowing a viewer to navigate, select and discover content by time, title, channel, and so on, by using a remote control, a keyboard, a touchpad, or even a phone keypad. Sometimes, EPG is also known as an Internet Media Guide (IMG), Interactive Program Guide (IPG), or Broadband Contents Guide (BCG) in an equivalent meaning. Generally, EPG can be displayed in several types such as Mosaic EPG, Box EPG, Text EPG, Mini EPG, Tree EPG, and so forth in accordance with the service provider's business model.

- *PVR:* The PVR is a consumer electronics device service that records television programs to the hard drive-based digital storage in standalone set-top boxes or networks. This can provide the "time shifting," "trick modes," and complementary convenience functions such as recording onto DVDs, commercial skip, sharing of recordings over the Internet, programming and remote control by Personal Digital Assistants (PDAs), networked PCs, or Web browsers.
- *Business-to-business (B2B) hosting:* The B2B hosting service is an IPTV hosting service for the special group or business unit subscriber. It connects the channel, VoD, and portal service made by customer business unit to the IPTV platform to supply particular groups with real-time broadcasting channels, value-added interactive services, and T-community activity. Generally, it includes a regional community, religion unit, and small business company broadcasting and interactive data services.
- *Customer-to-customer (C2C) hosting:* The C2C hosting service is an IPTV hosting service for the customer as an individual channel type. This service allows subscribers to make their own channel by uploading A/V, application, and data that they produce and providing other subscribers what they create.
- *Multiangle service:* This service provides various different camera angles that one can view. The viewer can select angles which he/she likes to watch. For example, when the customer watches the baseball game in TV, he can see the first base, third base, or backfield according to viewer's wish not by the channel director's attention.

1.6.3 Interactive Data Services

- *T-information (news, weather, traffic and advertisement, etc.):* T-information is a television information service that supports a considerable amount of useful information such as news, weather, traffic, and advertisements. The viewer can choose contents on the overlapped linked program screen or an independent menu.
- *T-commerce (security, banking, stock, shopping, auction, and ordered delivery):* T-commerce is a television commercial service that allows viewers to purchase goods and use financial services such as banking, stock, auction, and so on. For example, personal banking services support the user to view account balances, review past account activity, pay bills, and transfer money between accounts.
- *T-communication (mail, messenger, SMS, channel chatting, VoIP, Web, multiple video conference, and video phone):* T-communication is a television communication service that enables people to exchange information such as voice, video, and data. Users can send (or receive) the mail and message while watching TV. In addition, viewers can simultaneously interact with other people via two-way video and audio transmissions using video conference.

T-communication is a key convergence service of telecommunication and broadcasting in the IPTV world.

- *T-entertainment (photo album, game, karaoke, and blog):* T-entertainment is a television entertainment service that contributes to viewer's amusement by providing exciting items such as powerful games, vivid karaoke, and friendly photo albums. Games can be subdivided into single player and multiplayer games according to the number of players. Moreover, games can be classified into network linked and stand alone, according to the network interaction type.
- *T-learning (education for children, elementary, middle and high school student, languages, and estate):* T-learning is a television learning service that educates viewers through lectures, tutorials, performance support systems, simulations, job aids, game, and more.

1.7 IPTV Deployment Challenges and Success Factors

Although IPTV technology is observing a great evolution, there are still many challenges that need to be overcome for IPTV technology to achieve a truly ubiquitous IPTV deployment. From the network point of view, one major challenge is the huge bandwidth required to carry IPTV services, which is generally far higher than the bandwidth required to support Voice over IP (VoIP) and Internet access services. In addition, VoD uses unicast transport to provide communications between the IPTV consumer devices and VoD servers, and in turn generates huge amounts of traffic in the network. Indeed, efficient networking solutions should allow the IPTV distribution networks to support high bandwidth carrying capacities and multicast transport mechanisms. Efficient caching mechanisms through caching video content nearest to the user (considering the location of video servers in the operator network) can reduce network bandwidth consumption.

Quality is also an important prerequisite and challenge in IPTV deployment. It is worth pointing out that the current quality of IPTV does not currently match that of cable TV services and still the triple-play offers fail to assure similar quality level during the simultaneous use of Internet and TV. There also exists a mobile version of IPTV called Mobile IPTV. Mobile IPTV extends IPTV services over IP-based networks to mobile networks. Mobile IPTV allows users to enjoy IPTV services anywhere and even while on the move. Quality is an obstacle for Mobile IPTV, where the two main drawbacks affecting the quality of mobile IPTV are the bandwidth and terminal capabilities [22]. This gap is expected to shrink as the bandwidth increases, especially with Fiber To The Home (FTTH) and Long-Term Evolution (LTE) technologies and the continuous improvements in video codecs. Platforms standardization is another challenge for the wide interoperable deployment of IPTV. Indeed, the great evolution in IPTV technologies led to the development of many different platforms, devices, and codecs. However, the lack of standards limits the interoperability and hence applications' innovations. Moreover, careful business models present a nonnegligible challenge in IPTV deployment. First, the coexistence of IPTV and the Internet TV should be carefully considered through a smart business model that would be beneficial for all parties involved in the deployment of Digital TV (whether Internet TV or IPTV). In addition, the existence of several actors in the IPTV value chain creates a tough competition among the different offers of IPTV services by pushing network operators and service providers to propose value-added services not only to attract new clients but also to reduce the churn. By developing well-thought out business models that work with the different actors, new business opportunities will be made available to all of them.

The main technical challenges faced by current IPTV systems are outlined below:

- *Architectural scalability and reliability:* With the growing number of IPTV subscribers, maintaining the stability of the IPTV architecture and providing efficient and reliable service delivery up to the last mile becomes a challenge. Poor scalability of the IPTV architecture can manifest itself in the form of excessive network demands (necessitating very expensive bandwidth upgrades), compromised QoS, and so on.

- *High-quality video:* IPTV has stringent QoS requirements, and meeting these requirements is a key success factor for service providers as they strive to build customer loyalty. Indeed, customers' QoE is the key value upon which service providers must differentiate themselves. The delivery of IPTV is a nascent and highly competitive area, and for customers to truly embrace IPTV, and the broader Multi-play offering, it must meet or exceed their QoE expectations.

- *Capacity of access networks:* The capability of access networks is a crucial issue for the successful deployment of IPTV services. Since most homes are currently equipped with multiple end-user devices (TVs, computers, etc.), there is a greater demand for high-speed access networks. Technologies offering higher bandwidth support (e.g., ADSL2+) help in supporting IPTV services. Furthermore, encryption technologies can help reduce bandwidth requirements and therefore counteract increasing bandwidth capacity demands.

- *Interoperability and standardization:* Standardization is imperative for the development of a worldwide mass market in low-cost IPTV services. Many entities are working on different aspects of IPTV. However, there is no well-defined coordination among them. To gain acceptance and reach technical and commercial success, it is essential to coordinate the efforts of all entities working in various parts of IPTV.

- *Robust and secure content protection:* Content Security is also one of the most critical issues for IPTV success. Without proper content security, subscribers will stay with their current TV service providers. The choice of appropriate content protection solution is imperative to complete negotiations successfully throughout the content acquisition process. Without IPTV security, service providers cannot live up to the expectations of both customers and content creators in terms of availability, level of quality, and exclusivity when a premium is requested.

■ *Compatibility and accessibility:* Compatibility which will facilitate multiple services is also expected to be one of the significant factors influencing IPTV adoption. Compatibility at the individual level means that consumers can enjoy various services on the IPTV format. This implies that IPTV will become a common platform for voice, video, and data resulting in triple and even quadruple services. IPTV will not just be another medium for telecasting; it will change the entire value–chain relationship. Instead of one-to-many, there will be new opportunities for one-to-one interaction on an unprecedented scale. Customers' focus on compatibility present some technical challenges to the providers as to integration, making sure the mixture of software, hardware and related gear that make up an IPTV network functions properly.

■ *Personalization of content and services:* IPTV providers will have to differentiate their service bundles from the video services that are already available on the market in various forms such as Web TV and mobile TV. The IPTV providers may develop interactive and proactive applications. Interactive functionality is the most promising feature for new IPTV services. The user wants more control over the contents especially for the rewind and pausing options. Users want the function to enable or disable on-screen advertising, the selection of languages, instant messaging boards, and so on. The new IPTV services should provide more interactive interfaces to attract and facilitate the end-users. Indeed, IPTV can satisfy the consumers' new demands if IPTV effectively brings TV media to the public who can serve as a producer as well as a consumer.

■ *Managing heterogeneity of access networks:* Clients may request IPTV services using varied local access networks including ADSL, Cable Modem, Universal Mobile Telecommunication System (UMTS), Wireless LAN (WLAN), Worldwide Interoperability for Microwave Access (WiMAX), and so on. All these network connections support distinct characteristics and vary for their uplink and downlink capacities. This variation in the bandwidth capacities can influence the offered video quality.

■ *Multiaccess devices:* End users may access IPTV services using a wide variety of devices such as TV, Laptop, PDAs, cellular phone, other handheld devices, and so on. These devices possess heterogeneous characteristics in terms of screen resolution, data rate, and so on. To be able to ensure smooth contents delivery and acceptable QoS levels, it is essential to consider the capabilities of all devices at the receiving end. Furthermore, it is also possible that a single home accesses different TV channels using different terminals at the same time. In this context, the service provider has the challenge of dealing with the available bandwidth and efficient content adaptation to meet the needs of all users.

■ *Dynamic subscriber behaviors:* In a realistic multiplay user environment, subscribers behave in a dynamic fashion. Household receiving Multi-play services from a single provider may be simultaneously initiating channel-change and new Internet-connection requests while having multiple VoIP telephone conversations. When scaled across the subscriber base, this dynamic behavior can be very demanding on the control plane of IPTV network elements, and can potentially jeopardize the QoE IPTV viewers receive.

■ *Dynamic network characteristics:* The condition of the access network can change dramatically during the process of video streaming. Therefore, it is important for IPTV providers to efficiently perform dynamicity management in order to monitor the current network conditions regularly and maximize the utilization of available resources while minimizing packet drop ratios. Congestion control mechanism may also be employed to ensure the smooth delivery of media content. This becomes even more essential when there are many customers.

■ *Efficient video coding:* Video encoding is considered as the most important aspect in the multimedia contents streaming applications and other IPTV services. Efficient video coding techniques lead to improved content portability and management. Highly efficient coding schemes can also help in improving access network throughput resulting in better QoS/QoE for IPTV users.

■ *Efficient routing schemes and bandwidth utilization:* Efficient routing plays a major role for the efficient deployment of the VoD and IPTV services. VoD unicast transmissions increase the traffic load on the networks compared to IPTV (multicast). The situation becomes worse while delivering the service to the millions of clients. Service providers must implement efficient routing to serve clients with efficient utilization of the available bandwidth and putting lower loads onto the networks.

■ *Overall QoS/QoE:* Successful deployment of IPTV services requires excellent QoS for video, voice, and data. QoS metrics for video include jitter, number of out-of-sequence packets, packet-loss probability, network fault probability, multicast join time, delay, and so on. QoS metrics for voice includes Mean Opinion Score (MOS), jitter, delay, voice packet loss rate, and so on. QoS metrics for IPTV services include channel availability, channel start time, channel change time, channel change failure rate, and so on. In order to gain wide acceptance, it is crucial for IPTV providers to be able to provide high-quality TV service, video, voice, as well as interactive services [23]. To efficiently provide clients with a satisfying IPTV service and to maintain a high QoE level, user data should be prioritized depending on their service category. However, it is hard to determine how much bandwidth should be reserved for a QoS-guaranteed flow especially when the video quality that the user perceives is actually dependent on individual human perception [24].

■ *User interface and multichannels view:* With the growing IPTV industry and the availability of hundreds of TV channels, users want to watch more than one channel at the same time. For example, someone might wish to have a look at some game score and at the stock market while watching a movie. We need to design special multichannel interfaces along with the picture-in-picture features that can allow the user to open different channels with their preferred settings.

■ *Billing issues:* To become competitive, profitable, and meet growing consumer expectations, IPTV service providers will need an end-to-end, flexible billing solution that can provide rapid roll-out and trials of new Broadcast, VoD Pay-Per-View, and other services. The solution will need to provide both customer-centric and partner-centric features and automatically flow and expand to support all IPTV components while communicating with other IPTV enabling systems.

1.8 Future of Digital TV

The future of Digital TV concerns mainly the huge increase in mobile TV and video streaming services, personalized services, and social evolution through opening the Digital TV model (including IPTV, Mobile TV, and Web TV) to social networking applications. TV services personalization is one of the essential pillars for new and rich service offers based on content individual adaptation: the domain of targeted advertisements is an important example enriching service offers. TV services personalization promises to open up new market opportunities and enable the creation of new services through advanced services personalization, and consequently, more digital content consumption and investments. In this context, new business models are expected involving different actors along the Digital TV chain, which has a socio-economic dimension. Users will become an active part in the content creation; the advertiser will be ready to pay more for the targeted ad space. The revenues can come from two sides—the users and the advertisers. A dynamic customer profile database will also allow taking a step beyond advertising—becoming a multi-sided marketplace where other kind of sellers and buyers can meet. Digital TV is expected to change the user experience via quality, personalization, new services (such as content personalization, recommendation, and personalized channels), mobility, user interactions, targeted services (advertisements), and multidevices usage. We note that TV services personalization requires the identification of watchers in a distinguished and individual manner. Several studies are being done on adequate and user-friendly identification techniques for TV services personalization (currently focusing on IPTV). These include, using a special PIN code for each user through remote controls that could be even personal remote controls (using for instance the users' cell phones). Near Field Communication (NFC), Radio Frequency Identifiers (RFID), and finger prints scanners are also being studied as advanced identification technologies. A decision on which identification technology can be a success depends on many factors, as the users' acceptability, market feasibility for introducing a new identification technology upon the Digital TV architecture (including IPTV, Web TV, and Mobile TV), the expected investments cost of deploying such identification technologies compared to the expected revenues from the Digital TV services personalization in general.

The IMS (IP Multimedia Subsystem) offers a standard unified infrastructure for management and control of services (through decoupling the control from the service layer). IMS appears to be a promising underlying architecture for Digital TV deployment allowing for interoperability and facilitating the deployment of new services fitting the services' convergence and personalization—which meets the social networking current trends and paves the way to an open TV converged model including (IPTV, Web TV, Mobile TV, and Social Network Services).

Several standardization bodies are studying IMS to develop solutions for the fixed and mobile TV services (examples are ETSI/TISPAN, Open IPTV Forum, 3GPP, and ITU), and several industrials (France Telecom, Ericsson, Huawei, Thomson, and ZTE) nowadays show interest to the IMS deployment for Digital TV.

IPTV via satellite is a market that has attracted a lot of attention over recent years. Compared to satellite TV, IPTV appears to be a compelling and highly competitive platform. However, true demand and opportunity for satellite-delivered IPTV is not entirely clear at this early stage, where many questions remain regarding the precise demand for satellite broadcasting in a fiber and copper-centric telco environment, particularly in the United States. In this early stage of the market cycle, extensive analysis on the potential of IPTV via satellite in terms of market and technology trends and the role that the satellite industry can play in the emerging IPTV industry has to be undertaken. In the future, a converged model is expected in which satellite technology's ability to provide cost-effective broadcast services can be positioned as a natural broadcast enabler and would seem a perfect fit to distribute and deliver TV programming to IPTV providers. In this context, satellite capacity providers can act as "head-ends-in-the-sky" and become infrastructure core components in enabling IPTV services, and could generate retail revenues through revenue-sharing arrangements with content providers and IPTV service providers.

1.9 Conclusion

This chapter presents a detailed review on Digital TV technologies showing the challenging transition in the TV world from analogue to digital TV as well as the history and current status of Digital TV and emerging trends for the digital TV market. IPTV evolution compared to Web TV is also presented with a particular emphasis on IPTV's different architectural components, its standardization activities and market status. Another important contribution of this chapter is the focus on the different competitors for IPTV technology and the current challenges associated with IPTV deployment. The chapter concludes by discussing some technical issues related to the future for Digital TV while considering different existing technologies (such as IPTV, Web TV, Mobile TV, Social Networking Services, and Satellite TV).

Acronyms

ADSL	Asymmetric Digital Subscriber Line
API	Application Programming Interfaces
ARPU	Average Revenue Per User
ATIS	Alliance for Telecommunications Industry Standard
B2B	Business to Business
BC	Broadcast
BCG	Broadband Contents Guide
C2C	Customer to Customer
DRM	Digital Rights Management
DVB	Digital Video Broadcast
EPG	Electronic Program Guide
FG	Focus Group
FiOS	Fiber Optic Service
FTTN	Fiber-To-The-Node
GEM	Globally Executable MHP
HD	High Definition
HDTV	High Definition Television
IIF	Interoperability Forum
IMG	Internet Media Guide
IMS	IP Multimedia Subsystem
IP	Internet Protocol
IPG	Interactive Program Guide
IPTV	Internet Protocol Television
iTV	Interactive TV
LTE	Long-Term Evolution
MHP	Multimedia Home Platform
MoD	Music on Demand
MOS	Mean Opinion Score
MPEG	Moving Pictures Experts Group
MPLS	Multi Protocol Label Switching
MRG	Multimedia Research Group
NFC	Near Field Communication
NGN	Next Generation Network
OTT	Over The Top

PDA	Personal Digital Assistants
PON	Passive Optical Networking
PVR	Personal Video Recording
QoE	Quality of Experience
QoS	Quality of Service
RFID	Radio Frequency Identifiers
SD	Standard Definition
SD&S	Service Discovery and Selection
SIP	Session Initiation Protocol
STB	Set Top Box
TMG	Telecommunication Management Group
UGC	User Generated Content
UMTS	Universal Mobile Telecommunication System
VDSL	Very-High-Speed DSL
VoD	Video on Demand
VoIP	Voice over IP
WiMAX	Worldwide Interoperability for Microwave Access
WLAN	Wireless LAN

References

1. O'Deiscoll G., *Next Generartion IPTV Services and Technologies*, John Wiley & Sons, Hoboken, NJ, 2008.
2. Global Insights on Digital Media, *Analysis—Forecasts and Comments on the Global Digital Media Market, Fact Book* January 2009. http://www.telecomsmarketresearch. com/Global_Insights_Mobile_Communications.shtml (last accessed July 22, 2011).
3. Multimedia Research Group, Inc., http://www.mrgco.com/index.html (last accessed July 22, 2011).
4. IPTV Global Forecast –2009 to 2013 Semi-annual IPTV Global Forecast Report, May 2009, http://www.mrgco.com/iptv/gf0509.html (last accessed July 22, 2011).
5. Cesar Bachelet, Personalization and better interfaces will drive monetization of content, *IPTV World Forum: Technical Symposium*, March 25, 2009. http://www.analysysmason. com/About-Us/News/Insight/IPTV-World-Forum-personalisation-and-better-interfaces-will-drive-monetisation-of-content/ (last accessed July 22, 2011).
6. Telecommunications and Internet converged Services and Protocols for Advanced Networking TISPAN), Service *Layer Requirements to Integrate NGN Services and IPTV*, ETSI TS 181 016 V3.3.1, July 2009.
7. Monpetit M.J., Your content, your networks, your devices: Social networks meet your TV experience, *ACM Computers in Entertainment*, 7(3), September 2009.
8. Micokzy E., Next generation of multimedia services—NGN based IPTV architecture, *Proceedings of the 15th IEEE International Conference on Systems, Signals, and Image Processing*, Bratislava, June 2008, pp. 523–526.
9. International Telecommunication Union—Telecommunication (ITU-T), http://www. itu.int/ITU-T

10. Telecoms & Internet converged Services & Protocols for Advanced Networks (TISPAN), http://www.etsi.org/tispan/
11. IPTV Forum, http://www.iptv-forum.com/ (last accessed July 22, 2011).
12. ATIS, http://www.atis.org/ (last accessed July 22, 2011).
13. 3GPP, http://www.3gpp.org/ (last accessed July 22, 2011).
14. DVB, Digital Vides Broadcast http://www.dvb.org/ (last accessed July 22, 2011).
15. DVB-IPTV, http://www.dvb.org/technology/standards/ (last accessed July 22, 2011).
16. ETSI TS 201 812 V1.1.1, *Digital Video Broadcasting (DVB); Multimedia Home Platform (MHP) Specification 1.0.3.*
17. MPEG, http://www.chiariglione.org/mpeg/ (last accessed July 22, 2011).
18. The MPEG Home Page, http://www.chiariglione.org/mpeg/ (last accessed July 22, 2011).
19. She J., Hou F., Ho P., Xie L., IPTV over WiMAX: Key success factors, challenges, and solutions, *IEEE Communications Magazine*, 45(8), Aug. 2007, 87–93.
20. The Broadband Forum, http://www.broadband-forum.org *IPTV Explained*, http://www.broadbandservicesforum.com/images/Pages/IPTV%20Explained.pdf (last accessed July 22, 2011).
21. ITU Telecommunications Standardization Sector: Focus Group on IPTV Draft, *Classifications of IPTV Service and its Meaning*, FG IPTV-ID-0026, 2006.
22. Tawfik A. et al., Adaptive packet video streaming over IP networks: A cross-layer approach, *IEEE Journal on Selected Areas in Communications*, 23(2), February 2005, pp. 385–401.
23. Xiao Y., Du X., Zhang J., Hu F., Guizani S., Internet Protocol Television (IPTV): The killer application for the next-generation internet, *IEEE Communications Magazine*, 45(11), November 2007, 126–134.
24. Lee K., Trong S., Lee B., Kim Y., QoS-guaranteed IPTV service provisioning in IEEE 802.11e WLAN-based home network, *Proceedings of IEEE Workshop on Network Operations and Management*, Salvador Da Bahia, April 2008, pp. 71–76.

Chapter 2

Open-IPTV Services and Architectures

Emad Abd-Elrahman and Hossam Afifi

Contents

2.1 Introduction 30
2.2 Definitions and Terminologies 31
    2.2.1 Definitions...31
    2.2.2 Open-IPTV Services..32
        2.2.2.1 Physical Set-Top-Box (P-STB) 32
        2.2.2.2 Software Set-Top-Box (S-STB) 33
        2.2.2.3 Virtual Set-Top-Box (V-STB) 33
        2.2.2.4 Open Set-Top-Box (O-STB) 33
    2.2.3 Some IPTV Terminologies 33
2.3 Open-IPTV Architectural Components...34
2.4 Content Transmission Forms (Managed and Unmanaged Networks)........36
    2.4.1 Unicast Transmission...36
    2.4.2 Multicast Transmission...37
    2.4.3 Validations-Based Testbed 38
    2.4.4 WEB-TV and VOD Scenarios...39
    2.4.5 Live-TV Scenario 39
2.5 Open IPTV Business Model 41
    2.5.1 Design Factors 41
    2.5.2 Collaborative Architecture..41
        2.5.2.1 Current Architecture Status.......................................42
        2.5.2.2 Collaborated Open Architecture................................43

2.6 Nomadic IPTV Services.. 44
    2.6.1 Nomadism and Roaming.. 44
        2.6.1.1 Nomadic Access Issues...45
    2.6.2 Nomadic Access..45
        2.6.2.1 Access under Managed Network 46
        2.6.2.2 Access under Unmanaged Network.................. 46
2.7 Collaborative Architecture for Open-IPTV Services 47
    2.7.1 Open-IPTV Life Cycle 47
    2.7.2 Collaboration Issues and Model Analysis...............................49
    2.7.3 Traditional Broadcast TV versus Web-TV 49
    2.7.4 Cloud Design Motivations..50
    2.7.5 Collaboration Culture and Benefits 51
    2.7.6 Making the Domestic Cloud Happen....................................51
    2.7.7 Cost Analysis: CAPEX vs. OPEX..52
2.8 Open-IPTV Use-Cases Analysis 53
2.9 Conclusion 55
References 55

2.1 Introduction

This chapter provides a brief description and new visions regarding the convergence network architecture for an Open-IPTV model. Also, it counts the benefits of this architecture to mobility and security issues. With the new Open-IPTV model, the migration toward converged networks between different Content Providers (CP) infrastructures becomes natural. The general management methodology will add value to the overall control between those providers. This optimization includes two cost factors that are important for a provider: Capital Expenditure (CAPEX) and Operational Expenditure (OPEX).

Recently, the aspect of cloud network has proliferated within the Internet Service Providers (ISPs). It enhances the multimedia business efficiency. Cloud infrastructure solutions have matured and provided reasonable interoperability under the current Internet regulations and standards [1]. While the cost of deploying delivery network solutions for IPTV has increased over the last several years, the operational expense of maintaining and managing the network also continues to rise. That is why we are searching for collaboration between IPTV service providers.

The rest of this chapter is organized as: Section 2.2 highlights some definitions and terminologies about IPTV. Section 2.3 introduces the components of the open IPTV architecture and its analysis in flow control way. Section 2.4 compares between different ways of content transmissions. Section 2.5 differentiates between the current model for business IPTV delivery and the proposed one for collaborative delivery based on cloud network for domestic region. Section 2.6 handles some issues relevant to access IPTV in Nomadic situations. The model analysis, motivations from cloud design, collaborations, and cost analysis are discussed in Section

2.7. Section 2.8 analyzes some use cases access in different nomadic situations. Section 2.9 concludes the study of this chapter.

2.2 Definitions and Terminologies

The current status of IPTV model can be summarized in Figure 2.1. We have three models: IMS-based standard model based on the IP Multimedia Subsystem (IMS) [2] core as a controller, NGN-based [3] standard model based on Next Generation Network architecture, and finally, the Internet model, Google TV [4]. We have combinations of these infrastructures like Digital Video Broadcasting DVB. The DVB project could use either IMS or NGN so we can call it a hybrid model. In DVB-IPI-based architecture, the DVB-IPTV service is the video service provided over IP like TV over IP or the Video-on-demand (VOD) over IP as specified in refs. [5–7].

We expect the future to have collaborative model (Cloud-Based) for IPTV delivery. It will have advantages over the other models in terms of low investments cost, better delivery performance, and converged system in design.

2.2.1 Definitions

The term Internet Protocol TeleVision (IPTV) has many definitions. But, the ITU-T definition [8:4] is the more general one: "IPTV is defined as multimedia services such as television/video/audio/text/graphics/data delivered over IP-based networks managed to provide the required level of QoS/QoE, security, interactivity and reliability."

Actually, there are two models for general IPTV management:

- *The managed model:* It concerns access and delivery of content services over an end-to-end managed network by the operator (like Orange or Free Triple-play operators in France).
- *The unmanaged model:* It concerns access and delivery of content services over an unmanaged network (e.g., the Internet) without any quality of service (QoS) guarantees. YouTube represents one such type of this model.

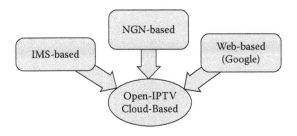

Figure 2.1   IPTV models collaboration.

2.2.2 Open-IPTV Services

The Open IPTV means offering TV over both managed and unmanaged networks. The terminal (TV) is also modified to accommodate built-in IP services.

To obtain an open IPTV service, we may need to reform the Set-Top-Box (STB). In order to propose a new aspect of STB that matches the new era of different access types and forms of videos, we need to highlight the characteristics of STBs. Moreover, we will define the advantages and disadvantages of adopting each type either for the operator or client. In Figure 2.2, we differentiate the current proposed services under the managed and unmanaged networks.

2.2.2.1 Physical Set-Top-Box (P-STB)

This system represents the actual implemented scenario. It mainly depends on physical Hardware of STB and leased connection between the consumer and content provider.

Advantages: The service security assurance and bandwidth guarantee are advantages of this model. Also, the good management of STB provided by the CP is naturally guaranteed.

Drawbacks: With the present model of IPTV, the delivery is based on physical STB restricted to specific location. But, as the consumers are increasingly becoming mobile, they demand bandwidth regardless of their locations to satisfy their entertainment. So, the lack in this model is the inability to support Nomadism (Mobility and Nomadic Access aspects).

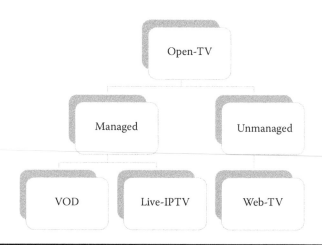

Figure 2.2   Open TV components.

2.2.2.2 Software Set-Top-Box (S-STB)

STB in this case depends on Software instead of Hardware for controlling the received channels and videos. It could be a USB disk carrying all necessary information.

Advantages: It will satisfy the consumer desires for enjoying all their subscription videos anywhere. So, this type is highly recommended for nomadic access services.

Drawbacks: The operators cannot be in control of everything. Management of user policies and authorization while changing the location is challenging.

As the future will probably adopt the new model of Set-Top-Box (STB) in a software form, we will no more need to restrict the physical location of IPTV services.

2.2.2.3 Virtual Set-Top-Box (V-STB)

It is a kind of S-STB but resides in the operator premises instead of user side. So, it is like an HTTP application accessed remotely by the client.

Advantages: The operator could control everything easily. For the client, no prerequisites are needed for one's system to start accessing the service except for a simple IP connectivity.

Drawbacks: Delay between the client application and the STB server is the big issue. Also, some security problems for user identification are produced in this type.

2.2.2.4 Open Set-Top-Box (O-STB)

It is a kind of hybrid STB that groups features from P-STB and S-STB. The O-STB can be accepted to run for any operator. Also, it can accept many kinds of video services like managed IPTV, Web TV, Social Networks (like YouTube), and VOD service.

Advantages: It will have an easy deployment manner and end of compatibility issues. Also, it is modern and suitable for new style of life.

Drawbacks: It may suffer from some complexity in design.

2.2.3 Some IPTV Terminologies

The common IPTV categories used could be classified as:

- *Pay-TV:* This service refers to the subscription-based TV service delivered in either traditional analog forms, digital or satellite. In different countries, we have similar terms referring to "Packs" and channels like Canal-Satellite, Showtime, ART, and so on.
- *TV-OTT:* TV Over-The-Top; it is one of the American TV services that provides a seamless consumer experience for accessing linear content through the

broadcast network on a TV set, as well as nonlinear services such as Catch-up TV and Video on Demand (VOD) through a broadband IP network. It is also designed to allow the provider to extend content and the consumer experience to additional platforms including PCs, mobile, gaming consoles, and connected TVs.

■ *IPTV "Follow-me:"* allows the user to continue to access one's IPTV service while moving and changing one's screen. (Content adaptation while Mobility is an issue.)

■ *Personal IPTV "My Personal Content Moves with Me:"* It allows the user to access one's personalized IPTV content in any place in one's domestic region with the reception of the bill on one's own home subscription (like Nomadic Access).

■ *TVA:* TV-Anytime is developed by a specific IPTV group [9] which is interesting in the interoperability and security for future TV.

■ *Open IPTV:* is a model of TV service that will be based on borderless technology. A hybrid model that merges the traditional Broadcasting TV with the Web-TV in one thing. OIPF (Open IPTV Forum) [10] is a well-known group in this field.

2.3 Open-IPTV Architectural Components

Open IPTV term could be defined as an integral solution for both managed and unmanaged IPTV services. It could be considered as a kind of TV anywhere/anytime but it goes beyond the subscriber domestic home region to include all possible access.

We have mainly two models of IPTV service delivery:

Managed service: This is the service provided by IPTV service providers based on standard STB and reserved link BW. The contents are mainly generated by the operators but in some situations the clients could generate and diffuse their contents like "Personalized TV."

UnManaged service: This may be called Internet IPTV which has no guarantee for QoS. YouTube is one of the famous representations of this model. The majority of contents in these methods could be generated by the customers themselves.

So, Open IPTV architecture means providing an integral solution for both managed and unmanaged IPTV services in one model. As a result, this will require searching new attaching point with the access network based on software Set-Top-Box (S-STB). Moreover, this kind of S-STB will add more scalability in the access. Video services can either be based on Managed or Unmanaged scenarios.

We can follow the state diagram in Figure 2.3 which explains the steps of accessing video service based on an open IPTV architecture as:

1. The client application regardless of the accessing device (PC, Mobile, or TV) initiates a request for the Web manager index. This manager is just a repository for indexing different videos, channels, or multimedia services. This manager also integrates the functionality of the "recommendation system" that recommends information items like movies, TV program/ show/episode, VOD, music, books, news, images, Web pages, and so on, that are likely to be of interest to the user. Finally, the recommender system gives its decisions based on comparing the user profile with some social reference characteristics.
2. By clicking on a specific choice from the previous manager, a request will pass to the back office manager asking about the client privacy.
3. The answer will either be positive or negative for accessing the client choice based on AAA decision and will forward a notification to index system.
4. The Web manager notifies the client application to start accessing the video but after accepting the DRM agreement.
5. The client application starts activating the client DRM engine.

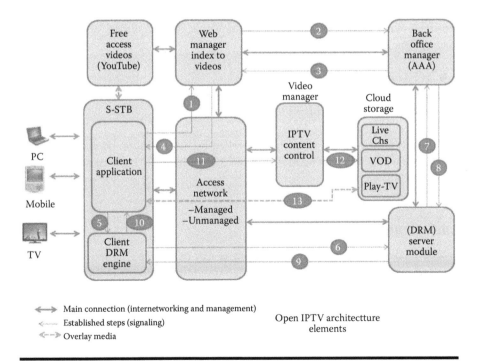

Figure 2.3   Flow control for open IPTV architecture.

6. The client DRM engine asks the DRM server module about the rights for this video.
7. The DRM server and the back office authorization center agree about this client rights.
8. The back office center gives the DRM server a resume about this client and file security information.
9. The DRM server notifies the client DRM engine the DRM policy and rights for the requesting video.
10. The DRM client engine sends those permissions to the client application to start playing the selected video.
11. The client application starts the access by passing a request to IPTV controller.
12. The Controller redirects this request to storage cloud system who is caching the videos and channels.
13. Finally the cloud management system builds an overlay session for accessing the video based on a specific resource optimization.

2.4 Content Transmission Forms (Managed and Unmanaged Networks)

The IPTV has different forms of transmission that depend mainly on the type of access (Managed or Unmanaged).

- HTTP/TCP (for Web-TV)
- Unicast/RTP/UDP (for VOD)
- Multicast/UDP (for Live-TV)

The first two scenarios could be controlled by using unicast transmission and the last one is controlled using multicast techniques in IPTV business model as the following.

2.4.1 Unicast Transmission

Unicast is point-to-point. It can be an effective way in light applications like e-mail. But, for heavy loaded applications like video streaming, where it will be inefficient if more than one user is streaming. If we have N clients that join the server (Streaming Server SS) then, this server must handle N sessions for the same copy of data packets as shown in Figure 2.4. Also, all routers (Multicast Nodes MCN) located near the traffic source will handle large numbers of sessions and increase the burden of those devices. HTTP and RTP/UDP are two different kinds of unicast. Although they are not efficient, half of the Internet traffic is today video streaming

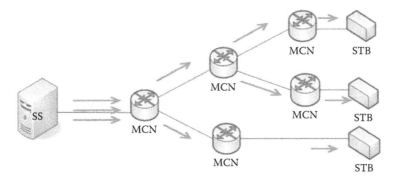

Figure 2.4 Unscalable UNICAST transmission.

coming from the main short video services such as YouTube. That is why unicast is representing the main mode of transmission over Internet.

2.4.2 Multicast Transmission

It is a kind of Broadcast (BCAST) transmission but it specifies a group of receivers not all like BCAST. As shown in Figure 2.5, only one copy of traffic is generated and passed to all devices inside the paths between source and destination. Also, there is no linear increase in traffic as the number of client increases.

For this type of access, we have two basic elements: Management protocol to manage users joining and leaving multicast groups like protocol IGMP [11] and MCAST routing protocols to forward MCAST packets like Protocol Independent Multicast (PIM) which has two popular modes: Sparse Mode PIM-SM [12] for using in light LANs and Dense Mode PIM-DM [13] for using in populated networks.

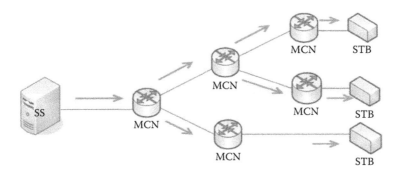

Figure 2.5 Scalable MCAST transmission.

2.4.3 Validations-Based Testbed

In this section, we focus on validating the open-TV aspects by implementing the three scenarios:

1. Web-TV (using HTTP/TCP as a reliable transmission protocol)
2. VOD (using HTTP/TCP as a reliable transmission protocol)
3. Live-TV (using RTSP and UDP)

We highlight the important aspects that are needed to be present in a typical prototype. This implementation depends on Android smart phones as clients. Also, all scenarios are simulated based on unicast mode. We built a simple architecture (like a Testbed shown in Figure 2.6) that consists of an integrated local area network of our sub-network inside our campus (Telecom SudParis) as follows:

1. One Server to stream either via VLC or LIVE555 under Linux
2. A client computer using Android emulator or VLC client
3. An access point to test Mobile IP clients (Wifi (IEEE 802.11) via the access point with an SSID configured as: Open IPTV)
4. Two Mobile clients as: Sony Ericsson Xperia—Android 2.1 and Samsung S Galaxy—Android 2.2 that are interconnected to WLAN by the previous access point Wifi (IEEE 802.11) via the access point SSID preconfigured

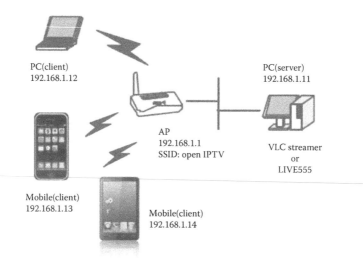

Figure 2.6   Testbed architecture for accessing IPTV.

2.4.4 Web-TV and VOD Scenarios

We conducted this test using HTTP/TCP as a reliable transmission protocol that can match transmission over the unreliable networks like Internet. The server side streams the videos via VLC server. Clients use Web browser application built by Android under Linux. One of the advantages of this scenario is its capability to support VOD which permits clients to fast-forward or rewind the video sequence.

We used Wireshark tools [14] to analyze the control messages and observed the following sequence of queries between the client and server as shown in Figure 2.7.

- *HEAD:* Query the header of the media
- *GET:* Query to the media content
- *Partial Content Information:* Response from the server response about the type of media which is (Video MP4)
- *GET(2):* The second request following the translation of the navigation bar on the video to an area not stored in memory buffer
- *Continuation of Traffic:* Responses of video content in continuation

2.4.5 Live-TV Scenario

We simulate this scenario by using VLC [15] as client and LIVE555 [16] as a streaming server.

In this part, we used the same Testbed in Figure 2.6 but with LIVE555 server [16] under Linux as a streamer. The client uses VLC as a client to view the stream in an on-line mode without any options of fast-forward or rewind. Then, we analyzed the sequence of messages between the two parties as the following in Figure 2.8.

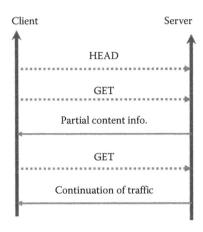

Figure 2.7   Flow control messages in accessing Web-TV by Mobile nodes under Android.

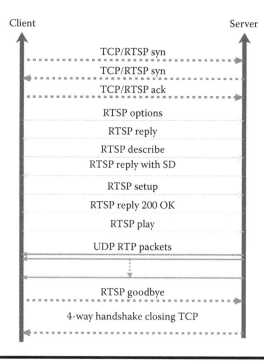

Figure 2.8 Flow control messages in accessing Live-TV by VLC clients under Unicast mode for firewall issues that prevent MCAST.

The control part is done using standard Real Time Streaming Protocol (RTSP) commands [17] as follows:

■ *Three-way handshaking of TCP/RTSP:* There are three packets used normally to open any TCP connection with the sequence control bits SYN/SYN/ACK.

■ *RTSP options:* It is an RTSP query sent by the client to the server so as to tell him about the client software used (VLC) and asking about the server software as (LIVE555).

■ *RTSP reply:* This is the server confirmation for the previous message from the server side to the client side and inside it the server proposed some public options like DESCRIB, OPTIONS, SETUP, TEARDOWN, and PLAY.

■ *RTSP describe:* The client tried to start by describing and accepting the server application used as a streamer.

■ *RTSP reply with SD:* The server answers using session description protocol all information about the media session and descriptions.

■ *RTSP setup:* The client initiated request for session establishment with the server based on the previous descriptions.

■ *RTSP reply 200 ok:* The server confirms the request of RTSP setup by sending back ok.

- *RTSP play:* The client uses the method play to start viewing the video.
- *Bundle of UDP RTP packets:* This bundle of packets represent the video packets encoded by MPEG-TS with as each RTP packet contains seven chunks with 188 Bytes as a Packet Elementary Stream PES for each one.
- *RTSP goodbye:* It is a control report sent by the receiver based on RTCP message to close a connection.
- *Four-way handshaking closing TCP:* Those are the ordinary four-way handshaking for closing any TCP connection with the sequence control bits ACK/FIN-ACK/FIN-ACK/ACK.

2.5 Open IPTV Business Model

The trend of cloud computing is a concept that fits well into an IPTV architecture. The concept of collaborative resource has been discussed in different works such as the one presented in ref. [18]. The authors proposed the concept of Alliance as a general aspect of Virtual Organizations. Their Alliance concept is based on integration and collaboration between clients' requests or demands and providers for resources. Moreover, they study the motivations from reforming the distinction in the current situation of organizations that will lead to good business model. The work mainly discussed the collaboration problems and some security aspects toward virtual organizations. Also, the work in ref. [19] proposed the idea of on-demand cloud service within IPTV-based servers virtualization. But, this work did not touch the area of domestic collaboration between different providers. So, what are the factors that affect design of an open model?

2.5.1 Design Factors

We believe that two factors press on the providers decisions while taking a new infrastructure investment:

- *Capital Expenditure (CAPEX):* It is representing the cost of network foundation and all nonconsumable system devices and infrastructure.
- *Operational Expenditure (OPEX):* It is representing the running cost for provider network including all cost of operation and maintenance.

For the long-term investments, the operators will reduce those costs in the domestic cloud. Moreover, the new added services related to quality and interactivity will be costless.

2.5.2 Collaborative Architecture

The competitive space between different IPTV operators pushes them to implement high similarity in clients' services. This means that, the majority of VoD and

IPTV channels are the same corresponding to the culture and social interests of each country. Thus, if we make some convergence between the different providers will not affect the overall policy of these operators. Moreover, it will enhance the service delivery and reduce incremental cost for future service investments.

2.5.2.1 Current Architecture Status

The current architecture of CP (as shown in Figure 2.9 as France case study) has mainly three layers:

Layer 1: Interconnection layer (infrastructure providers, like France Telecom in France)
Layer 2: Control layer (CP Islands, like Orange, Free …)
Layer 3: Management layer (The 3rd party hosting and cashing videos and channel servers, like Akamai [20])

The isolations and different islands are the main features of this model. Each operator has a huge investment and local management for the same services that was provided by all the other competitors in many cases.

The open models require an open infrastructure design and also an open management policy. So, the providers must avoid their selfishness and think about two important things:

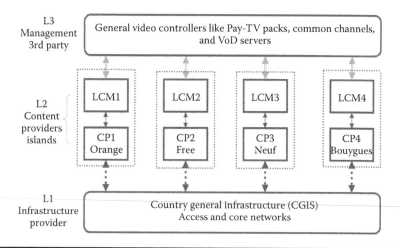

Figure 2.9 The current model of different islands CP. CP: Content Provider; LCM: Local Content Management; GCM: General Content Management; arrows with dot lines: the interface between CPs access network and core network; arrows with double lines: the interface between LCMs and GCM.

- The great benefits from the collaboration through cloud computing for example
- The satisfactions of consumers with new services

Thus, the Open-IPTV model needs a lot of cooperation between different partners for achieving success. The next section highlights the collaborated architecture.

2.5.2.2 Collaborated Open Architecture

The collaborative model design is illustrated in Figure 2.10. This model proposes more interactions and collaborations between different operators. As mentioned, the S-STB is the best match for IPTV delivery; we use it as the point interface to multiscreen access client.

UAR: User Authorization Request
UAA: User Authorization Answer
CAI: Content Adaptations Interactions

The collaboration exists in the form of common access to CCI and DGCM layers by the client access layer. This case will lead us to a new methodology of

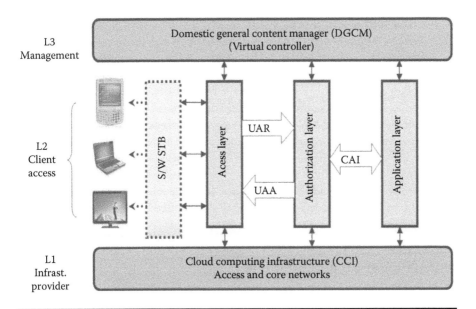

Figure 2.10 The proposed collaborative model for domestic cloud network to new IPTV system.

accessing called Resource-On-Demand (ROD) that will save the time and cost for service configuration. Moreover, UAR and UAA processes for clients services authentication and authorizations pass mutually and independently of user access network. The content adaptation for different screen has two aspects; one based on the capability of S-STB and the other on the access device specifications. The CA process for the domestic sphere for the client is mainly done by STB.

2.6 Nomadic IPTV Services

Most of the IPTV operators either in triple play or quadruple play services provide the video services-based physical Set-Top-Box (STB) device. This box mainly restricts the users to be inside their homes for having the service. Some operators give their clients the rights to access their IPTV service based on PC or Mobile phones devices [21] inside their own managed networks using something like VLC media player [15]. This access mainly based on RTP/UDP stack because the UDP is fast although unreliable. The operators in this scenario could guarantee the network reliability because the network is managed (end-to-end). But, the customers still need to have the same access outside the home network which means anywhere (Nomadic Access).

The Nomadic Access to Nomadic Service (NA-to-NS) aspect has four elements that need to be clarified:

1. *Nomadic Access:* The Nomadic Access is the way of accessing a home service outside the home network.
2. *Nomadic Network:* The Nomadic Network is the network that provides network access for roaming and remote users to the home network or the home services. Remote users are those with a portable, desktop PC or DTV at home that is already connected to the Internet (e.g., via a domestic ISP) and need to access the same subscribed service in nomadic situation.
3. *Nomadic Service:* The Nomadic Service is the service that can be obtained from a nomadic network and can be accessed independently of the home service provider. The nomadic service may be a service provided by the nomadic device to the network.
4. *Nomadic Device:* Traditionally, it was mainly the mobile devices but it could be extended to include any digital or IP devices like DTV, PC, or any device that can obtain the service behind an anonymous IP-STB.

2.6.1 Nomadism and Roaming

"Nomadism" is an equivalent term to "Roaming" but under some conditions. So, it is a kind of mobility called "Personal Mobility" which means the user could access one's service in any place under "any access network" by using the same credential or accounts. If the service is obtained outside the operator network based on mobility of Mobile phones it will be considered as Roaming. But, if the service is obtained either from same operator or from different operator based on any access device, this can be considered as Nomadism. So, Nomadism can be defined as the global definition for Roaming services and we can call it service "Portability."

2.6.1.1 Nomadic Access Issues

1. *Network Address Translation (NAT):* The main issue in the all-IP-based home network is the NAT problem. Because in the nomadic situation the user may have a private IP address and this adds more complexity to build VPN. Also, if it is possible to build the VPN with NAT we may have some problem relevant to the service itself and the ports numbers used.
2. *Privacy Problem:* The privacy is a big issue in the nomadic network. All users need to use the minimum sensitive information while they are in nomadic situation. This is to keep and guarantee your privacy and prevents others from knowing your sensitive information.
3. *Digital Rights Management (DRM):* All operators guarantee the DRM issues while providing their IPTV service by depending on the physical STB settled at the user home. But, for nomadic situations, they have some troubles for controlling the DRM and on which criteria they will build their guarantees.
4. *Cross Domain Authentication (CDA):* Obtaining a nomadic access rights is similar to the roaming service available in a mobile. So, the authentication can be regulated by the home network AAA servers or by the visited networks according to the roaming policy. Cross authentication was provided by the roaming group in refs. [22–24].
5. *Multicast Tree Structure:* As the life TV is diffused based on multicast technology like using PIM-SM [12] multicast routing protocol and managing the clients joining and leaving by IGMP [11], this tree could be changed and updated so as to optimize the operator network performance. This will require adopting a new way of rearranging the nodes and the MCAST trees.

2.6.2 Nomadic Access

For accessing nomadic services from any place we mainly have two ways:

1. *All Based Home Network:* In this scenario, both service and content could be provided by the home network. So, the nomadic network uses specific software to connect to the home network and to forward the service. This connection establishment has two phases. In the first phase, the client connects either to one's ordinary Internet Provider (from home) or to the nomadic network using any dial-in point-to-point connection like PPPoE (PPP over Ethernet). Then, the PPPoE makes a connection to a LAN, and provides authentication, authorization, and accountability issues for this client from

one's home network. After the user passed this verification phase over multiple Internet providers and had the settings to specific service, he could make a VPN (Virtual Private Network) connection using for example, PPTP (Point-to-Point Tunneling Protocol). This will encrypt the traffic by passing it through a "tunnel." This tunnel is secure so that other people cannot intercept the traffic and read as it passes across the Internet or over different wireless connections.

2. *Service-Based Home Network:* Service is provided by the home network while content is provided by the visited network or nomadic network. In this scenario, the client has to pass a secure authentication and authorization connection with one's home service providers. Then inter policy between providers will give the content requested from the very near point to the visited network.

Those two ways could be supported by an access from managed network or unmanaged network. Each type has its pros and cons as the following.

2.6.2.1 Access under Managed Network

By using a specific infrastructure built and managed by the operators.
Pros:

- *QoS Guaranteed:* As the Internet adopts Best Effort (BE) delivery, it cannot guarantee the Bandwidth requirements for video. But for managed networks it could.
- *Security:* High degree of security and data integrity. Also, high guarantee for DRM can be achieved.
- *Transmission:* It can apply scalable Multicast transmission which reduces the burden rates on the video sources.

Cons:

- *High cost:* The monthly subscription for some countries could be raised to 90 Euros for triple play service which is considered very high to many customers.
- *Locality:* It is not easy to access the same service outside your home because the majority of services are strictly to physical STB.

2.6.2.2 Access under Unmanaged Network

The Unmanaged network includes any Internet access like Internet cafe or any shared DSL connection.

Pros:

■ *Low cost:* There is no comparison in cost if we compare a dedicated line (ADSL or FTTH) with a specific Bandwidth (bit rate) and free Internet connection shared between different clients.
■ *Portability:* It is easy to access Internet from any place which complies with our objective (Nomadic Access).

Cons:

■ *QoS unguaranteed:* As the Internet adopts Best Effort (BE) delivery, it cannot guarantee the Bandwidth requirements for video.
■ *Security issue:* The DRM and security access plus privacy are very low in this type of networks.
■ *Transmission issue:* The routers over Internet cannot permit Multicast routing protocols. So, it is a big problem when the number of users increases and cause high burden for the source of video to handle too many sessions at the same time (Unicast). Hence, this solution is unscalable or at least inefficient.

2.7 Collaborative Architecture for Open-IPTV Services

To have collaboration, we need to understand what is convergence? As previously, convergence by telecom means having one network serves data and voice or videos. The rapid evolution in networks proposes another type of convergence which is the service convergence. Figure 2.11 illustrates the relations between converged networks "Infrastructure" with converged service "Management Tools." The access networks and storage or caching systems are integrated very fast with the converged infrastructure. But, the media servers and management tools or software are still needed to move toward convergence. This is due to the isolation islands for each operator systems.

2.7.1 Open-IPTV Life Cycle

Open-IPTV Model and Relation Aspects: To support a correct model, we need to explain the relations between the four actors that lead to a successful Open-IPTV model (see Figure 2.12) as follows:

■ *CP:* They must study between them the content convergence so as to facilitate the consumer access methods and delivery and also the content adaptation to match different screens.
■ *Operators:* They must convince the delivery of Open-IPTV before missing the dominant and control of Web delivery because the service will come in the near future.

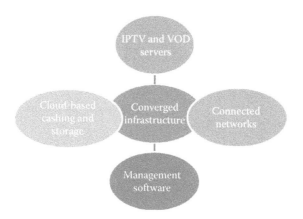

Figure 2.11   Future network-service convergence.

- *Infrastructure Providers:* It is the time for convergence infrastructure and Cloud Computing design to appear so as to enhance the user access methods and ameliorate the mobility and security user issues.
- *Consumers:* No way for the consumers from integrating themselves with the new technology and new methodologies of future Internet services. If the

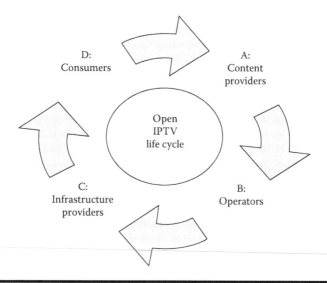

Figure 2.12   The Open-IPTV service actors where A is the first point of the cycle and responsible for contents generation, B is the service provider who is responsible all services management, C is the network access part and the point of attachment for the client, D is the last point of the cycle which represents the end user. (In some cases A, B, and C could be represented by one provider.)

consumer does not go with this development then we will have a missing part in the ring or the cycle will not be completed. Client culture and motivations must be changed so as to help for the model success.

2.7.2 Collaboration Issues and Model Analysis

This part provides some aspect and study analysis relevant to the new Open-IPTV model and the impacts on all members of the new life cycle as the following:

- The impacts of infrastructure integration on workflows and privacy policy.
- The impacts of applying and integrating convergence model between different CP on:
  - Consumer privacy protection measures
  - Business operational cycle
  - Financial management performance
- The optimization of resources under new collaborative conditions.
- The new behavior of consumers under new methodologies of services accessing with different screens is shown in Figure 2.13.

So, the content adaptation and the QoS assurance are the two factors that affect studying multiscreen IPTV delivery and band width optimization. Moreover, these two factors play an important role toward cloud migration. They are the turning point in the design and collaboration between different providers.

2.7.3 Traditional Broadcast TV versus Web-TV

The future perspective for the relationship between traditional TV watching and Web-TV in terms of average number of hours/month is shown in Figure 2.14. Some

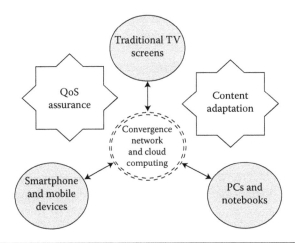

Figure 2.13 Open-IPTV model and multiscreens consumer issues.

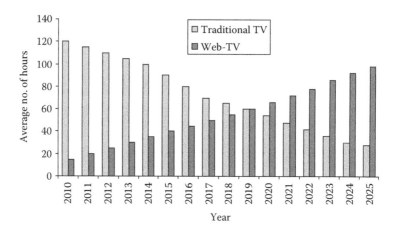

Figure 2.14 The expectation watching for Web-TV will exceed the amount of watching traditional TV by the year 2020.

IPTV monitor sites on a weekly basis [25] and specialists in technical and industrial reports [26,27] estimate that, any viewer in advanced countries like United States and Europe watches traditional TV for 120 hours per month and Web-TV for 18 h per month in 2010. But, this scenario will be reversed after the year 2019. They expected that, the Web-TV watching will exceed the traditional TV in 2020 and this excess will continue as shown in Figure 2.14.

New research shows that while TV broadcasts still dominate initial viewing, more users are turning to the new services like (TVE) TV everywhere to watch TV when they have free time.

For some segments of the industry TVE is being able to watch any television show, any movie, any online video content anywhere you would like on any screen, television, PC, mobile phone and, of course, the Smartphone at any time.

2.7.4 Cloud Design Motivations

CP businesses rely on the network (infrastructure) and the data center (hosting servers). They are the key factors in providing a successful IPTV service. Without them, achieving business goals like increasing market share, customer satisfaction, or operating margins and profitability is nearly impossible. However, over time, both networks and the data center have matured and, in the process, become more complex than ever before. The adoption of new trends, applications, and services over the years has brought with it inflexible designs, point product solutions, and a plethora of operating systems and management applications. This complexity leads to increased operating expenses in both the delivery and data center networks and limits their potential. So, the traditional ways for enhancing the current state by hierarchal way are not suitable. They lead to more sophisticated network and high cost especially the Capital Expenditure (CAPEX) one. Also, by default, as a result of complexity the Operational expenditure (OPEX) will be augmented. So, we expect that the curve of cost will reach a high peak.

From all causes mentioned above, we can conclude that the two main reasons for adopting cloud design are the reducing of high costs of investments and consumer quality assurance for satisfying the service. But, the migration toward cloud is not easy from the providers' perspective.

So, we suggest the first step to move IPTV service to the cloud: adopting the convergence mechanism between the current providers. Then the cloud will be the Internet available one. We categorize the IPTV service cloud into two aspects as:

- *Domestic cloud:* It resulted from collaborative providers in the same country.
- *International cloud:* It is the cooperation between different domestic clouds over Internet.

2.7.5 Collaboration Culture and Benefits

The collaboration model as a first step to cloud design is to consider a new value-added culture between different providers. It will reduce the whole cost for new services deployments and increasing the total revenue from the network services. As, the future IPTV services goes to future generation based on multiscreen multiservices, the development of content and service providers must take parallel path for achieving customers satisfactions which are

- High availability
- Good scalability
- Minimum OPEX cost and moderate CAPEX cost
- Good interoperability between different providers
- More facilitations to clients access services like mobility issues

On the other hand, the realization part of Figure 2.13 is not practically simple. This is because the huge cost for content provider implementations to adopt multiscreen systems flow adaptations. But, we think that the collaborative model will achieve the most cost reduction for this scenario.

2.7.6 Making the Domestic Cloud Happen

One main problem facing all CP toward migration to cloud computing networks is the lack in security. If the content is very sensitive, the cloud cannot guarantee its security as much as required.

The reliability and troubleshooting also represent a big difference between the cloud providers and traditional ones. But, the new release of "Domestic Cloud" can adopt the convergence mode.

So, the cooperation between the existing providers for obtaining a convergence network could lead them to the future cloud computing network as a suitable solution in this time. The trusting in sharing contents and resources is a way toward full migration to the concept of cloud. We suggest the collaborative solution as a "Domestic Cloud" network for all types of videos and IPTV service.

As mentioned before, in France, the majority of IPTV providers provide the same group of VoD, channels, Pay-TV packs, and other types of videos. They almost have 90% of the contents common. Parts of these videos are hosted by another party or cashing systems like AKAMAI system [20] and small parts of videos are hosted by the providers themselves. So, if they share their resources, they will provide good services and they can overcome the bandwidth bottleneck problems. The benefits from this convergence are

- *Interoperability:* get unified management for different providers' networks.
- *Portability:* achieve some degree of service mobility between operators "Nomadic" principle.
- *Integration:* obtain complete service "Customer Satisfactions."
- *Quality of Services (QoS):* service assurance without restriction.

The successful model of the domestic cloud will lead to host international services which are the corner stone of cloud networks and International Cloud (IC). The third-party cloud service is the great objective but the route we follow must be taken step-by-step.

2.7.7 Cost Analysis: CAPEX vs. OPEX

All direct and indirect infrastructures and servers costs represent the whole part of CAPEX cost. We estimate that, for the cloud design, it can have more reduction of the essential costs till 80% of the total cost. Moreover, the reduction mainly depends on the degree of collaboration between the providers in the same domestic region.

Consider the following assumptions:

N: the number of providers' links in the domestic region
C: the total cost for each provider

Then, the total CAPEX for all providers $= N * C$
For collaboration model it will equal: $N * c/N * C$
where $c \approx L * C^2$ (c: the cost of link between two operators as a result from N-Square problem)
So,

$$CAPEXc = (L * C^2)/C = L * C$$

where

CAPEXc: collaboration CAPEX cost
L: the infrastructure Link foundation between providers after collaboration.

So, the profit (*Revenue R*) from collaboration as a reduction in CAPEX is

$$R = N * C - L * C = (N - L) * C$$

If we have a third parity (Cloud provider), then the CAPEX for the current operators will be zero and all costs will just be OPEX costs.

Therefore, the typical cost optimization regarding traditional data center design versus cloud design is really remarkable. The long-term costs will be reduced. Also, the benefits from adoption of cloud are the utilization principles of servers-based needs. This means that, the infrastructure is used only when there is a real need and it is released otherwise.

2.8 Open-IPTV Use-Cases Analysis

This part mainly discusses the analysis about two use cases for accessing nomadic IPTV service as the following:

- The user tries to access one's subscribed channels and videos from another place but relevant to one's operator. Using the same URI host part in this address (userA@operatorA.com) for operatorA network but from userB access device STB.

  This scenario gives the clients with the same IPTV operator, the same home privileges for accessing their own subscribed or personalized channels outside their home networks. As the main issue of access is the operator guarantee that the user requests the nomadic access is the home user itself. This problem could be solved by using additional hardware H/W like USB Key or Dongle that can be used as a personal badge to give the client one's rights as soon as it is accepted by the visited network STB as shown in Figure 2.15. So, after starting the access by the nomadic user (userA), the Visited Network (VN) and Home Network (HN) are considered as the same. After the validation of this user H/W the service triggering (ST) could be started for initiating the service from the (VN) easily through Visited Application Server (VAS).

- The user tries to access one's subscribed channels and videos from another place but irrelevant to one's operator. Using different URI host part. userA@ operatorA.com in operatorB network.

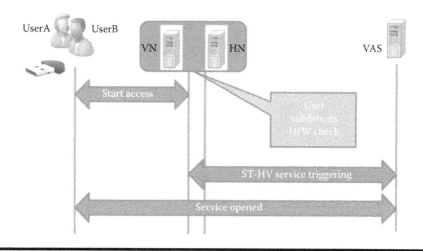

Figure 2.15 Flow control for the user who accesses one's service outside one's home by the same operator.

This scenario is equal to roaming issue in mobile operators. So, we propose collaborative access between different providers to enable their customers having the right access of their personal services in any operator network. It is suitable to apply either Federation Identity (FI) or Multi-Identity (MI) according to the policy agreement between operators. The software-based solutions are more practical in this case or we can apply the hardware-based

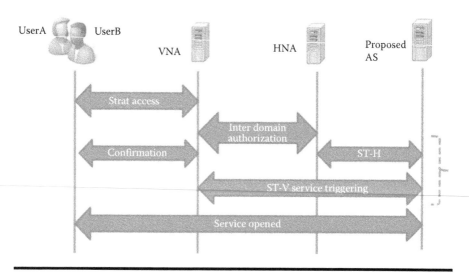

Figure 2.16 Flow control for the user who accesses one's service outside one's home by a different operator.

solution but on condition of compatibility solved. So, after passing the access verification by the visited network (VN), the service could be proposed by the home network or the visited network according to the inter-domain policy between the two operators as shown in Figure 2.16.

2.9 Conclusion

Today, collaboration is considered as a vital solution in IPTV business models. It has high impacts on IPTV market share, seamless mobility between different operators, common way of security, cost reduction in investments and good network performance. This chapter contributions were to: present state-of-the-art for new IPTV terminologies like OTT, Pay-TV, TVA, and Open TV; illustrate the benefits in convergence networks design and their impacts on CAPEX and OPEX costs. Also, demonstrate the performance of future IPTV service against traditional TV and introduce the multiscreen idea and the bandwidth optimization for multimedia delivery to different consumer's screens and then analyze the impacts of using cloud computing infrastructure in the converged network to different providers. Moreover, a simple testbed for open TV was conducted and analyzed in terms of control messages exchanged between server and clients. Apart from normal access, we also introduced the aspect of Nomadism and its access problems, ways and some use cases for IPTV delivery.

Finally, we expect the demise of isolated CPs' islands over the Internet at least in the domestic regions.

References

1. M. Armbrust, A. Fox, R. Griffith, A.D. Joseph, R. Katz, A. Konwinski, G. Lee et al. A view of cloud computing, *Communications of the ACM* 53(4), 50–58, 2010.
2. Draft ETSI TS 182 027 V0.0.9 (2007-04); Telecommunications and Internet converged Services and Protocols for Advanced Networking (TISPAN); IPTV Architecture; IPTV functions supported by the IMS subsystem, ETSI Technical Specification Draft, 2007.
3. Telecommunications and Internet converged Services and Protocols for Advanced Networking TISPAN); Service Layer Requirements to integrate NGN Services and IPTV, ETSI TS 181 016 V3.3.1, 2009.
4. Google TV: http://www.google.com/tv/
5. Digital Video Broadcasting (DVB); *Transport of MPEG-2 TS Based DVB Services over IP Based Networks*, ETSI TS 102 034 V1.4.1, 2009.
6. Digital Video Broadcasting (DVB); *Carriage of Broadband Content Guide (BCG) Information over Internet Protocol (IP)*, ETSI TS 102 539 V1.3.1, 2010.
7. Digital Video Broadcasting (DVB); *Remote Management and Firmware Update System for DVB IPTV Services (Phase 2)*, ETSI TS 102 824, 2010.
8. ITU-T Recommendation Y.1901 (01/2009)—Requirements for the support of IPTV services, clause 3.2.15, p. 4.

9. TV-Anytime Forum: http://www.tv-anytime.org/
10. OIPF (Open IPTV Forum): http://www.openiptvforum.org/
11. Internet Engineering Task Force; *Internet Group Management Protocol, Version 3*, RFC 3376, October 2002: http://www.rfc-editor.org/rfc/rfc3376.txt
12. Internet Engineering Task Force; *Protocol Independent Multicast (PIM) Sparse Mode*, 2006: http://www.ietf.org/rfc/rfc4601.txt
13. Internet Engineering Task Force; *Protocol Independent Multicast (PIM) Dense Mode*, 2005: http://www.ietf.org/rfc/rfc3973.txt
14. Wireshark Tools: http://www.wireshark.org/
15. Video Lan Client VLC: http://www.videolan.org/
16. LIVE555 Media Server: http://www.live555.com/mediaServer/
17. Internet Engineering Task Force; *Real Time Streaming Protocol (RTSP)*, 1998: http://www.ietf.org/rfc/rfc2326.txt
18. J.M. Brooke and M.S. Parkin; Enabling scientific collaboration on the grid; *Original Research Article Future Generation Computer Systems*, 26(3), 521–530, March 2010.
19. P. Yee Lau, S. Park, J. Yoon and J. Lee; Pay-as-you-use on-demand cloud service: An IPTV case; *International Conference on Electronics and Information Engineering (ICEIE 2010)*; pp. V1-272–V1-276.
20. AKAMAI: http://www.akamai.com/
21. Free Box TV: http://www.free.fr/assistance/2236-freebox-multiposte-executer-vlc-media-player-sur-votre-ordinateur.html
22. B. Aboba, J. Lu, J. Alsop, J. Ding, W. Wang; *Review of Roaming Implementations*, RFC 2194, September 1997: http://tools.ietf.org/html/rfc2194
23. B. Aboba, J. Vollbrecht; *Proxy Chaining and Policy Implementation in Roaming*, RFC 2607, June 1999: http://tools.ietf.org/html/rfc2607
24. B. Aboba, G. Zorn; *Criteria for Evaluating Roaming Protocols*, RFC 2477, January 1999: http://tools.ietf.org/html/rfc2477
25. FierceIPTV: http://www.fierceiptv.com/
26. ReportLinker: http://www.reportlinker.com/
27. Parks Associates' Report; *TV Everywhere: Growth, Solutions, and Strategies*, February 2011: http://www.parksassociates.com/report/tv-everywhere-report2011

Chapter 3

Mobile TV

Sherali Zeadally, Hassnaa Moustafa,
Nicolas Bihannic, and Farhan Siddiqui

Contents

3.1 Introduction 58
    3.1.1 Evolution of Mobile TV 58
3.2 Mobile TV Standards 60
    3.2.1 Digital Multimedia Broadcasting 60
    3.2.2 Satellite Digital Multimedia Broadcast 60
    3.2.3 Terrestrial DMB 61
    3.2.4 Digital Video Broadcasting-Handheld 61
    3.2.5 Media Forward Link Only 61
    3.2.6 Integrated Service Digital Broadcasting-Terrestrial 62
3.3 Mobile TV Distribution 63
3.4 Business Models and Opportunities for Mobile TV 64
    3.4.1 Mobile Device Support for New Services Based on TV
        and Video 65
    3.4.2 Business Models 66
3.5 Open Issues and Challenges 67
    3.5.1 Heterogeneous Characteristics of Terminals and Devices 68
3.6 Conclusion 72
Acronyms 73
References 74

3.1 Introduction

Recent advances of digital technologies have led to an explosive growth in the development and emergence of applications rich in multimedia content. Portable TV has been around since the 1980s but suffers many drawbacks which primarily stem from the analog technology it uses. Mobile TV services emerged in the mid-1990s after the introduction of digital TV for terrestrial standards. Mobile TV refers to the Audio–Video (AV) transmission or signal distribution, point-to-point and point-to-multipoint, and moving receivers. In short, mobile TV extends many TV services to mobile users (Park and Jeong, 2009).

Figure 3.1 shows an overall mobile TV architecture. In the first stage, a wireless interface enables communication between the access network and the mobile receiver. In a mobile TV architecture various wireless access networks, such as wireless LAN (WLAN), WiMAX, and cellular networks can exist. Each wireless technology has its own characteristics which service providers should carefully consider when deploying mobile TV. In the second stage, the wireless section extends to the sender so that both the sender's and receiver's devices can be mobile. Moreover, user-created content is becoming more popular in the Internet community; any mobile user can create TV content and provide it to other mobile TV users.

3.1.1 Evolution of Mobile TV

Interests in mobile TV started in the 1990s with the emergence of digital terrestrial TV along with the proliferation of mobile terminals and technological improvements in transmission and compression technologies. Mobile TV is considered as an extension of television outside the home sphere. In 1996 and 1997, the German broadcaster RTL carried out Digital Video Broadcasting—Terrestrial (DVB-T) service trials on mobile terminals with low mobility. In 1998, the Japanese digital

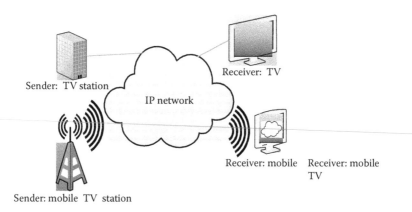

Figure 3.1   Delivery of traditional TV and mobile TV.

broadcaster experts group approved the Integrated Services Digital Broadcasting-Terrestrial (ISDB-T) as a digital terrestrial standard supporting mobile terminal access to the transmitted signal for receiving video and other data services. In 1999, mobile TV was introduced by the Singapore Broadcast Corporation (SBC) as the world's first service for TV broadcasting in public transport. Since 1999, standardization efforts, mainly by Digital Video Broadcasting (DVB), have continued to enhance mobile broadcast services with efficient power consumption and video compression solution based on IP for video and data delivery. Following these standardization efforts, the Digital Video Broadcasting—Handheld (DVB-H) became the standard for DVB on handheld devices and was approved in November 2004 by the European Telecommunications Standards Institute (ETSI). DVB-H enables the delivery of mobile television to handheld devices such as cell phones. Crown Castle, a US-owned company, began operating pilot DVB-H services in 2004, and in the United Kingdom, the mobile network O2 started using DVB-H in 2005 (Jacklin, 2005).

Nowadays, South Korea offers mobile broadcast TV services using the Korean Digital Multimedia Broadcasting (DMB) standard through terrestrial airwaves (i.e., Terrestrial DMB (T-DMB)) and satellites to mobile devices (through Satellite DMB (S-DMB)). South Korea is also promoting a new Broadband Wireless Access standard called "WiBro" based on Orthogonal Frequency Division Multiplexing (OFDM) for highly mobile users.

The mobile TV ecosystem has recently changed to the way we access TV-oriented content including TV streaming, Electronic Program Guide (EPG) related information and additional services provided by devices such as scheduling of Personal Video Recorders (PVRs). This revolution has mainly been initiated by companies such as Apple through their App Store applications. App Store has motivated developers to create downloadable applications thanks to revenue sharing and now content-oriented applications are no longer specified and distributed by mobile operators only. As a result, any content provider is now able to distribute its own contents over any IP network without the control of mobile operators as content aggregators. To address this strong competition, mobile operators are delivering an enhanced Quality of Experience (QoE) to end users who have subscribed to a paid mobile TV offer.

End users are now expecting ubiquitous access to TV content anywhere, anytime, from any device. The first trend of mobile TV is related to the frequent access of TV content fueled by the explosive growth of electronic messaging and communication services such as Instant Messaging (IM) and social networking technologies. Indeed, TV content has become a medium that is largely shared or forwarded when chatting or publishing personal comments on social network websites. The second trend of mobile TV is about disconnected mode for content consumption which is becoming more increasingly popular in urban contexts where network performances can quickly degrade in some environments (such as public transport) and affect time-sensitive services such as video streaming. The use of disconnected mode is also being largely motivated by the increasing storage capacities on personal devices as well as the decreasing cost of storage for those devices.

3.2 Mobile TV Standards

Different mobile TV standards have been developed by various countries. Some of these standards include DVB-H, DMB, S-DMB, MediaFLO, and so on. In this section, we compare different mobile TV standards taking into account features such as audio/video codecs, transmission schemes, power saving techniques, frequency bands, and so on.

3.2.1 Digital Multimedia Broadcasting

Digital Multimedia Broadcasting (DMB) (Gerami, 2010) is a digital radio transmission technology that is seen as the next-generation digital broadcasting service for indoor and outdoor users. This technology enables people on the go to enjoy high-quality video, theater-quality audio, and data through handheld devices such as handsets or in-car terminals while moving at a speed of up to 124 miles/h. DMB was developed by the Electronics and Telecommunications Research Institute of Korea in 2004 based on Digital Audio Broadcasting (DAB) standard. DMB uses the same video encoding scheme as DVB-H namely, H.264. However, DMB uses a different audio encoding technology, MPEG-4 Bit Sliced Arithmetic Coding (BSAC). It also uses MPEG-2 Transport Stream (TS) as the transport format. While DMB does not have the function of time slicing, it deals with a lower bandwidth of 1.54 MHz for keeping the tuner power low.

3.2.2 Satellite Digital Multimedia Broadcast

Satellite Digital Multimedia Broadcast (S-DMB) is a mobile satellite system which provides a communication medium in areas where terrestrial coverage is not available. S-DMB is suitable where terrestrial networks have not been deployed. S-DMB provides a high data rate broadcast service to any 3rd Generation Partnership Project (3GPP) standardized Class 3 handsets (250 mW). The system operates in the International Mobile Telecommunication 2000 (IMT-2000) frequency band allocated to Mobile Satellite Services and is based on a dedicated high-power geostationary satellite (Gerami, 2010). In the S-DMB system, a content server sends the live TV feed through an encoder and transmits the data to an S-DMB satellite in the frequency range of 13.824–13.883 GHz. The geostationary satellite rebroadcasts the signals directly to terrestrial repeaters at 12.214–12.239 GHz and directly to cell phones on the S-band (2.630–2.655 GHz). The terrestrial repeaters fill in the gaps where satellite signals get disrupted, such as in a city surrounded by tall buildings or in the subway. The dual broadcasts are coordinated so that if a subscriber happens to be within the range of the satellite and a tower at the same time, the subscriber receives both broadcasts and ends up selecting the reception with a stronger signal. An S-DMB system can reach data rates of 128 kbps. S-DMB adopts MPEG-2 Advanced Audio Coding (AAC) and H.264/ Advanced Video Coding (AVC) for audio and video compression. As with T-DMB, S-DMB is also based on the DAB transmission standard. However, unlike T-DMB, S-DMB uses Code Division Multiple Access (CDMA) for channel access instead of OFDM and has a wider channel bandwidth of 25 MHz (Zhou et al., 2009).

3.2.3 Terrestrial DMB

Terrestrial DMB (T-DMB) is a Terrestrial Digital Radio Transmission System for broadcasting multimedia to mobile units. T-DMB adopts H.264 for video coding and MPEG-4 Part 3 for audio coding and multiplexes them into the MPEG-2 transport stream. In T-DMB, Differential Quadrature Phase Shift Keying (DQPSK) modulation is used over OFDM channel access, which alleviates channel effects such as fading and shadowing.

3.2.4 Digital Video Broadcasting-Handheld

Digital Video Broadcasting-Handheld (DVB-H) is a European standard designed to use terrestrial TV infrastructure to deliver multimedia service to mobile devices. Since DVB-H only specifies the behavior below the IP layer of a mobile TV system, other service layer standards over DVB-H are required for describing, signaling, delivering, and protecting IP-based mobile TV services (e.g., streaming and file-downloading services). Currently, two similar service-layer standards are available over DVB-H, and they are the Open Mobile Alliance (OMA) Mobile Broadcast Services Enabler Suite (BCAST) standard and the DVB IP DataCasting (DVB-IPDC) standard. In DVB-H mobile TV systems, an IP platform is a harmonized address space that is available in one or several DVB-H transport streams. In addition, an IP platform may contain one or several multicast IP flows where there is no IP address conflict. In this case, each IP flow is an IP packet stream where every IP packet contains the same source and destination IP addresses. Consequently, an IP flow is identified by its source and destination IP addresses. In DVB-H mobile TV systems, a mobile TV service such as streaming or file downloading is carried in one or several IP flows. Hence, a DVB-H to DVB-H handover basically involves the service continuation of the currently receiving IP flows on a DVB-H terminal when the DVB-H terminal is moving from a DVB-H broadcast cell to another DVB-H broadcast cell (Chiao, 2009).

3.2.5 Media Forward Link Only

Media Forward Link Only (MediaFLO) (Yeon, 2007) is a technology proposed by Qualcomm that uses a limited number of high-power transmission towers.

MediaFLO uses OFDM transmission with approximately 4K carriers with either Quadrature Phase-Shift Keying (QPSK), or 16-Quadrature Amplitude Modulation (16-QAM) for the modulation of the carriers. MediaFLO provides multiple types of encoding schemes such as H.264, MPEG-4, and RealVideo. MediaFLO supports a multiple-level error correction system and an efficient coding which provides efficiencies of 2 bps/Hz to allow 12 Mbps with 6 MHz. MediaFLO uses the OFDM to simplify the reception from multiple cells, and the handsets receive the same packet from multiple cells which can improve the reception. In addition to saving power consumption by only accessing the part of the signal being watched, MediaFLO handsets can provide fast channel switching time which is less than 1.5 s. MediaFLO can be characterized as follows:

- *Turbo code algorithms:* They allow more aggressive error correction than is possible using the traditional Viterbi coding present in other systems.
- *Cyclic Redundancy Check (CRC):* If the packet is received perfectly, then the receiver does not have to calculate the Reed Solomon outer code, allowing additional receiver power savings.
- *Time interleaving data:* It is a good way to reduce the impact of impulse noise and changing multi-path, but it can also increase channel acquisition time, making surfing difficult.

3.2.6 *Integrated Service Digital Broadcasting-Terrestrial*

Integrated Service Digital Broadcasting-Terrestrial (ISDB-T) is similar to DVB-T and is used in Japan to provide digital service to TV sets and handheld mobile devices. It systematically integrates various kinds of digital content, each of which may include High Definition Television (HDTV), Standard Definition Television (SDTV), sound, graphics, text, and so forth. ISDB consists of various services. In an ISDB-T network, signals for both fixed and mobile reception services can be combined in one transmission via the use of hierarchical layers. Transmission occurs as a continuous flow by minimizing delays in signal acquisition when the user switches from one channel to another. However, this also means that the receiver must be powered on continuously while programs are being viewed. Consequently, power consumption becomes a significant design challenge for handset manufacturers. ISDB-T systems are characterized as follows (Yeon, 2007):

- *MPEG-2 interface:* The input signals to the system and output signals from the system conform to the MPEC-2 TS specifications.
- *Flexible use of modulation schemes:* Different types of digital content can be simultaneously transmitted with the appropriate modulation schemes and appropriate bit-rates for each type of content can be integrated into the ISDB stream.

- *Use of control signal:* Informs the receiver of the multiplexing and modulation configuration.
- *Partial reception:* Some of the services can be received by a lightweight, inexpensive narrowband receiver.

3.3 Mobile TV Distribution

The most common distribution means of mobile TV are cellular networks and more recently WiFi networks. Mobile TV distribution was initially largely focused on cellular networks. Network capacities are upgraded when required to sustain the TV data growth in order to maintain a high network Quality of Service (QoS). It typically consists of upgrading mobile network nodes such as the Gateway GPRS Support Node (GGSN), mobile transmission links, and service platform capacity to handle all the simultaneous TV sessions. Since cellular networks are conceptually delivering unicast-based transport flows, the content is transmitted to each receiver individually making mobile TV distribution through cellular networks similar to the distribution of any other data through cellular networks. Furthermore, the increase of video consumption on the Internet, proliferation of online games, and other Internet applications over mobile devices makes cellular networks even more congested. Consequently, these cellular networks often do not have sufficient bandwidth to provide live mobile TV services with a similar user experience and quality as classical digital TV (Klein, 2010).

Some technologies have emerged for broadcasting content over cellular networks, including Multimedia Broadcast Multicast Service (MBMS), Integrated Mobile Broadcast (IMB), and Evolved MBMS (e-MBMS) for Long-Term Evolution (LTE) networks. These technologies are currently either not mature yet or not widely deployed.

Multimedia Broadcast Multicast Service (MBMS) is a point-to-multipoint service in which data are transmitted from a single source entity to multiple destinations. MBMS allows network resource sharing; however, it suffers from high-power requirements from the base station limiting the adoption of this technology by cellular operators. In contrast, IMB is intended for delivering point-to-multipoint services over the Time Division Duplexing (TDD) spectrum of current 3G networks, without affecting services using Frequency Division Duplexing (FDD). TDD uses one frequency for transmission in one direction and another frequency for transmission in the other direction. TDD can also use a single frequency by allocating timeslots, transmitting in one direction during a particular timeslot and in the other direction during another timeslot. FDD transmissions occur in one direction using one frequency, and transmissions in the other direction take place on another frequency. The difference in frequencies minimizes the interference between the transmitter and receiver (Poole, 2006). IMB is only suitable for 3G networks with unused TDD spectrum and is not relevant to LTE networks. IMB requires a completely separate dedicated end-to-end infrastructure and a dedicated receiver making it still an immature technology. Evolved MBMS (e-MBMS) has emerged to support multimedia broadcasting over LTE networks and has a similar concept and high-power requirements as MBMS.

As with cellular networks, WiFi networks also require high bandwidth for Live Mobile TV distribution. However, this bandwidth requirement is less severe with WiFi networks which have higher bandwidth than cellular networks. WiFi networks also incorporate QoS mechanisms that are already in use for continuous media (such as Internet Protocol TV (IPTV)) applications. The main problem with WiFi networks is their limited coverage. Nevertheless, WiFi networks could provide a complementary service to cellular networks for Live Mobile TV distribution, wherever a WiFi network is available (Klein, 2010).

Access to mobile TV services from WiFi technology is largely promoted by end users for the following reasons:

■ The connection kit for WiFi access is becoming more and more user friendly for WiFi capable Smartphones.
■ Unlike cellular networks which limit access to Subscriber Identity Module (SIM)-enabled devices, access to mobile TV services from WiFi access points can be based on non-SIM methods for user's authentication. This flexibility in network attachment enabled by WiFi access is of great benefit for personal devices such as non-SIM equipped tablets that represent a large part of tablet sales.
■ WiFi access generally means free of charge access for the end user compared to cellular networks which require a data subscription plus an active fair usage to limit data consumption by cellular users.
■ WiFi access is also able to deliver high throughput sometimes similar to a fixed broadband user experience. This advantage is reinforced with new adaptive streaming technologies where streaming profiles can be updated by the device itself leading to better QoE. Adaptive streaming technologies over Hypertext Transfer Protocol (HTTP) differ from legacy Real-Time Streaming Protocol (RTSP)-based mechanisms where the bandwidth allocated for a given streaming session is under the full control of the network operator who may apply rate shaping on TV sessions due to mobile network dimensioning constraints.
■ WiFi coverage is also increasing rapidly because both mobile and fixed operators are responding to user demands by launching offers generally called "Community WiFi." These offers allow a fixed broadband subscriber to open and share its WiFi connection to other customers.

3.4 Business Models and Opportunities for Mobile TV

The current market for mobile services and multimedia services involves several actors including Mobile Operators, Content and Application Providers, Devices Vendors, and Platforms vendors. Mobile Operators aim to increase their market share through: (i) increase in the number of customers and reducing churn, and (ii) increasing revenues coming from nonvoice content services (multimedia services, TV, and Video). Content Providers are looking at ways to increase the use of mobile phones as a managed channel for content delivery, while Application Providers (Developers) are looking at developing, protecting, and monetizing mobile content assets.

Devices Vendors aim to increase their mobile devices market share, monetizing devices as a content delivery/consumption platform, and increasing their nondevice services revenues (as in the case of Apple services through iTune). Platforms Vendors want to increase their content distribution services and aggregation platforms.

3.4.1 Mobile Device Support for New Services Based on TV and Video

For mobile TV customers, the number of mobile devices per household is rapidly increasing, where on average each household is equipped with more than one mobile device according to a study done by Orange Labs involving 2000 mobile and TV end users in April 2010. This study has also shown that 46% of the end users rely on their mobile devices for TV and video content consumption with different consumption rates for different age categories (ranging from once a day "especially for young end-users" to once a week).

Considering the prominent role of mobile devices in the daily life of end users, different actors in the mobile TV and IPTV domains are working to provide new TV and entertainment services involving mobile devices, which in turn will fuel new business opportunities among the different actors. Examples of such new services include:

- *Mobile set-top-box (STB)*: A service in which the mobile device is used as an STB (i.e., software STB) allowing TV to be watched using whatever the access network is available. A high QoE can be achieved with a content displayed on the TV set after been decoded by the mobile STB, similar to IPTV QoE when network resources are available. This service is expected to open up a new market opportunity among mobile operators, device vendors, content and application providers. Mobile operators will increase the range of their service packages for mobile end users. Device vendors will need to consider the integration of an STB-like module during the manufacturing process, and application providers will have the opportunity to develop and sell new applications similar to the STB-like module.
- *Mobile user location*: A service in which the mobile device is used to locate the user based on the Global Positioning System (GPS) technology integrated into the mobile device, and detects the mobile user's proximity to centers of interest (for instance, theaters, shopping malls, restaurants, etc.) to distribute advertisements in the form of short videos related to the user's actual location

(including user location such as the user's residential home). By making use of technologies such as Bluetooth, the user's presence can be detected and depending on the type of the user's terminal, the TV content can be adapted to match the terminal requirements. Such a service can also open up new market opportunities for mobile operators, device vendors, application providers, and third-party sellers (theaters, shopping malls, etc.). Mobile operators could sell the localization information of users to content providers, but devices vendors will need to consider more enhanced localization technologies during the manufacturing process. Application providers will have the opportunity to develop and sell new applications through targeted advertisements (such as short videos). Third parties (such as theaters, shopping malls, restaurants) will pay the application providers and mobile operators to distribute their advertisements related to their businesses.

■ *Edutainment*: A service in which the mobile device is used to display interactive educational materials for an online educational module to perform online quizzes, and to evaluate the user's educational level in different modules (for instance by using online questionnaires) to know the prerequisites that should be followed. This service is also expected to open a new market opportunity between mobile operators, devices vendors, application providers, and third parties (mainly educational institutes). Mobile operators could sell the E-learning facility service to their mobile clients as part of their offered packages. Device vendors will need to consider more enhanced interactive technologies (for instance, based on tactile screens, sound facilities, etc.) during the manufacturing process. Application providers will have the opportunity to develop and sell new applications related to E-learning based on interactive videos, and finally third parties (mainly educational institutes) will pay mobile operators for providing this learning facility service to mobile end users.

3.4.2 Business Models

Business models can increase the revenue of mobile operators by deploying various new Application Programming Interfaces (APIs) that include: Short Message Service (SMS) transmission API, SMS reception API, Multimedia Messaging Service (MMS) transmission API, MMS reception API, Advertising API, User Context API, Location API, and Publishing to App store API (BlueVia 2011) which are described below.

■ *SMS transmission:* The mobile client can configure its application to receive inbound text messages from end users ensuring that messages are correctly routed to the application and the mobile client in turn shares a percentage of the price for the inbound text message with the end users. For instance, this SMS could contain the Uniform Resource Locator (URL) for a recommended video content from a friend or an announcement of a new movie.

■ *SMS reception:* The mobile client can configure its application to send outbound text messages to end users, while the end user must authorize the reception of these messages by providing a token to the mobile client application. This SMS could contain the URL for a recommended video content from a friend or an announcement of a new film. The price of the outbound SMS is shared between the mobile client and the end user.

■ *MMS transmission:* The mobile client can configure its application to receive inbound MMS messages from end users ensuring that messages are correctly routed to the application while sharing a percentage of the end-user price of the inbound MMS message. For example, this MMS could contain a daily weather map picture showing the traffic conditions for a given area in real time while giving the URL to follow more details on the Internet.

■ *MMS reception:* The mobile client can configure its application to send outbound MMS messages to end users who must authorize the reception of these MMS messages by providing a token to the mobile client application. The price of the outbound MMS is shared between the mobile client and the end user.

■ *Advertising:* The mobile client could incorporate video- and graphic-based advertising in its application to generate a revenue stream while receiving a percentage of the net advertising revenue received by the service provider.

■ *User context:* The mobile client could send information about its context (presence information, region, activity, preferences) directly to the service provider to receive a list of the available videos matching the user context with the possibility to watch snapshots of these videos or to buy them. The mobile client pays extra for having the customized list of available videos.

■ *Publishing to app store:* The mobile client could publish one's own application (that could be a "User Generated Content" (UGC)) at the available App store. This allows the mobile client to sell one's application directly to the customer while sharing the revenues with the App store provider.

3.5 Open Issues and Challenges

The advent of mobile TV identifies a new TV era in which the core elements are personalization, unique programming, and interactivity. In contrast to traditional TV, watching TV on a mobile phone is a more personal experience and the interactivity controls are immediately available. The success of mobile TV depends on the availability of desirable, popular content to the end user.

The extent to which the mobile operators succeed in reaping the benefits of mobile TV will depend on consumer satisfaction and response. The success of mobile TV in the market will also rely on factors such as display size and resolution of image content, transmission quality, and the price structure. Currently, there are several issues that need to be considered to enable a successful mobile

TV consumer market. A list of technical challenges (Mushtak and Ahmed, 2008; Park et al., 2008; Park and Jeong, 2009) related to mobile TV deployment is discussed below.

3.5.1 Heterogeneous Characteristics of Terminals and Devices

Mobile clients connect to the network to access TV services using a variety of devices and terminals such as mobile phone, Personal Digital Assistants (PDAs), and other handheld devices. These devices have heterogeneous characteristics such as screen resolution, data rate, and resource capabilities. All these heterogeneous factors lead to a small video display, low-power processor (because of a small battery), limited storage, and so on. Being lightweight is also an important requirement of a mobile terminal. These capability limitations mean that only a restricted set of technologies is possible for mobile TV solutions. For example, the content-providing server should consider the mobile terminal's screen size when sending a video stream to ensure smooth content delivery and acceptable QoS.

Although these limitations are persistent for low and mid-range mobile devices, smartphones devices (classified as high-end devices) now overcome most of those technical weaknesses as mentioned previously.

Over the last few years, research effort by Penhoat et al. (2011) takes into account this heterogeneity when services are broadcasted. Figure 3.2 highlights the need to define context-aware broadcast architectures.

Furthermore, it is possible that members of residential home access different TV channels using various terminals at the same time. In this context, service providers

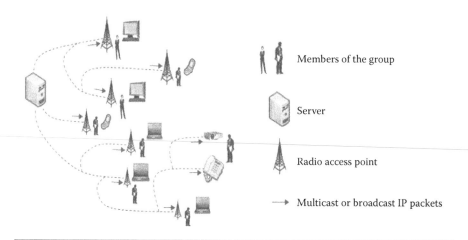

Figure 3.2   Broadcast transmission in a heterogeneous environment.

must deal with the available bandwidth and efficient content adaptation to meet various needs of all members of that home.

- *Bandwidth requirements:* Although the wireless link's effective bandwidth is growing rapidly, it will not be sufficient for mobile TV until the Fourth Generation (4G) wireless network is fully deployed when concurrent greedy data sessions are ongoing in a given cell. Even when the 4G wireless network is available, the bandwidth might not be sufficient if bandwidth-greedy services such as Ultra Definition (UD) video emerge and the number of users increases rapidly. Wireless links' bandwidth are significantly lower than wired ones, and the number of new high-bandwidth applications continues to increase. Therefore, bandwidth-aware solutions are always desirable for mobile TV services in the wireless environment.
- *Wireless mobility:* Mobile TV solutions need to be adaptive to user's preferences and the possible limited capabilities of mobile devices. Compared to fixed wired network environment, this task is especially challenging in wireless networks because of the error-prone radio channels and frequent Non-Line-Of-Sight (NLOS) transmissions (Singh et al., 2008). When the mobile TV terminal moves around, packets can suffer quality degradation caused by factors as shadowing and fading as they travel through wireless channels. Even when mobile TV terminals are stationary, temporal reflectors and obstacles in the wireless environment can affect the received signal quality and cause bursty packet losses. Such quality degradation is intrinsic to wireless communication links. Therefore, mobile TV servers and terminals should react adaptively to the wireless link's varying conditions. Dynamic management of the wireless link is crucial for the smooth playback rate during a streaming session. To prevent service interruption due to peers entering and/or leaving the system, we need a robust, adaptive mechanism to manage such link changes (Park et al., 2008).
- *Service coverage:* The purpose of mobile TV devices is to provide anytime, anywhere, any mobile device access to TV services. However, it is virtually impossible to deploy a wireless network that covers all geographical areas with no dead spots. Enabling vertical handover between heterogeneous wireless networks (Wireless Local Area Networks (WLANs), 4G networks, etc.) can help address the service-coverage limitations. Vertical handover occurs when a mobile terminal changes the type of connectivity it uses, usually to support its mobility. But, solutions for seamless mobility with vertical handover support still need further investigation (such as the ability to sustain uninterrupted mobile TV services while moving across different wireless networks without any significant performance degradation and how to select the best network among them for mobile TV). When the mobile terminal performs its access network selection, the mobile operator must still be able to deliver valuable information on available resources per access network. In this

context, the 3GPP defined Access Network Discovery and Selection Function (ANDSF) can be very useful. Indeed, the ANDSF database delivers operator policies to assist the mobile terminal during the access network selection.

■ *Scalable video coding:* The Scalable Video Coding (SVC) technology lets the system consider the network terminal types and available bandwidth. Although SVC enables scalable representation of video content with high coding efficiency, it is difficult to perform real-time encoding because of the SVC encoders' complexity. Further study is needed on how to best control the SVC rate according to network resource availability.

■ *QoS and QoE:* For high-quality mobile TV services, supporting key QoS characteristics, such as packet loss, bandwidth, delay and jitter, and packet–error ratio, are important. Mobile TV delivery systems must be able to handle QoS requirements through careful system design (e.g., over-provisioning or the use of Next Generation Networks (NGNs) implementing dynamic resource reservation procedures), careful traffic control in the network (such as traffic engineering and service differentiation), and optimized buffering and error-correction at the receiver. In particular, reacting quickly to varying conditions of wireless links is critical. Supporting user-perceived QoE by providing a resource-aware mobile TV service is also important (e.g., by increasing or decreasing the transmission rate according to the user's expectation) (Park et al., 2008).

  – *Middleware support:* Middleware is one of the key functions of mobile TV services. By deploying middleware, a service provider can control the usage of TV service remotely. Moreover, middleware acts as a transparent solution that enables mobile TV support on heterogeneous platforms. So far, there are several well-known middleware solutions that work with Set-Top-Boxes (STBs). However, these middleware technologies incur high overheads to be implemented in mobile device. We need to design lightweight middleware for mobile TV that can be easily deployed and supported by portable devices.

  – *Power constraints:* A special problem for mobile terminals is the limited battery capacity. Battery constraint in mobile devices has also been identified as one of the barriers in widespread usage of mobile TV. To save energy, time-slicing for DVB-H transmissions has been developed. The use of a technique called time-slicing, where bursts of data are received periodically, allows the receiver to power off when it is inactive leading to significant power savings. DVB-H is largely based on the successful DVB-T specification for digital terrestrial television, enhanced with additional features designed to take account of the limited battery life of small handheld devices. DVB-H also employs additional forward error correction to further improve the already excellent mobile performance of DVB-T (dvb-h, 2009).

  – *User interface issues:* The immediacy of content delivery is a significant factor, especially in the case of viewing live events. Consider the case that there is a significant time delay of several seconds between the transmission of standard TV and mobile TV. An audience seeing the scoring of a goal in an important championship with noticeable delay is likely to be severely annoyed, even if this delay is as low as 15 s. In such a situation, viewers of the delayed presentation may hear cheering from their neighbors, while they themselves wait for the goal.

The same issue exists in the case of channel switching. People do not want to wait too long for the presentation of the new channel. Providing low response times for channel hopping, being at least comparable to the ones known from standard TV, still represents an important open issue for mobile TV (Buchinger et al., 2009).

Moreover, when converting the video format from standard TV to mobile TV, the output text size is generally too small. As a result, the perceived QoS can be significantly affected (Buchinger et al., 2009). Since text appearance is also an influential factor for mobile TV quality it is crucial to study and assess the effect of text legibility and quality on overall perceived video quality. Studies have shown that people expect mobile TV to be easy to use. People are not willing to navigate through menus and therefore the need for special TV buttons on the phone has been expressed (Buchinger et al., 2009). Added functionality must focus on making the handset easy-to-use for mobile TV and should include: easy navigation and fast channel change, personalized list of TV stations, program reminders, and recording capability. In this context, mobile TV service providers need to incorporate all these design features efficiently in a dedicated application. Extra features should not be added at the expense of losing existing functionality (Mobile-TV_IBM, 2011).

– *Digital rights management:* As mobile TV services are profitable and large-scale services, the need for piracy countermeasures and business profit realization is intensified in order to protect the business rights for all those involved in content generation and distribution. Digital Right Management (DRM) is a shield for content owners and service providers protecting their legal rights, and it introduces a series of technical methods and policies managing access to software, music, movies, clips, or other protected content. Another reason for implementing DRM solutions is to control multi-viewing of a given content in disconnected mode. The promise around multiviewing lies in the ability for the end user to copy and display a purchased VoD (also called "e-sell Through") on multiple devices at the premises of the user. The major issue related to the use of DRM is the selection of a DRM technology that is convenient for all actors in the mobile TV ecosystem. Those main actors involved are content owners, mobile operators as service providers, and device makers. The selected DRM for mobile TV purpose is also expected to be compliant with other TV streams namely WebTV and IPTV. Since piracy and hacking are unavoidable, DRM technologies become increasingly important as a key component for developing sustainable business models (Zhang et al., 2009).

– *Content personalization:* The need for content personalization is very high in the case of mobile TV systems. The main challenge in this area is to identify mobile user preferences and usage patterns in order to enable this personalization. Specific content personalization types include the following: (a) *Filtering massive quantities* of content to identify items that best match a user's tastes or needs followed by recommendations of new content that may be of potential interest to a user; (b) *Proactively discovering* content with certain attributes (e.g., a certain language soundtrack or multiple audio channels). There are two dimensions to personalization (Uribe et al., 2009): (a) *Structure personalization* focuses on the content configuration itself (e.g., font size). This personalization shows the content the way users want to watch it, not of the content itself; (b) *Content personalization* focuses on the content itself, based on many factors (such as users' preferences). There are also personalization levels (Uribe et al., 2009) that dictate where personalization can be done: User-side personalization is done on the user device. In this way, the structure personalization can be more precise, just because users can do it each time. Server-side personalization is done on the server side, according to user characteristics (such as preferences, content consumption, etc.). Personalization can also be implemented at both, user and server sides.

■ *Consumer adoption:* The success of mobile TV depends on consumer adoption. Mobile TV provides for the customer a new and desirable service that is simple to subscribe and use. But consumers will only be willing to pay for mobile TV if they are satisfied with the content, quality, price, and user friendliness.

■ *Business issues:* The major business concern with mobile TV is the possibility of low consumer demand for mobile TV viewed on tiny screens. Wide adoption requires a business model for making mobile TV services attractive to users. As we mentioned previously, user interface is another obstacle to a successful mobile TV business. The small mobile device form hinders development of a fancy user interface. Mobile TV growth will require a highly creative and innovative human–machine interface suitable for the mobile device. Watching live TV while the user is mobile is one of mobile TV's most attractive features. So, access to popular real-time TV programs and rich content should be provided.

3.6 Conclusion

Mobile computing technologies have completely revolutionized the way we live, work, and entertain ourselves. People are now empowered with different types of technologies that enable them access to information from anywhere, at anytime using computing devices that are becoming increasingly more powerful and affordable. Mobile TV is one of the services that has been attracting a lot of attention recently because of the tremendous revenue potential for all the players involved in its generation, distribution, management, and use. We described several standards that have emerged over the last couple of years to promote mobile TV in various parts of the world to speed up its deployment and adoption. Despite the increasing demand of such a service from mobile users, the speed of adoption has been slow. This is because many technical challenges still need to be solved. We reviewed many of these challenges in this chapter and discussed some of the issues that still need to be addressed to enable the ubiquitous deployment and adoption of mobile TV. We also pointed out the importance of appropriate business models for mobile TV that need to be in place to ensure service profitability by the various parties involved.

Acronyms

16-QAM	16-Quadrature Amplitude Modulation
3GPP	Third-Generation Partnership Project
4G	Fourth Generation
AAC	Advanced Audio Coding
ANDSF	Access Network Discovery and Selection Function
API	Application Programming Interface
AV	Audio Visual
AVC	Advanced Video Coding
BSAC	Bit Sliced Arithmetic Coding
CDMA	Code Division Multiple Access
CRC	Cyclic Redundancy Check
DAB	Digital Audio Broadcasting
DMB	Digital Multimedia Broadcasting
DQPSK	Differential Quadrature Phase Shift Keying
DRM	Digital Right Management
DVB	Digital Video Broadcasting
DVB-H	Digital Video Broadcasting—Handheld
DVB-IPDC	DVB IP DataCasting
DVB-T	Digital Video Broadcasting—Terrestrial
E-Learning	Electronic Learning
E-MBMS	Evolved MBMS
EPG	Electronic Program Guide
ETSI	European Telecommunications Standards Institute
FDD	Frequency Division Duplexing
GGSN	Gateway GPRS Support Node
GPS	Global Positioning System
HDTV	High Definition Television
HTTP	Hypertext Transfer Protocol

IM	Instant Messaging
IMB	Integrated Mobile Broadcast
IMT 2000	International Mobile Telecommunication 2000
IPTV	Internet Protocol TV
ISDB-T	Integrated Services Digital Broadcasting-Terrestrial
LAN	Local Area Network
LTE	Long-Term Evolution
MBMS	Multimedia Broadcast Multicast Service
Media FLO	Media Forward Link Only
MMS	Multimedia Messaging Service
NGN	Next-Generation Network
NLOS	Non-Line-Of-Sight
OFDM	Orthogonal Frequency Division Multiplexing
OMA	Open Mobile Alliance
OMA BCAST	OMA Mobile Broadcast Services Enabler Suite
PDA	Personal Digital Assistant
PVR	Personal Video Recorders
QPSK	Quadrature Phase Shift Keying
QoE	Quality of Experience
QoS	Quality of Service
RTSP	Real-Time Streaming Protocol
SBC	Singapore Broadcast Corporation
S-DMB	Satellite-DMB
SDTV	Standard Definition Television
SIM	Subscriber Identity Module
SMS	Short Message Service
STB	Set-Top-Box
SVC	Scalable Video Coding
TDD	Time Division Duplex
T-DMB	Terrestrial-DMB
TS	Transport Stream
UD	Ultra Definition
UGC	User Generated Content
URL	Uniform Resource Locator
WLAN	Wireless LAN

References

Blue Via, Business Models, May 2011, https://bluevia.com/en/page/view/menupath/main.
gotomarket.sell.businessModels (last accessed July 29, 2011).

Buchinger S., Kriglstein S., and Hlavacs H., A Comprehensive view on user studies: Survey
and open issues for mobile TV, in *Proceedings of the 7th ACM European Conference on
European Interactive Television*, Leuven, Belgium, 2009.

Chiao H., Advances in mobility management of DVB-H mobile TV systems, in *Proceedings of IEEE Region 10 Conference TENCON*, Singapore, November 2009, pp. 1–6.

dvb-h, DVB, DVB-H, DVB-SH and DVB-IPDC are the key enabling technologies for mobile television, 2009, http://www.dvb-h.org/technology.htm (last accessed July 29, 2011).

Gerami M., Policies and economics of digital multimedia transmission, *International Journal of Computer Science Issues*, 7, Issue 2(4), March 2010, 21–30.

Jacklin M., Mobile TV and Data, *Broadcast Engineering Magazine*, February 2005. http://broadcastengineering.com/mag/broadcasting_mobile_tv_data/ (last accessed July 29, 2011).

Klein I., *What is Mobile DTV (MDTV)? Which Networks can Deliver True DTV Experience to Mobile Devices?*, White Paper, Enabling TV Everywhere, Siano Mobile Silicon, October 2010, http://www.siano-ms.com/CN/images/White_papers/Siano_What_is_MobileDTV.pdf (last accessed July 29, 2011).

Mobile-TV_IBM,IBM Corporation—Global Business Services, *Primetime for Mobile Television*, 2011, http://www-935.ibm.com/services/us/gbs/bus/pdf/ibv-ge510–6275-02.pdf (last accessed July 29, 2011).

Mushtak M. and Ahmed T., P2P-based mobile IPTV: Challenges and opportunities, *Proceedings of the IEEE/ACS International Conference on Computer Systems and Applications*, Washington, DC, 2008, pp. 975–980.

Park S. and Jeong S., Mobile IPTV approaches, challenges, standards, and QoS support, *IEEE Internet Computing*, 13(3), June 2009, 23–31.

Park S., Jeong S., and Hwang C., Mobile IPTV expanding the value of IPTV, in *Proceedings of the 7th IEEE International Conference on Networking*, Cancun, Mexico, April 2008, pp. 296–301.

Penhoat J., Guillouard K., Lemlouma T., and Salaun M., Analysis of the implementation of utility functions to define an optimal partition of a multicast group, in *Proceedings of the 10th International Conference on Networks*, St. Maarten, The Netherlands, January 2011.

Poole I., What exactly is UMTS TDD, *IET Communications Engineer*, 4(4), August-September 2006, 46–47.

Singh H., Kvvon C., Kim S., and Ngo C., IPTV over wireless LAN: Promises and challenges, *Proceedings of the 5th IEEE Conference on Consumer Communications and Networking Conference*, Las Vegas, January 2008, pp. 626–631.

Uribe, S., Fernandez-Cedron, I., Alvarez, F., Menendez, J., and Nuez, J., Mobile TV targeted advertisement and content personalization, in *Proceedings of 16th IEEE International Conference on Systems, Signals, and Image Processing*, Chalkida, Greece, June 2009, pp. 1–4.

Yeon C., Mobile TV technologies, in *Proceedings of 5th IEEE International Conference on Information and Communications Technology*, Giza, Egypt, December 2007, pp. 2–9.

Zhang Y., Yang C., Liu J., and Tian J., A smart-card based DRM authentication scheme for mobile TV system, in *Proceedings of IEEE International Conference on Management and Service Science*, Wuhan/Beijing, China, September 2009, pp. 1–4.

Zhou J., Ou Z., Rautiainen M., Koskela T., and Ylianttila M., Digital television for mobile devices, *IEEE Multimedia*, 16(1), 2009, pp. 60–71.

Chapter 4

Connected TV: The Next Revolution?

Erwan Nédellec

Contents

4.1   What Is a Connected TV? 78
4.2   Different TV Set Manufacturers and Their Solutions............................79
        4.2.1   Samsung Electronics...79
        4.2.2   LG Electronics..80
        4.2.3   Philips...80
        4.2.4   Sony..81
        4.2.5   Panasonic...81
        4.2.6   Google TV...81
4.3   Technical Fragmentation..82
        4.3.1   Website vs. Widget vs. Native User Interface 82
        4.3.2   Web Technologies..83
        4.3.3   Graphic Resolutions..83
        4.3.4   Fonts...83
        4.3.5   Cookie Support 84
        4.3.6   Media Player...84
        4.3.7   Decoding Capabilities 84
        4.3.8   Remote Control 84
        4.3.9   Content Protection...85
        4.3.10 Transport Protocol...85
        4.3.11 Performance..85

4.3.12 Technical Fragmentation Also Exists within
        a Same Manufacturer...85
4.3.13 Technical Fragmentation Also Exists within a Connected TV...... 85
4.4 How to Cope with the Technical Fragmentation? 86
4.5 Standardization 86
    4.5.1 CEA 2014 Specifications..86
    4.5.2 DLNA 87
    4.5.3 Open IP TV Forum..87
    4.5.4 HBBTV..88
    4.5.5 HTML 5 89
    4.5.6 Youview 89
4.6 So at the End, Is It Really a Revolution?...89
Acronyms.. 90

Revolution embraces two notions, the first one is the fact that something changed dramatically, and the second one is that it occurred suddenly. If we consider the latter, the connected TV has not appeared suddenly in the households. There have been several attempts in the last decade, but they all have failed until now. The most markworthy was the creation between Microsoft and Thomson Multimédia in 1999 of a joint venture named TAK. The TAK company developed a line-up of connected TV which were able to render interactive services (like e-mail, EPG, news) and also to navigate on the Internet. A keyboard specially designed was provided with the TV set. At that time, the issue was not the connected TV itself, but the low bit rate provided by the Internet connection through the PSTN which induced a poor user experience. But times have changed and the broadband access penetration changed the habits of everyone dramatically, the revolution occurred for surfing on the Internet, even if the pace of that revolution has not gone as fast as everyone could have hoped! So let us see if it will be the same for the TV sets in the future, let us see the usages, the technical trends, and we may say in 10 years that we are not using our TV sets the same way!

4.1 What Is a Connected TV?

A connected TV is a TV set which provides facilities to be connected on an IP network. The network interface is most commonly Ethernet, but the manufacturers usually offer for an extra cost Wi-Fi USB dongles in order to extend the capabilities of the TV set. The built-in Wi-Fi support is only present on high-end models today.

In the early stages of the connected TV, the use cases were mainly around sharing personal contents within the home network relying on the DLNA* specifications, and so it did not imply the access to the Internet. But nowadays, each connected TV provides its own solution and ecosystem to access the Internet.

---

* Digital Living Network Alliance (cf. http://www.dlna.org/).

4.2 Different TV Set Manufacturers and Their Solutions

4.2.1 Samsung Electronics

In 2007, Samsung Electronics started by launching its InfoLink solution. At that time, the solution was very basic by providing simple widgets which allowed the user to enjoy weather forecast, news, and stocks information through widgets displayed on the top of the video. There were no video widgets at this stage, the widgets pulled RSS feeds only.

In 2009, Samsung Electronics (like some other TV set manufacturers) signed an agreement with Yahoo! in order to embed Yahoo!'s solution in its line-up. The Yahoo! solutions were more accomplished as they provided an SDK for third-party developers, allowing rendering of video widgets, offering monetization, and user-identification facilities based on Yahoo!'s back-end solution. The Yahoo! solution* relies on their Konfabulator widget engine and on XML, Javascript, and CSS technologies. Today, the most popular widgets are Facebook, Twitter, Ebay, Pandora, Amazon, USA Today, Flickr, CBS, and so on.

Since 2010, Samsung Electronics has been accelerating the pace on connected TV by investigating a lot in its own proprietary solution named Internet@TV which was renamed as Smart TV in 2011.

As the global leader on the market, Samsung Electronics invested more aggressively than the others in marketing campaigns; they also launched contests in several areas of the world (like Europe, United States, Brazil, and China) with many prizes to win (e.g., €500,000 in Europe and $500,000 in the United States). In May 2011, Samsung Electronics claimed to have published more than 550 applications on their application store and to having reached over 5 million applications downloaded since the store first opened in February 2010. The most popular applications worldwide are Youtube, Google maps, AccuWeather, Vimeo, and Texas Hold'em, but in practice, the ranking is different in different countries of the world as many applications are more regional than global (like Dailymotion, for example).

The functional scope for the user is similar to the solution designed by Yahoo! The user can download and install widgets on their TV set. Samsung Electronics provides facilities for identification, and may provide facilities for monetization in the future. Technically, the developer can either develop widgets relying on Web technologies such as Javascript, HTML, XML, CSS, and DOM, or develop widgets relying on the Adobe Flash technology.†

The Samsung Electronics Smart TV solution also supports some protected content solution and some adaptive streaming protocols in order to render premium contents.

---

* More details on http://connectedtv.yahoo.com/ and on http://developer.yahoo.com/connectedtv/
† More details on http://www.samsungdforum.com/ and on http://www.samsungsmarttvchallenge.eu/

An interesting point to note is that as Samsung Electronics is a multiscreen player, they also developed an application which benefits from the convergence with their mobile devices. For example, an application on the TV can stream videos to their Galaxy tablet, or the Galaxy tablet can act as an enhanced remote control.*

4.2.2 LG Electronics

LG Electronics, the other major Korean manufacturer also initially signed an agreement with Yahoo! and then launched its proprietary solution named Netcast in 2010, and renamed Smart TV in 2011. In 2010, the user was able to access websites specially designed to be properly rendered on the LG TV sets. At that time, LG signed several partnerships with the main actors in the world in order to provide services on LG TV sets (like Youtube, AccuWeather, Orange, Maxdome, CinemaNow, NetFlix, etc.). Each time, LG mixed the global and regional actors in order to provide relevant contents. In 2011, they extended their technical specifications by introducing the widgets support, so it means that the user can download applications as well. Technically wise, the developer can either develop websites relying on Web technologies, or widgets relying on Web technologies or the Adobe Flash technology. Like many other CE manufacturers, they also provide content protection facilities and adaptive streaming solutions in order to enjoy premium contents.

4.2.3 Philips

Philips launched its Net TV offer in 2009 by providing access to the websites designed for their TV sets. They signed several agreements with global and regional partners such as Youtube, Ebay, Arte, Bild.de, Tagesschau, and so on. Philips chose to be as close as possible to the standard by relying on the CEA 2014 specifications which profile a set of Web specifications (XHTML 1.0, CSS TV Profile 1.0, ECMAscript 262, DOM level 2) and which define an API in order to control the media player.† More recently, Sharp selected the Philips solution for its own TV sets.

Even if Philips prefers to render websites specially designed for their TV sets, they also offer the possibility to browse the full Internet by entering a URL on the TV set. In practice, the user experience is very bad with a standard remote control, because you are navigating laboriously from one link to another using the four arrows and the OK button. More recently, some other manufacturers like Sony (with its Google TV), Samsung Electronics, and LG Electronics also provided the browsing on the full Internet. The user experience is a little better with a nonstandard remote control (especially when a cursor is displayed on the TV set), but the TV set

---

* More information on https://market.android.com/details?id=com.samsung.smartview/
† More information on http://www.philips.co.uk/c/about-philips-nettv-partnerships/22183/cat/

is definitively not the most accurate device for surfing on the Internet today. It may make sense in the future if there is a revolution in the way of browsing the full Internet on a TV set, like Apple which created a break in the mobile ecosystem for surfing on the Web from a mobile.

4.2.4 Sony

Sony launched its Bravia Internet Video solution in 2009. They focused on providing catch-up videos to their customers by signing several regional partnerships like ARD in Germany, M6 in France, Mediaset in Italy, RTVE, Antena3 and Lasexta in Spain, and Five in the United Kingdom. The user experience is consistent with the cross media bar navigation already present on the Sony PlayStation game console. The user interface is very similar from one content provider to another. Besides that strategy, Sony is also acting as a content provider after launching in late 2010 their own service named Qriocity for premium music and premium video on demand.

4.2.5 Panasonic

Panasonic launched its Vieracast solution in 2008, which became Viera Connect in 2011. In 2011, Panasonic also released an SDK in order to allow third-party developers to create new applications,* but a developer must pay an annual membership fee of $129 for a basic account, or $599 for a premium account. Like many other manufacturers, Panasonic signed agreements with many global and regional actors in order to provide relevant contents (such as Youtube, Bloomberg, Eurosport, Picasa, BBC, Acetrax, Dailymotion, etc.).

4.2.6 Google TV

Google is not a CE manufacturer, but as they often succeed in what they undertake, all the TV ecosystems were paying utmost attention to that new comer. The Google TV products have been co-developed by Google, Intel, Sony, and Logitech. Logitech launched its STB, and Sony launched its TV set in October 2010. Both products were based on the Intel Sodaville chipset, the Android operating system, and the Google Chrome browser. Keyboards were specially designed by Sony and Logitech for browsing on the full Internet. Besides the navigation on the Internet, Google signed agreements with partners in the United States in order to provide websites specifically designed for the Google TV set. At the time of launch, they were no application stores. Unfortunately for Google and its partners, the product was not mature at launch and the user interface suffered from the lack of consistency in ergonomic choices. Moreover, the Hulu entertainment

---

* More information on http://developer.vieraconnect.com/

website and the U.S. television networks like NBC, CBS, FOX, and ABC have blocked people from watching full-length shows on their websites using their Google TV-enabled devices. So, a few days after having announced the acquisition of Widevine Technologies in December 2010, which will provide to Google a solution to securely deliver video to connected devices, and just prior to the Consumer Electronics Show in January, Google asked manufacturers to hold off making any new product announcements at the show. So, Google TV 1.0 failed, but all the TV ecosystems are now paying utmost attention to the upcoming Google TV 2.0.

4.3 Technical Fragmentation

As one may have noticed, each manufacturer made its own technical solution which is a nightmare for content providers as it is often not possible to reuse the development made for one connected TV and port it on another one. The multiscreens paradigm for a content provider implies to adapt each time its service to a new connected TV. At mid-term, content providers can hope to rely on standardized solutions, but in the interim period, the content provider must cope with that technical fragmentation, and must design its architecture in order to limit the adaptation needed each time.

In order to have a good view of the technical fragmentation, we are going to review the major technical differences between the different solutions.

4.3.1 Website vs. Widget vs. Native User Interface

In some cases, if the service corresponds to a website specially designed for the connected TV, it means that when the user wants to access the service, the TV set instantiates the embedded browser by specifying internally the URL to render. The TV set accesses websites, like a computer does, except that the website must fit technical requirements to be properly rendered on the TV set.

In some other cases, if the strategy of the manufacturer is not to allow the rendering of websites, they prefer that the content provider develops widgets. A widget in that context usually corresponds to a ZIP file containing all the layout presentation (HTML, Javascript, CSS files) and a Manifest file which allows the widget engine to instantiate the widget. The widget engines are often based on the same layout engine as browsers, which is why the rendering layer relies on the same family of technologies in both cases. As the widget embeds the presentation layer only, it usually communicates with the back-end of the content provider through XML HTTP Requests (aka XHR).

The last case corresponds to native applications already embedded in the TV set. They are more like templates that a content provider might have slightly modified, but the level of customization is very low. Those applications may communicate with the back-end of the content provider through XML HTTP Requests (aka XHR), which a widget or a webpage could do as well. The only advantage of this approach is that a content provider will invest less time as his job will mainly be to provide metadata, as the user interface is already defined.

4.3.2 Web Technologies

Even if many TV set manufacturers support Web technologies, there are so many different flavors that it is impossible to have one solution for all. It is partly due to the fact that the W3C standardization does not mandate technical requirements in the specifications, and so TV set manufacturers are always partially compliant with the specifications (as it is the case for the different Web browsers on a PC anyway). It is also due to the fact that some manufacturers prefer to have a full proprietary solution, and they do not rely on standardized specifications at all. For example, only some TV set manufacturers implemented the famous API defined in the CEA 2014 specification which allows the control of the media player (that specification is referenced by DLNA, Open IP TV Forum, HBBTV). Even if that API is not perfect, it minimized the burden of porting a service to a new connected TV for a content provider.

As the standardization is not yet ready to address all the use cases in the field today, a content provider must first pay attention to the components present in the TV set, and then to the standardized specifications. In practice, it is easier today to develop a service which is rendered, thanks to the same layout engine (e.g., webkit) on two different TV platforms than porting a service on two different TV platforms which are compliant with the same standardized specifications (e.g., CEA 2014).

4.3.3 Graphic Resolutions

Almost all TV set manufacturers selected the 1280×720 pixel resolution which corresponds to the UI HD profile defined in the CEA 2014 specification, but some selected a different resolution which is painful for the content provider as it needs to rescale the size of the whole user interface. When the ratio between the different resolutions is the same, the impact is hopefully less serious.

4.3.4 Fonts

Even if in standardization, the Tiresias font is popular, TV set manufacturers are often reluctant to embed third-party fonts. So, they are always pushing the content provider to use their own fonts which are optimized for rendering on their TV sets, and of course, that proprietary fonts cost nothing to the TV set manufacturer. But as a consequence, the content provider may have to adapt its service to optimize the rendering of the user interface on the TV set.

4.3.5 Cookie Support

Even if it is very useful for content providers to use cookies, the permanent cookies are unfortunately not always supported on TV sets offering a Web technology-based solution. So, the content provider must take that into account in its design.

4.3.6 Media Player

The media player is always the sensitive component which never behaves in the same way in different TV platforms.

For example:

■ The MMS protocol implementation is never perfect (fall back directly in HTTP in order to avoid implementing the MMS protocol, no support of the MMS redirection, etc.).

■ The policy for the management of the buffer is designed first for files, and as a consequence, live streams are not always properly rendered.

■ Some media player performs several HEAD and GET HTTP requests before rendering contents which does not work with protected HTTP URL which are requestable only once.

■ Trick modes are not always working properly, especially for the protected contents, so it is always safer to use seek mode instead.

4.3.7 Decoding Capabilities

The decoding capabilities of the TV sets are pretty similar. Almost all TV platforms are able to decode MP4 and ASF containers, H.264 and WMV video codecs, MP3, AAC, and WMA audio codecs. The differences are mainly on AAC support because some TV platforms support the AAC-LC only, while others support AAC-HE.

Regarding the image decoding, all the TV platforms support the JPEG decoding, but only some benefit from the JPEG hardware decoding capabilities provided by the TV set chipset.

4.3.8 Remote Control

The key codes for the remote control are not always the same on different TV platforms, except on those which are compliant with the CEA 2014 specification. But adapting the values of the different key codes is not a big burden. The content provider might have more impact on its user interface if the user interface must be controllable through a pointing device (like the magic RCU provided by LG Electronics), and not only on the standard remote control. With a pointing device, the user moves a pointer on the screen, and clicks on the icons. The way of designing the user interface is very different as the clickable zone must be big enough;

moreover, there are only a couple of buttons on the remote control which have the consequence to display more buttons on the user interface. And as the user may stop to use his standard remote control to switch to the pointing device, the service must work with both navigation modes at the same time. So, it must be taken into account since the beginning of the design, and not at the end.

4.3.9 Content Protection

As usual, there is no universal solution on content protection, the most popular solutions are PlayReady (Microsoft), Windows Media DRM 10 Portable Device (Microsoft), Widevine, Flash Access 2.0 (Adobe), Marlin (Intertrust), but there is almost no manufacturer implementing all five solutions at the same time.

4.3.10 Transport Protocol

The manufacturers usually implement HTTP, HTTPS, and MMS. They have also started to implement adaptive streaming solutions like HLS (Apple), Smooth Streaming (Microsoft), or Widevine.

4.3.11 Performance

Even in a perfect world where all the TV set manufacturers will be compliant with the same standardized specifications, the rendering of the services will be different depending on the performance of the TV sets. Today, we can notice big differences among the manufacturers. We can only hope that differences will be less important in the future as the chipsets are getting more and more powerful.

4.3.12 Technical Fragmentation Also Exists within a Same Manufacturer

All TV set manufacturers are changing their line-up every year, and as a consequence, the technical specification. So, the content provider might have to adapt each year his service to be properly rendered on the new TV set models.

Moreover, TV sets are upgraded periodically with firmware upgrades. Sometimes, regressions may occur with implications to adapt to the content provider service as well.

4.3.13 Technical Fragmentation Also Exists within a Connected TV

For example, on a Samsung Electronics TV set, the HBBTV specifications and the Smart TV specifications are very different, so it means that for an actor like TF1 in

France, if they want to be accessible from both the red button displayed on the top of their live channel, and from the Smart TV hub, they must develop one HBBTV-compliant website, and one Smart TV widget.

4.4 How to Cope with the Technical Fragmentation?

The standardization is always a good answer to mitigate the technical fragmentation. Unfortunately, the standardized solutions are not currently fully ready, so content providers must cope with during the interim period.

As it is not possible to mutualize the presentation layer for all connected TV sets at this stage, the content provider must first focus on the data layer. It must design and roll out perfect Web services which allow several different connected TV sets to access the contents and the metadata. For example, the API for the back-end must be thought as a multidevice API and it must provide facilities to manage the statistics for the different devices (which is different from the different manufacturers). The content provider must also select the delivery technologies which better fit their strategy (transport protocol, encoding, content protection) depending on the TV set manufacturers that it is targeting.

The content provider may mutualize the presentation layer in some specific cases, but must mutualize the data layer in all cases (back-end, delivery platforms).

4.5 Standardization

Several fora or standardization bodies may address the connected TV ecosystem in some way, but we are going to focus on those which seem to be the most relevant at mid-term.

4.5.1 CEA 2014 Specifications

The Consumer Electronics Association created the CE-HTML in the CEA 2014 specifications in order to profile a flavor of Web technologies for developing user interface for CE devices, and more specifically for the TV sets.

The CEA selected:

■ The XHTML 1.0 specifications (to make it short, the XML format equivalent to HTML 4.01)
■ CSS TV Profile 1.0 (which is based on CSS 2.1 and provides some extensions, but in practice, the connected TVs are more based on CSS 2.1 today)
■ DOM Level 2

- ECMAScript 262, 3rd edition (the standardized Javascript)
- XMLHttpRequest object

Other interesting aspects are the facts that the CEA defined:

- An API for controlling the media player
- Extensions to support spatial navigation and multitap navigation

The CEA 2014 also referenced UPnP Remote UI for rendering remotely the user interface on a device, defined a mechanism to display third-party notifications, and another for matching the capabilities of the device with those of the server, but in practice, it is not implemented in current products at this stage.

4.5.2 DLNA

The Digital Living Network Alliance forum began in 2003 with the goal of securing the interoperability between CE devices for sharing personal contents within the home network. DLNA provides specifications but they also have a certification program which allows the compliant devices to display the DLNA-certified logo.

The DLNA forum announced in May 2011* that it is working on new interoperability guidelines for the playback of high-quality, premium commercial video. Those new guidelines refine the DLNA 2.0 specification and leverage on DTCP-IP-protected streaming to make service provider content more easily available for playback on CE devices. As DLNA 2.0 referenced the CEA 2014 specifications, the guidelines may also refer to the CE HTML specifications and the remote UI facilities.

4.5.3 Open IPTV Forum

The Open IPTV Forum was created in March 2007 as a pan-industry initiative with the purpose of producing specifications for IPTV. Unlike DLNA which focuses on the home network, OIPF defines an end-to-end solution for both managed and unmanaged networks. The quality of service between the two may differ with the managed network which is able to offer a more consistent and reliable quality of service by its nature, while the unmanaged network corresponds to the open Internet which has a best-effort quality of service due to the unpredictable moment-to-moment bandwidth available.

The OIPF forum will deliver specifications, profiles, testing, interoperability, and certification tools. They already released two versions of the specifications, but the certification program is not yet finalized.

---

* More information on http://www.dlna.org/news/pr/view?item_key=91b712addabcc5ff9ba833 8bb988ef83d5ccfe46

The OIPF specification is based on existing standards and Web technologies including CEA, DVB, ETSI, and W3C.

For the presentation layer, the specification leverages on the CEA 2014 specifications, and for the content protection, Marlin has been selected for the release 1, but alternative options like CI+ and DTCP-IP may be used.

OIPF defines profiles in order to achieve a complete interoperability for equipments and services related to the same profile. Any implementation based on Open IPTV Forum specifications that does not follow the profile specification cannot claim Open IPTV Forum compliancy.

For the release 1, three profiles have been defined:

- OIP: open Internet profile
- BMP: baseline managed profile
- EMP: enhanced managed profile (IMS-based network)

4.5.4 HBBTV

Hybrid Broadcast Broadband TV is a pan-European initiative aimed at harmonizing the broadcast and broadband delivery of entertainment to the end consumer through connected TV and set-top boxes. They focus on providing access to entertainment contents like catch-up TV, video on demand, interactive advertising, personalization, voting, games, and social networking, as well as program-related services such as digital text and EPG.

The HbbTV specification is based on existing standards and Web technologies, including OIPF (Open IPTV Forum), CEA, DVB, and W3C.

Version 1.1.1 of this specification has been approved by ETSI as ETSI TS 102 796 in June 2010.

The major (and simple) use case is the so-called "red button" use case. A broadcaster controls the display of a pop-up on the top of the video, and if the user clicks on the red button of the remote control, the TV set may display the website of the broadcaster which allows him to access the catch-up of the broadcaster.

HBBTV is technically ready to start as on one hand, broadcasters in France, Germany, and Spain started to add HBBTV-compliant information in their DVB-T and DVB-S streams, and as on the other hand, many TV set manufacturers implemented the HBBTV stacks in their 2011 line-up. In the meantime, the business side is less clear. Some TV set manufacturers are not so happy to develop a regional solution while they are investigating a lot worldwide in marketing campaign for their own proprietary solution. Some broadcasters are not ready to provide their catch-up for free at this stage. So, even if technically wise, the solution is ready, it will take some time for the different stakeholders to secure their business.

4.5.5 HTML 5

HTML 5 specifications are not addressing particularly the connected TV devices, but many stakeholders have their eyes on those specifications which will be (in a couple of years) more compelling than the old CEA 2014 specifications (released in 2007). It is also understandable as many TV set manufacturers selected webkit as the layout engine, and that open-source layout engine implements many HTML 5 features. HTML 5 specifications have been mainly driven by Google and Apple, and both Safari (the Apple's browser) and Chrome (the Google's browser) are based on webkit. Moreover, the W3C launched in February 2011, the Web and TV Interest Group in order to identify requirements regarding the relationship between services on the Web and TV services, to review existing work and to ensure that the Web will function well with TV.

4.5.6 Youview

Youview (formerly named Project Canvas) has roughly the same functional scope than HBBTV, but the technical solution is different. Youview has been initially driven by the BBC and they focus on a solution for the United Kingdom. In addition to the BBC, three other broadcasters (Channel 4, Channel 5, and ITV plc) and three communication companies (Arqiva, BT, and TalkTalk) are onboard. It must be noted that Sky Broadcasting and Virgin Media have both major concerns on that approach, and are doing their best to not make it a success. Youview is targeting STB at short- and mid-term, and not connected TV.

4.6 So at the End, Is It Really a Revolution?

It is, at least, not yet a revolution. Currently, people are buying connected TV or more precisely connectable TV, but only between 10% and 25% (it depends on the country) of the connectable TV were really connected on the home network in 2010. Those bad figures are twofold. First, it may be still complex for nontechnical savvy people to connect a device on a home network. Second, a customer buys a TV first, to enjoy live TV either through a satellite, a cable, or a terrestrial tuner. The connected TV feature is more the icing on the cake today. In the coming years, usages may change as the customers will be able to enjoy very good VOD and catch-up contents directly on their TV sets. An interesting trend to follow will be the popularity of the Arte HBBTV-compliant service. Arte is a broadcaster based in France and Germany, and contrary to the other broadcasters, they are less interested in earning money on their catch-up as Arte benefits from the funding by the European Union to sustain Arte's cultural missions. So in the Arte case, there is no business issue. And as the broadcasters are in a very good position for educating their televiewers, those may later enjoy watching catch-up contents which could accelerate the pace of the penetration of connected TV in the households.

At this stage, connected TV are still devices dedicated for audio and video contents and are not multipurpose devices like smart phones are in the mobile ecosystem. Today, the connected TV provides mainly interactive services, and a glimpse of VOD and catch-up, but in the future they may integrate videotelephony solution in the TV set. So, it will also be interesting to follow that trend, because the consequence will be that companies like Apple, Google, and Microsoft could be the key players as they are today on smart phones. Google already tried to put a foot in the door with the Google TV 1.0; there are many rumours on the Web that Apple may launch a connected TV in the coming months, and it will make sense if Microsoft also launches a market place on its game console as a first step. Regarding the videotelephony, Microsoft bought Skype in May 2011, and Apple has a very good solution with Facetime.

The CE manufacturers (specially the strong brand name) are really afraid of becoming commodities for those software companies, but hopefully for them, the generalization of the tablets in the households may really back off the trend of changing the TV set into a multipurpose device.

So at this stage, we can only note that we are just in the predawn years of a revolution, or it will never be a revolution.

Acronyms

API	Application Programming Interface
CE	Consumer Electronics
CEA	Consumer Electronics Association
CSS	Cascading Style Sheets
DLNA	Digital Living Network Alliance
DOM	Document Object Model
EPG	Electronic Program Guide
HBBTV	Hybrid Broadcast Broadband TV
OIPF	Open IPTV Forum
PSTN	Public Switched Telephony Network
RSS	Rich Site Summary
SDK	Software Development Kit
STB	Set Top Box
UI	User Interface
VOD	Video On Demand
XHR	XML HTTP Request

Chapter 5

3DTV Technology and Standardization

Gilles Teniou

Contents

5.1 Stereoscopic 3DTV 92
    5.1.1 Introduction 92
    5.1.2 Rendering Technologies...92
    5.1.3 Frame-Compatible Formats 93
    5.1.4 Full Resolution Per View Formats..95
5.2 3DTV Standardization 95
    5.2.1 Introduction 95
    5.2.2 Contribution Network...95
    5.2.3 Distribution Network 96
    5.2.4 Home Network...97
5.3 3DTV Challenges 97
    5.3.1 Enhanced Quality of Experience 97
    5.3.2 Multiview Video Formats 98
    5.3.3 Mobile 3DTV...98
5.4 Conclusion 99

5.1 Stereoscopic 3DTV

5.1.1 Introduction

Although the first 3D movies were produced approximately 60 years ago, it is only in the late 2000s that 3DTV was considered for introduction into the home. The significant progress in the Consumer Electronics area in providing 3DTV sets and the success of the 3D movie *Avatar* both contributed in making 3D the next step after the transition from SD to HD.

The Japanese cable channel BS 11 has been providing 3D programming since 2008, and many content providers and broadcasters such as ESPN, DirecTv, Orange, and BSkyB launched 3D programs in 2010 taking the opportunity to use premium events such as the FIFA football world cup.

In the display area, TV-set manufacturers such as Panasonic, LG, Samsung, and Sony, after presenting 3D-enabled prototypes during the Consumer Electronic Show (CES) in Las Vegas (US) in January 2010, also released 3DTV-sets into the market during spring 2010.

Finally, 2010 was the year of the first Blu-ray 3D movies. Available with 3D Blu-ray players, the Sony PlayStation 3 received a firmware upgrade to enable 3D Blu-ray Disc playback.

With the availability of 3D content and 3D-capable devices, 3DTV standards are needed in order to guarantee interoperability between different equipment and, for the industry, the interoperability between elements of the audiovisual delivery chain.

5.1.2 Rendering Technologies

Stereoscopy is the method of combining two single-plane pictures in order to produce a relief effect perceived by the human brain. With each eye being shown a different angle of a same scene, the human visual system—with subjective assessments—is able to recreate depth information.

With glasses-free systems not being mature enough for 3DTV, special glasses are needed in order to select the appropriate view for each eye. Two technologies exist:

- Active shutter glasses which are synchronized with a 3DTV set displaying alternatively the left and right views of a video. Active glasses require batteries.
- Passive glasses which use a polarized filter placed on both the screen and the glasses. For example, the current 3DTV can interlace the left and right views in a single image on the screen whereas the filters on the glasses only allow the left eye to see the odd lines and the right eye to see the even lines of the screen. In this case, image resolution is halved when compared with active systems but new systems such as active retarder will attempt to solve this problem.

More recently, 3D-capable mobile devices were introduced to the market. These glasses-free systems are often called auto-stereoscopic displays although in this case,

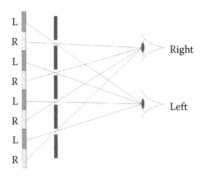

Figure 5.1   Parallax barrier.

they are based on a stereoscopic video format. The technology generally implemented is a grid placed over the screen, called the parallax barrier. When activated,
this barrier prevents the eyes of the user from viewing all the pixels of the display
such as depicted in Figure 5.1.

The second main 3D-rendering technology is based on a lens sheet consisting of
a series of vertical hemi-cylindrical lenses placed so as to direct light for different
viewing angles. When correctly placed, each eye can receive a different view from
the other. Although the rendering quality remains poor for stereoscopic 3DTV, this
physical layer can provide better results by increasing the number of views (i.e.,
beyond stereoscopy) (see Figure 5.2).

5.1.3 Frame-Compatible Formats

The stereoscopic 3DTV needs to deliver two views of a video content (left and right).
When considering the current 2D HDTV infrastructure that transports stereoscopic content to the TV through a live channel or an on-demand session, each view
has to be downsized in order to fit in an HD frame. Commonly, interlace contents

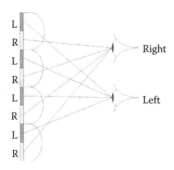

Figure 5.2   Lenticular lens sheet.

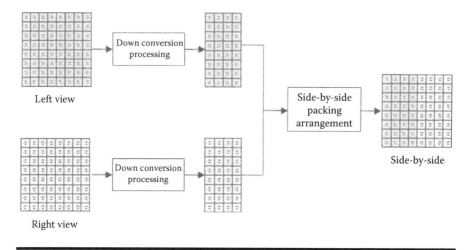

Figure 5.3   Side-by-side arrangement.

are horizontally downsized and placed "Side-by-Side" whereas progressive formats are usually placed in a "Top-and-Bottom" manner such as depicted in Figures 5.3 and 5.4.

Another pixel arrangement combining the left and right views inside an HD frame is called the "quincunx" format for which each view is down-sampled following a checkerboard pattern.

These arrangements are called "frame-compatible" due to their compatibility with the 2D HD infrastructure in which both views are conveyed.

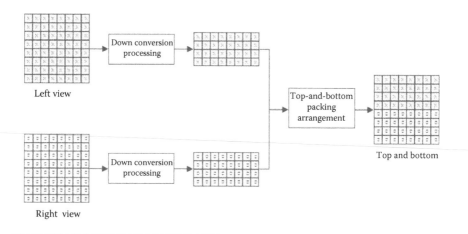

Figure 5.4   Top-and-bottom arrangement.

5.1.4 Full Resolution Per View Formats

In order to avoid the lack of definition introduced by the frame-compatible formats, it is necessary to transmit both views at full resolution. In this case, a dedicated infrastructure is required since the amount of data is twice as much as the frame-compatible formats. This representation format is often called "frame packing" because the left and right views are packed together.

5.2 3DTV Standardization

5.2.1 Introduction

From production to display, the 3D video signal is transmitted through different networks. For each of them, standardization groups took the responsibility to define specific signaling in order to be 3D compatible. Figure 5.5 illustrates such a delivery chain and the associated standards.

5.2.2 Contribution Network

3D video contents need a minimum of signaling prior to use at the TV head-end encoders or the Video on Demand (VOD) encoders. The Society of Motion Picture and Television Engineers (SMPTE) is standardizing 3D HOME Master requirements. The 3D task force was initiated in 2008. One year later, the SMPTE published recommendations on the 3D Home Master format (recommending the use of the 1920 × 1080 formats at 60 Hz for each view). Late 2009, a dedicated working group was created to define specifications in the 3D Home Master for the transport of 3D content between mastering equipment and ingest to

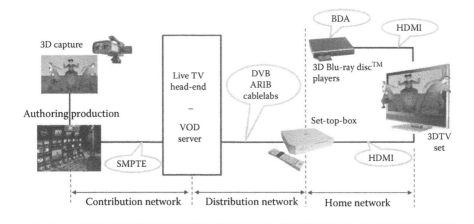

Figure 5.5   3DTV delivery chain.

the distribution system. The scope of the working group covers the image format, the associated metadata (indication of the left and right view, the depth map, 3D subtitles transport...). This working activity is still ongoing.

5.2.3 Distribution Network

In order to facilitate a rapid launch of the 3DTV services, the video format used to convey the 3D signal had to be compatible with the common HDTV infrastructure. This limitation had a significant cost impact when deploying 3DTV for a limited number of early adopters. This is the reason why frame-compatible formats were chosen. The video compression format remained the same as conventional HDTV, MPEG-2, and/or H.264/AVC. Consequently, the main goal of the standardization groups in this area was to indicate the 3D nature of the content to the decoder in order to allow the automatic switch of the display into the correct 3D mode.

The Digital Video Broadcasting (DVB) group launched two 3DTV working groups in 2010 (commercial and technical modules) after the conclusions of a dedicated study mission which stated that stereoscopic 3DTV was mature enough to be standardized.

DVB defined frame-compatible formats as being part of phase1 (whereas phase2 was identified for full resolution per view experience). A major requirement of the frame-compatible-based specifications was that there was no need to replace the decoder (a Set-Top-box or an Integrated Decoder TV). Only a firmware upgrade was accepted for dealing with the 3D-specific signaling.

In February 2011, DVB Steering Board approved the first 3DTV specifications. The transport layer has been updated to include a flag in the *AVC_Video_Descriptor* inside the Program Map Table (PMT) which indicates if the video stream contains 3D or 2D material. The video layer, only based on H.264/AVC, includes requirements on the use of *frame_packing_arrangement* SEI messages to inform which 3D layout is used (side-by-side or top-and-bottom). With such signaling the decoder is able to tell the TV-set which format must be rendered. Additional specifications were made available within the service information, particularly the declaration of a 3D service type and descriptors to convey the video depth range and disparity information. An extension of the DVB subtitles has been defined in order to appropriately render 3D subtitles and graphics.

In Japan, in August 2010, the association for Promotion of Digital Broadcasting (Dpa) produced requirements for 3D signaling of video content. The use of this signaling is limited to 1080i/60Hz side-by-side contents for which a 3D flag is embedded into the *user_data* field of the video picture layer for MPEG-2 video content, and the *frame_packing_arrangement* SEI message for the H.264/AVC video format. Also, the event information table (EIT) of the transport stream was updated to identify a 3D program, for example, to be used by the electronic program guide (EPG). The Japanese association of radio industries and businesses (ARIB) included these requirements into a technical report on Operational Guidelines for Digital Satellite Broadcasting, in March 2011.

In the United States, in January 2010, CableLabs, a consortium of cable opera-tors, published a new release of its Content Encoding Profiles specification which includes the requirements for formatting 3D content into a frame-compatible for-mat for use by cable television systems. The use of the *user_data* field for MPEG-2 video and the *frame_packing_arrangement* SEI message for H.264/AVC is required such as ARIB and DVB (for H.264/AVC only). Unlike DVB, the signaling at the transport layer in the PMT is based on specific metadata called the *3d_MPEG2_descriptor*. This descriptor contains a syntax element which indicates if the content is 3D or 2D encoded.

5.2.4 Home Network

In the home network, the local playback of 3D video content on Blu-ray discs and the digital connection between a decoder and the 3D TV-set are considered.

First, the HDMI LLC specifies the signaling of the digital link now widely adopted to connect different devices such as Set-Top-Boxes, game consoles, or Blu-ray players to the TV set. The HDMI 1.4a release includes 3D format signaling with a list of mandatory 3D formats to be supported by the HDMI sinks.

Specifically for the Broadcast contents, HDMI LLC defined three mandatory formats which are 1080i @ 50 or 59.94/60 Hz side-by-side, the 720p @ 50 or 59.94/60 Hz top-and-bottom, and the 1080p @ 23.97/24 Hz top-and-bottom. The two first 3D formats are mainly dedicated to broadcast TV signals whereas the last is more appropriate for movies (e.g., VoD services).

It is also important to note the mandatory format dedicated to the movie con-tents (e.g., for Blu-ray) which is 1080p @ 23.97/24 Hz frame packing, which means that both the left and right views are transmitted through the HDMI link together in full resolution.

The Blu-ray Disc Association (BDA) is an industrial consortium in charge of developing the Blu-ray Disc technology. Late 2009, the BDA released 3D specifica-tions in which the 3D extension of H.264/AVC, called multiview video coding (MVC), was promoted in order to offer a full resolution per view experience in 1080p (there is no support of interlace content). The required "Stereo High Profile" provides the backward compatibility with a 2D video which is the main commer-cial feature of MVC. In this case legacy "2D" Blu-ray players can still decode one of the two views (the one which is H.264/AVC encoded) and ignore the additional data specific for 3D (H.264/MVC encoded).

5.3 3DTV Challenges

5.3.1 Enhanced Quality of Experience

The first 3DTV DVB specifications were focused on frame-compatible formats for which the definition per eye cannot be better than half-HD. The so-called phase2 aims at providing specifications which enable a full HD per view 3DTV experience. This has the promise of delivering a higher quality of experience, similar to that of the 3D Blu-ray discs in terms of definition.

In this context H.264/MVC remains a strong candidate technology to fulfill such commercial requirements. Moreover, as presented in the BDA section, H.264/MVC is 2D backward compatible. This codec feature is very important for free-to-air broadcasters who address horizontal markets such as Digital Terrestrial TV (DTT). This 2D backward compatibility, also called (2D) service compatible, allows progressive introduction of 3DTV with the guarantee of offering a 2D version of the video to those who are not (yet) equipped with 3D-capable devices, even if for a lot of content shot in stereoscopic 3D, the production rules and grammar are different between 2D and 3D (i.e., there is no guarantee that a good 3D content can translate automatically into a good 2D content).

On the other hand, 3DTV frame-compatible compatible (aka FCC) technologies might be of interest to broadcasters and content providers who already deployed frame-compatible 3DTV services. The recent Motion Picture Expert Group (MPEG) activity on MPEG Frame Compatible (MFC) aims at defining the best coding schema for providing frame compatible and full-resolution per view versions of a stereoscopic 3D content.

5.3.2 Multiview Video Formats

Another activity at MPEG is called Free Viewpoint TV (FTV). Its goal is to define a new coding schema based on multiple views (more than two) and potentially the use of auxiliary data such as depth maps to aid in reconstructing interpolated views between the reference ones. With such a data model, it is possible to envision services of 3DTV based on multiple views on auto-stereoscopic displays. The depth control, adapted according to the screen size, for example, and using interpolated views is another potential use case which could take advantage of such a technology. Finally FTV could help stereoscopic systems with eye-tracking capabilities to render a better quality of experience with the use of additional views. A standardization activity could start in 2012.

5.3.3 Mobile 3DTV

Although glasses-free displays are not mature enough to be considered for 3DTV at home, their constraints on the placement of a unique viewer fits with the mobile TV experience. Mobile devices came into the market at the beginning of 2011 with displays based on parallax barriers which provide a good-enough quality of experience.

Since 2009, the 3rd Generation Partnership Project (3GPP) group has been involved in a study on mobile video formats for transmitting 3D video content to mobile devices. The Codec working group (SA4) is now working on mobile 3DTV use cases in order to evaluate the potential impact on the 3GPP specifications. This study aims at taking into account the subjective quality of a mobile 3DTV service in order to identify the appropriate video format and the signaling to be used. If 3DTV specifications are drafted at the end of this study, they will impact on 3GPP release 11 standards.

5.4 Conclusion

From production to the display, many standardization groups have been efficient in producing specific standards or new releases which provide dedicated guidelines to enable the implementation of an end-to-end 3DTV delivery chain. All of these groups' efforts participate in avoiding the fragmentation of technical solutions and ensure the interoperability of 3D-compatible devices available in the marketplace with the 3DTV services provided by broadcasters, network operators, and content providers.

The next challenge for 3DTV standards is to provide a better quality of experience, particularly for the 3DTV early adopters who will be the first to accommodate the actual rendering quality of the 3DTV services deployed today.

Chapter 6

Digital TV Architecture Standardization

Olivier Le Grand

Contents

6.1 Introduction 102
6.2 IPTV Standardization 102
    6.2.1 ETSI TISPAN 102
        6.2.1.1 IMS-Based IPTV 102
        6.2.1.2 NGN-Integrated IPTV 104
        6.2.1.3 Content Delivery Network Architecture 105
    6.2.2 Other SDOs Activities on IPTV 107
        6.2.2.1 Open IPTV Forum 107
        6.2.2.2 ITU-T IPTV-GSI 108
        6.2.2.3 ATIS IIF 109
6.3 3GPP Standards for the Support of Audiovisual Services 109
    6.3.1 Non-IMS-Based Architecture 110
        6.3.1.1 PSS Architecture 110
        6.3.1.2 MBMS User Service Architecture 111
    6.3.2 IMS-Based Architecture 111
6.4 Managed P2P Content Distribution 112
    6.4.1 ITU-T Distributed Services Networking 112
    6.4.2 3GPP IMS-Based P2P Content Distribution Systems 113
    6.4.3 Relevant IETF Protocol Design Activities 113
6.5 Harmonization with Web-Based Technologies 114

6.6 The Emergence of Cloud Computing... 115
Acronyms.. 115
References 117

6.1 Introduction

The scope of this chapter is to describe network architecture solutions developed or in the process of development regarding digital TV standardization, more specifically IPTV in Section 6.2, 3GPP mobile TV systems in Section 6.3 as well as managed Peer-to-Peer (P2P) solutions for which standardization started in Section 6.4.

Detailed aspects related to end-user devices and home digital aspects are not addressed here since they have been covered in Chapter 4.

6.2 IPTV Standardization

As IPTV was becoming an increasingly important emerging service in the telecommunication market, activities were started in standardization related organizations in order to develop end-to-end solutions for the delivery of IPTV services.

Given that the standardization of Next Generation Networks (NGN) was already well advanced in Standard Developing Organizations (SDOs) such as ETSI and ITU-T, the delivery of IPTV services using NGN appeared as the natural way forward leading to the standardization of the so-called "managed" IPTV architectures.

The objective of this section is to provide an overview of the IPTV end-to-end standardized solutions developed by ETSI (www.etsi.org), Open IPTV Forum (www.openiptvforum.com), ITU-T (http://www.itu.int/ITU-T), and ATIS (www.atis.org).

It has to be noted that although the IPTV standardization landscape may appear fragmented, solutions defined in these SDOs have many commonalities in terms of architectural principles though the level and depth of the produced specifications sometimes differ. In this chapter, the prime focus is made on the specifications produced by ETSI TISPAN since going at the level of stage 3 (i.e., protocol definition).

6.2.1 ETSI TISPAN

6.2.1.1 IMS-Based IPTV

ETSI TISPAN started work on IPTV standardization in NGN release 2 building upon the standards developed for NGN release 1 which were mainly addressing the support of conversational services such as Voice over IP by making use of the 3GPP IP Multimedia Subsystem (IMS) adapted for the context of fixed networks.

This led to the specification of an IMS-based solution for the support of IPTV whose architecture is shown in Figure 6.1.

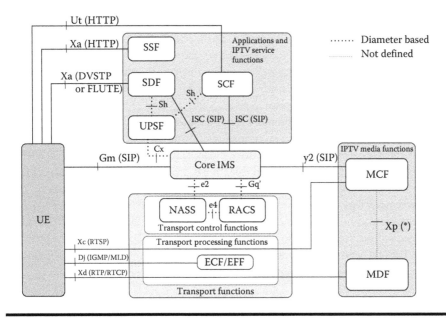

Figure 6.1 IMS-based IPTV architecture.

6.2.1.1.1 Reuse of NGN Components

The IMS-based IPTV architecture builds upon the ETSI TISPAN NGN architecture components developed for NGN Release 1 [ETSI ES 282 001], that is, the Core IMS [ETSI ES 282 007], the Network Attachment Sub-System (NASS) [ETSI ES 282 004], and the Resource Admission Control Subsystem (RACS) [ETSI ES 282 003]. These components were updated in NGN Release 2 (and Release 3 subsequently) in order to support IPTV services.

The "Core IMS" supports functions related to authentication, authorization, and signaling for the setup of the service provisioning and content delivery. The "Core IMS" routes signaling messages to the appropriate application server (AS) or triggers the applications based on settings maintained in the User Profile Server Function (UPSF), that is, the IMS user profile and possibly IPTV-specific profile data.

RACS includes functions for policy control, resource reservation, and admission control while NASS provides IP address provisioning, network level user authentication and access network configuration as defined in TISPAN. For resource reservation and admission control the "Core IMS" interacts with the RACS using the Diameter protocol [RFC 3588].

6.2.1.1.2 New Components for the Support of IPTV

The Service Discovery Function (SDF) and Service Selection Function (SSF) are functions which provide information necessary to the user equipment (UE) to select an IPTV service.

The IPTV Service Control Functions (SCF) performs service authorization during session initiation and session modification, which includes checking IPTV users' profiles in order to allow or deny access to the service. The IPTV SCF also selects the relevant IPTV media functions as well as performing credit control by interacting with the online charging functions. In practice, the SCF corresponds to a SIP Application Server (AS) performing control of services such as Content on Demand (CoD), Broadcast (BC), or Network Personal Video Recorder (N-PVR) services.

IPTV Media Functions are in charge of controlling and delivering the media flows to the UE. They are split into Media Control Functions (MCF) and Media Delivery Functions (MDF). The main roles of MCF are managing the media control flows and monitoring the control of MDFs. When receiving a request for content, the MCF selects the best MDF to deliver the content. The MDF is in charge of the delivery of the media and the storage of the media for CoD or N-PVR. The MDF also stores media up streamed or uploaded by the UE for User Generated Content (UGC) service. The IPTV Media Functions can be replaced by a Content Delivery Network in order to provide a more scalable delivery architecture (see Chapter 6.2.1.3).

6.2.1.1.3 Protocol Specifications

Based on the framework architecture and reference points described in [ETSI TS 182 027], ETSI TISPAN developed the corresponding protocol specifications in [ETSI TS 183 063].

Figure 6.1 illustrates the type of protocols used for the reference points defined in the IMS-based IPTV solution.

6.2.1.2 NGN-Integrated IPTV

As an alternative approach to IMS-based IPTV, ETSI TISPAN also developed a solution named the NGN-integrated IPTV subsystem architecture [ETSI TS 182 028]. Although also reusing NGN components such as NASS and RACS, this solution does not make use of the "IMS Core" but rather makes use of RTSP [IETF RFC 2326] instead of SIP [IETF RFC 3261] for the control of IPTV sessions. Figure 6.2 illustrates the different protocols used in the NGN-integrated IPTV solution.

6.2.1.2.1 Reuse of NGN Components

The NGN-integrated IPTV architecture reuses the Network Attachment Sub-System (NASS) [ETSI ES 282 004] and the Resource Admission Control Subsystem (RACS) [ETSI ES 282 003].

New components for the support of IPTV: In addition to the UPSF, MCF and MDF already used in the IMS based IPTV solution (cf. Figure 6.1), the following functional entities are introduced:

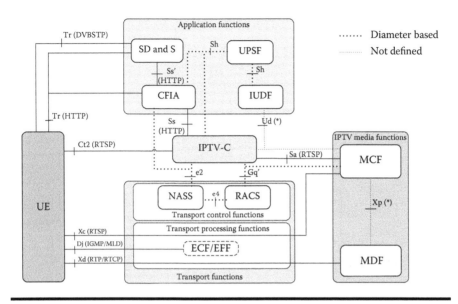

Figure 6.2 Protocols mapped to NGN-integrated IPTV functional architecture.

- The Service Discovery and Selection (SD&S) provides description information for discovery of the service as well as information required for description and selection of IPTV services.
- The Customer Facing IPTV Application (CFIA) provides IPTV service provisioning, selection, and authorization.
- The IPTV Control (IPTV-C) checks if the UE has been authorized by the Customer Facing IPTV Application to use IPTV Service Control and Delivery Functions. The IPTV-C provides selection of the Media Control Function (MCF).
- The IPTV user data (IUDF) is responsible for handling NGN-integrated IPTV user data.

[ETSI TS 183 064] provides detailed specifications of interfaces identified in Figure 6.2.

6.2.1.3 Content Delivery Network Architecture

Following the specifications of the IMS-based and NGN-integrated IPTV subsystems, ETSI TISPAN decided to develop in more detail the functions which are responsible for content delivery towards the end-user, that is, the IPTV media functions as shown in Figure 6.1. The objective was to introduce the possible use of Content Delivery Networks (CDN) in the overall architecture given that CDNs were becoming more and more popular for performing transparent and effective delivery of content to end users (see Chapters 10 and 11 on CDN).

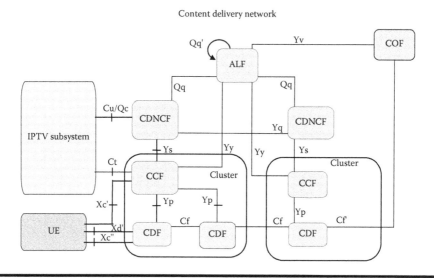

Figure 6.3   Overview of the CDN architecture.

The CDN functional architecture as specified in [ETSI TS 182 019] is shown in Figure 6.3. This functional architecture allows for deployment of CDNs in several architectural topologies, including flat and hierarchical architectures as well as hybrid topologies. Note that the CDN specified in [ETSI TS 182 019] can be utilized where appropriate for other services than IPTV.

6.2.1.3.1 CDN Functional Entities

The Content Delivery Network contains one or more Content Delivery Functions (CDF) grouped geographically or administratively in clusters. The CDFs in one cluster are controlled by a specific Cluster Controller Function (CCF). One or more CCFs are controlled by the Content Delivery Network Controller Function (CDNCF). The CDF and CCF are directly involved in delivering content to the UE. The CDNCF controls content by interacting with the IPTV subsystem. The CDNCF obtain content-related information (availability, location...) from the Asset Location Function (ALF).

The ALF contains knowledge of content available to the CDN and receives content-related data from CCFs for deployed content and from the Content Origin Function (COF) from new content. The COF provides content ingestion to the CDN and is used as an interface toward the Content Providers.

6.2.1.3.2 Relationship between CDN and IPTV Subsystems

In Figure 6.3, the IPTV subsystem may be an IMS-based or NGN Integrated subsystem as described in previous sections, the Content Delivery Network (CDN) taking the place of the IPTV Media Functions shown in Figure 6.1 for the IMS-based IPTV architecture and Figure 6.2 for the NGN-integrated IPTV architecture. In these figures, the CDNCF and the CCF take the place of the MCF while the CDF in charge of content delivery takes the place of the MDF.

There are two approaches to mapping the CDN to the IPTV Subsystem. The difference between the two approaches is that realization of the "Service Approach" is entirely through the Cu reference point to the CDNCF, while the "Query Approach" uses a separate reference point, Ct, for communicating with the CCF, in addition to the Qc (or Cu) reference point toward the CDNCF.

Regarding the reference points connecting the CDN to the IPTV subsystems, the Cu and Ct reference points perform the same tasks as the reference point y2 in the case of an IMS IPTV subsystem (cf. Figure 6.1) or the same tasks as the reference point Sa in the case of NGN-Integrated IPTV subsystem (cf. Figure 6.2). The difference is that y2 and Sa are connected to the MCF while Cu is connected to a CDNCF and Ct is connected to a CCF.

6.2.1.3.3 Future Activities

Following the CDN architecture specification, ETSI TISPAN is now in the process of developing the specification of CDN interfaces and protocol details. Furthermore, although [ETSI TS 182 019] provides informative scenarios for CDN interconnection, ETSI TISPAN is working on a new specification for the CDN interconnection architecture which should take into account the progress of the IETF in this area (cf. Chapter 10 for further information on IETF CDNI activities).

6.2.2 Other SDOs Activities on IPTV

6.2.2.1 Open IPTV Forum

The Open IPTV Forum (OIPF) (www.openiptvforum.com) provides the specifications for an end-to-end solution for the deployment of IPTV Services. The OIPF is mostly focussing on standardizing a common user-to-network interface (UNI) both for a managed and nonmanaged network (also called "Open Internet").

Managed Network IPTV Services are provided from within an operator's core network, enabling the Service Provider to make use of service enhancement facilities like multicast delivery and QoS provision.

Open Internet IPTV Services are accessed via an independently operated access network, with or without QoS guarantees. Open Internet IPTV services may be accessed via a service platform (e.g., a portal) that provides supporting facilities for multiple Service Providers.

Volume 4 ("Protocols") [OIPF_VOL4] brings together the specification of the complete set of protocols for the IPTV Solution, covering the reference point interfaces defined in the Release 2 Architecture [OIPF_ARCH2]. These reference points are classified as:

■ The User Network Interfaces (UNI), between the network or service provider domains and the consumer domain
■ The Home Network Interfaces (HNI), between the functional entities in the consumer network domain
■ The Network Provider Interfaces (NPI), between the functional entities in the network and service provider domains
■ Interfaces to external systems, for example, the DLNA home network

The NPI interfaces correspond largely to interfaces specified by ETSI TISPAN. Regarding the managed network model, two approaches, that is, based on the use of SIP (IMS-based approach) and based on the use of RTSP (non-IMS-based approach) are possible. The OIPF architecture also includes CDN functional entities, that is, CDN controller, Cluster Controller, and Content Delivery Function, making the OIPF CDN architecture compatible with the ETSI TISPAN CDN architecture defined in [ETSI TS 182 019].

6.2.2.2 ITU-T IPTV-GSI

Standardization for IPTV is also being conducted at ITU-T (International Telecommunication Union, Telecommunication Standardization Sector). ITU has Global Standards Initiatives (GSIs), which are groups intended to deliberate on discussions spanning multiple Study Groups (SGs). IPTV-GSI was established in 2008 to take over the work already done by FG IPTV (Focus Group on IPTV) in the 2006–2007 timeframe.

The ITU-T IPTV functional architecture specified in [ITU-T Rec. Y.1910] identifies three IPTV architecture approaches that enable service providers to deliver IPTV services:

■ "Non-NGN IPTV functional architecture" (Non-NGN IPTV) which is based on existing network components and protocols/interfaces which are not NGN based.
■ "NGN IMS-based IPTV functional architecture" (NGN-IMS-IPTV): The NGN-IMS based IPTV architecture utilizes components of the NGN architecture as identified in [ITU-T Y.2012] including the core IMS component. It is comparable to the ETSI TISPAN IMS-based IPTV architecture.
■ "NGN-based non-IMS IPTV functional architecture" (NGN-Non-IMS IPTV): The NGN Non-IMS IPTV architecture is also using components of the NGN architecture but without making use of the core IMS component. The "core IMS" functions are replaced by the IPTV service control functional block which plays a similar role as the control functions in the case of the ETSI TISPAN NGN-integrated IPTV architecture (cf. Figure 6.2).

The NGN-based IPTV architecture uses network attachment control functions (NACF) defined in [ITU-T Y.2014] to provide functions such as authentication and IP configuration. NACF is the ITU-T NGN equivalent of ETSI NASS [ES 282 004].

The NGN-based IPTV architecture uses resource and admission control functions (RACF) defined in [ITU-T Y.2111] to provide resource and admission control functions. RACF is the ITU-T NGN equivalent of ETSI RACS [ES 282 003].

ITU-T also specified in more detail the functional architecture of content delivery functions in NGN and the related procedures. Built upon the IPTV functional architecture described in [ITU-T Y.1910], [ITU-T Y.2019] provides an overview of content delivery and distribution functions, requirements for content delivery and distribution, a hierarchical architecture of content delivery and distribution functions including the definition of related reference points, and associated procedures.

Till now, no detailed stage 3 specification defining interfaces and protocols has been produced by ITU-T. However, ITU-T is involved in the development of IPTV Recommendations related to terminal, home network, middleware, QoS/QoE, and security aspects.

6.2.2.3 ATIS IIF

The North American organization called the Alliance for Telecommunications Industry Solutions (ATIS), through the IPTV Interoperability Forum (IIF) http://www.atis.org/iif/index.asp, is developing an industry's end-to-end solution for IPTV—a suite of globally acceptable standards and specifications that drive delivery of IPTV from the core of the network, to the end-user device. The IIF is working on the definition of necessary standards and specifications for IPTV network architecture, QoS and QoE, Security, and Interoperability. ATIS IIF and ITU-T IPTV-GSI are closely working together by cooperating on issues such as requirements and functional overviews.

ATIS specification ("IPTV high level architecture") [ATIS-0800007] provides a high-level architectural framework for end-to-end systems' implementation and interoperability for the supporting network design. Two "managed network" IPTV architectural approaches relying on the ITU-T NGN framework are provided, that is, a "Core IMS" approach and a non-IMS approach.

ATIS specification ("Content on demand service") [ATIS-0800042] further describes the use of the relevant functions identified in [ATIS-0800007] for delivery of an IPTV Content on Demand (CoD) Service.

6.3 3GPP Standards for the Support of Audiovisual Services

The Third-Generation Partnership Project (3GPP), http://www.3gpp.org/index.php, is developing standards for the support of audiovisual services in mobile systems called "Transparent end-to-end packet-switched streaming service" (PSS). PSS allows unicast streaming of content from a server called PSS server to a PSS client located in the mobile UE. In addition to PSS, 3GPP developed the Multimedia Broadcast/Multicast Service (MBMS) which is a unidirectional point to multipoint bearer service in which data is transmitted from a single-source entity to multiple recipients. MBMS is therefore a useful capability that can decrease the amount of data within the network and use resources more efficiently for the support of audiovisual services.

Similar to other SDOs such as ETIS TISPAN, 3GPP has defined two versions of PSS, that is, a non-IMS-based PSS and MBMS user service architecture as well as an IMS-based PSS and MBMS user service architecture.

6.3.1 Non-IMS-Based Architecture

Figure 6.4 illustrates the non-IMS-based PSS and MBMS user service architecture. This architecture was the first one to be defined by 3GPP in Release 4.

The mobile network which covers both the Packet Switched (PS) core network and the radio access network (RAN) enables the mobility and provides IP connectivity over unicast, multicast broadcast bearers between the servers (PSS server and BM-SC) and the clients (PSS client and MBMS client).

6.3.1.1 PSS Architecture

The PSS client and the MBMS client located in the UE perform service selection and initiation, receive and present the content to the user. The PSS client can discover the PSS services via multiple means like, for example, browsing. The PSS client interfaces to the PSS server transparently through the mobile network.

The PSS server performs control and streaming delivery functions on a unicast access type.

PSS allows for conventional streaming to be used, that is, the session control protocol used between the PSS client and PSS server is RTSP while the transport protocol is RTP, cf. [ETSI TS 126 234] for further details. In addition, Progressive download over HTTP as well as Dynamic Adaptive Streaming over HTTP (DASH) can be used as alternatives methods to conventional streaming [ETSI TS 126 247].

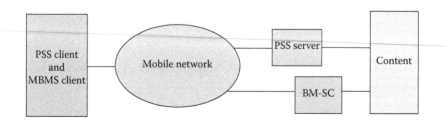

Figure 6.4   Non-IMS-based PSS and MBMS user service architecture.

6.3.1.2 MBMS User Service Architecture

The MBMS User Service architecture is based on an MBMS client on the UE side and a Broadcast/Multicast Service Centre (BM-SC) on the network side. The BM-SC performs control and streaming/download delivery functions in a hybrid unicast/multicast/broadcast access type.

The MBMS client interfaces to the BM-SC via layer 3 protocols defined between the UE and the GGSN (located in the mobile network) and the GGSN with the BM-SC (Gmb). The use of the Gmb and Gi interfaces in providing IP multicast traffic and managing MBMS bearer sessions is described in detail in [ETSI TS 123 246].

6.3.2 IMS-Based Architecture

Figure 6.5 describes the IMS-based PSS functional architecture. In addition to PSS functions, the IMS core and various functions are added. For simplicity, the MBMS user service-related functions are omitted.

The architecture shown in Figure 6.5 is similar to the ETSI TISPAN IMS-based IPTV architecture (cf. clause 6.2.1.1) where SDF, SSF, SCF, HSS, and IMS Core functions are also used. However, the following new functions are introduced:

■ The PSS Adapter performs protocol translation between SIP and RTSP to offer control of PSS servers. It proxies RTSP messages received from the UE and implements SIP/RTSP translation toward the PSS server.

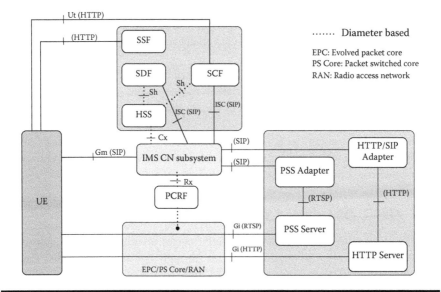

Figure 6.5   IMS-based PSS architecture.

- The HTTP/SIP Adapter provides for correlation of SIP sessions with HTTP incoming requests.
- The PSS server functionally contains media control and media delivery functions similar to ETSI TISPAN MCF and MDF.
- The HTTP server delivers content to the UE using progressive download over HTTP or Dynamic Adaptive Streaming over HTTP (DASH).
- The Policy and Charging Rules Function (PCRF) controls the charging and the establishment of resources in the mobile network, that is, in the RAN and PS core network.

Refer to [ETSI TS 126 237] for further details on IMS protocols and procedures for the control of PSS- and MBMS-based streaming and download applications.

6.4 Managed P2P Content Distribution

P2P refers to the concept that in a network of equals (peers) using appropriate information and communication systems, two or more participants are able to collaborate to make a portion of their resources (e.g., network bandwidth, storage capacity, computing power) available to the other P2P network participants without necessarily requiring central coordination. P2P has led to the implementation of numerous popular applications for internet users, for example, file sharing P2P applications, live media streaming, real-time communications.

In traditional telecommunication networks, service nodes are organized in a centralized manner which leads to some drawbacks, for example, load imbalance between different service nodes and single point of failure. It is also a problem for operators to extend capacities of service nodes continuously and in a timely fashion.

Many of these problems can be resolved or eased by using distributed approaches allowing load distribution between different service nodes, reallocation of service node's responsibility in case of failure and greater flexibility for system capacities by exploiting physical resources.

Distributed technology such as P2P may be used in managed networks to create an overlay network which has an ability to share resources between different nodes and links within a managed network deployment.

6.4.1 ITU-T Distributed Services Networking

ITU-T SG13 is currently developing architectures for distributed services networking (DSN) which aims to support services through the adoption of distributed technologies (e.g., P2P) in order to achieve a more robust system at a relatively low cost. In particular, DSN also aims at providing content-based services via distributed mechanisms (e.g., linear streaming, video on demand, content sharing).

In order to share content in a scalable and reliable manner, DSN provides an overlay network built on top of an underlying IP-based infrastructure (e.g., a NGN). This overlay network is constituted of nodes deployed in a given service provider's domain and nodes deployed in end-user domains. Nodes in the DSN overlay network can play different roles such as holding content, caching content, forwarding content to other nodes, controlling services, and their transport. Using the DSN overlay network, a given user equipment (UE) can find one or more nodes which hold or cache the content that the UE is looking for. Then by establishing communications with these nodes, the UE can then receive the corresponding content.

ITU-T is currently developing a new Recommendation [ITU-T Y.dsnarch] that will provide the functional architecture of DSN including the description of functional entities and reference points. In addition, Recommendation [ITU-T Y.dsncdf] will provide further details regarding the support of content delivery services. DSN protocol specification work may also be started once the DSN architecture is becoming mature enough.

6.4.2 3GPP IMS-Based P2P Content Distribution Systems

3GPP has initiated studies in Release 11 on the support of IMS-based P2P content distribution systems. The objectives are very similar to the ITU-T DSN ones since the intent is to use P2P technology between UEs and servers at the edge of the network. [3GPP TR 22.906] provides the overview, use cases, and other aspects (e.g., Mobility, Charging, Security, etc.) of IMS-based P2P Content Distribution Services (CDS). 3GPP is now studying the related architectural aspects in order to propose solutions in order to fulfill the use cases and requirements as defined in TR 22.906 while avoiding the duplication of work performed in other SDOs, in particular IETF where related protocol solutions are under development.

6.4.3 Relevant IETF Protocol Design Activities

Several IETF (www.ietf.org) working groups (WGs) are in the process of developing protocol solutions that may fulfill the architectural requirements of both ITU-T DSN and 3GPP IMS-based P2P CDS solutions.

The Peer-to-Peer Streaming Protocol (PPSP) Working Group is specifying signaling and control protocols for a P2P streaming system for transmitting live and time-shifted media content with near real-time delivery requirements. The goal of PPSP is to serve as an enabling technology, building on the development experiences of existing P2P streaming systems. Its design will allow it to integrate with IETF protocols on distributed resource location, traffic localization, and streaming control, and data transfer mechanisms for building a complete streaming system or a streaming delivery infrastructure.

The PPSP WG has identified in [IETF PPSP] two types of basic entities in P2P streaming, that is, the Tracker and the Peer. A Tracker is seen as a directory server which maintains a list of peers storing chunks for a specific channel or streaming file and answers queries from Peers for peer lists, while a Peer refers to a participant in a P2P streaming system that not only receives streaming content, but also stores and uploads streaming content to other participants.

The PPSP WG is designing a protocol for signaling between Trackers and Peers (the PPSP "tracker protocol") and a signaling protocol for communication among the peers (the PPSP "peer protocol").

The Application Layer Traffic Optimization (ALTO) Working Group is specifying a service that will provide applications with abstracted network information (e.g., operator's policies, the topological distance between two peers, the transmission costs of data to/from a peer...). With ALTO, applications such as P2P and CDNs have the potential to become more network-efficient (e.g., reduce network resource consumption) and achieve better application performance (e.g., accelerated download rate). The kind of information that is meaningful to convey to applications via an ALTO service is any information that applications cannot easily obtain themselves and that changes on a much longer timescale than the instantaneous information used for congestion control on the transport layer.

The ALTO WG is specifying a client–server protocol that could also be introduced in ITU-T DSN and 3GPP IMS-based P2P systems. An ALTO Client could be introduced in Peers residing in the UEs acting as User Peers and in network entities (Caching Nodes or Servers) acting as Network Peers. The ALTO protocol [IETF ALTO] would allow the ALTO clients to interrogate an ALTO server for a network map (e.g., list of peer entities) and/or a cost map (e.g., ranking available peers for a requested content in a specific order according to cost metrics) in order to receive back an optimal set of peers for the service requested.

6.5 Harmonization with Web-Based Technologies

This chapter has given a panorama of already standardized managed network solutions for IPTV in ETSI, Mobile TV in 3GPP as well as on-going standardization of managed P2P approaches for content delivery in 3GPP and ITU-T. Part of these solutions already relies on the use of Web-based technologies and HTTP [IETF RFC 2616], for example, for service navigation or adaptive streaming for content delivery.

In parallel to the standardization of these solutions, the World Wide Web Consortium (W3C) created an internal group called "Web and TV Interest Group" for discussion on smarter integration of Web and TV, with the aim to identify requirements and potential solutions to ensure that the Web will operate well with TV. Activities include in particular requirements placed on the [W3C HTML5] video, audio, and media interfaces by media formats that will be used for Web and TV as well as proposed APIs that meet these requirements.

It is expected that the Web and TV Interest Group will coordinate and provide inputs to the W3C activities on Web Real-time Communications whose objective is to define client-side APIs to enable Real-Time Communications in Web browsers. The intent of these APIs is to enable building applications that can be executed inside a browser and that support real-time communications using audio, video and any additional media. Extensions to HTML5, for example, the <audio> and <video> elements may also be proposed as a result. Cooperation with the IETF Real-Time Communication in Web browsers (RTCWeb) Working Group will be ensured in order that application developers making use of the W3C APIs can control the components or the architecture for the selection and profiling of the IETF-specified protocols.

In the end, efforts in both W3C and IETF should facilitate the move toward harmonization of TV and Web.

6.6 The Emergence of Cloud Computing

With the significant advances in Information and Communications Technology (ICT) over the last few decades, computing is evolving toward a model consisting of services that are commoditized and delivered in a standard manner. In such a model, users access services based on their requirements without regard to where the services are hosted or how they are delivered. Several computing paradigms have promised to deliver this computing vision, of which the latest one is known as Cloud Computing. Cloud Computing is a model for enabling ubiquitous, convenient, on-demand network access to a shared pool of configurable computing resources (e.g., networks, servers, storage, applications, and services) that can be rapidly provisioned and released with minimal management effort or service provider interaction.

Using cloud computing platforms for offering IPTV services will provide new opportunities to service providers leading to the definition of cloud-based IPTV architectures that will possibly be addressed by standardization bodies and could require enhancements to emerging cloud-based standards.

Acronyms

3GPP	Third-Generation Partnership Project
ALF	Asset Location Function
ALTO	Application Layer Traffic Optimization
API	Application Programming Interface
AS	Application Server
ATIS	Alliance for Telecommunications Industry Solutions
BC	Broadcast
BM-SC	Broadcast Multicast Service Centre
CCF	Cluster Controller Function

CDF	Content Delivery Function
CDN	Content Delivery Network
CDNCF	CDN Controller Function
CDS	Content Distribution System
CFIA	Customer Facing IPTV Applications
CN	Core Network
CoD	Content on Demand
COF	Content Origin Function
DASH	Dynamic Adaptive Streaming over HTTP
DLNA	Digital Living Network Alliance
DSN	Distributed Services Networking
DVB	Digital Video Broadcasting Project
DVBSTP	Digital Video Broadcasting Service discovery and Selection Transport Protocol, see [DVB TS 102 034]
ECF	Elementary Control Function
EFF	Elementary Forwarding Function
EPC	Evolved Packet Core
ETSI	European Telecommunications Standards Institute
FLUTE	File Delivery over Unidirectional Transport, see [IETF RFC 3926]
GGSN	Gateway GPRS Support Node
GSI	Global Standard Initiative
HTML	Hypertext Markup Language
HTTP	Hyper Text Transfer Protocol
IETF	Internet Engineering Task Force
IGMP	Internet Group Management Protocol, see [IETF RFC 3376]
IIF	IPTV Interoperability Forum
IMS	IP Multimedia Subsystem
IPTV	Internet Protocol TV
IPTV-C	IPTV Control
ITU	International Telecommunication Union
IUDF	IPTV User Data Function
MBMS	Multimedia Broadcast Multicast Service
MCF	Media Control Functions
MDF	Media Delivery Functions
MLD	Multicast Listener Discovery, see [IETF RFC 3810]
NACF	Network Attachment Control Functions
NASS	Network Attachment Subsystem
NGN	Next-Generation Network
N-PVR	Network Personal Video Recorder
OIPF	Open IPTV Forum
P2P	Peer-to-Peer
PCRF	Policy and Charging Rules Function
PPSP	Peer-to-Peer Streaming Protocol

PS	Packet Switched
PSS	Packet-switched Streaming Service
QoE	Quality of Experience
QoS	Quality of Service
RACF	Resource and Admission Control Functions
RACS	Resource and Admission Control Subsystem
RAN	Radio Access Network
RTCP	Real-time Transport Control Protocol, see [IETF RFC 3550]
RTCWeb	Real-Time Communication in Web-browsers
RTP	Real-time Transport Protocol, see [IETF RFC 3550]
RTSP	Real-Time Streaming Protocol
SCF	Service Control Functions
SCP	Service and Content Protection
SDF	Service Discovery Function
SD&S	Service Discovery and Selection
SIP	Session Initiation Protocol
SSF	Service Selection Function
TISPAN	Telecoms & Internet converged Services & Protocols for Advanced Networks
UE	User Equipment
UPSF	User Profile Server Function
W3C	World Wide Web Consortium

References

[3GPP TR 22.906]3GPP TR 22.906 v11.0.0, (2011-06); *Study on IMS Based Peer-to-Peer Content Distribution Services* (**Release 11**).

[ATIS-0800007]ATIS-0800007 (March 2007), *IPTV High Level Architecture*.

[ATIS-0800042]ATIS-0800042 (December 2010), *IPTV Content on Demand Service*.

[DVB TS 102 034]ETSI TS 102 034 (V1.4.1), Digital Video Broadcasting (DVB); *Transport of MPEG-2 TS Based DVB Services over IP Based Networks*.

[ETSI TS 123 246]ETSI TS 123 246 v10.1.0 (2011-06); *Multimedia Broadcast/Multicast Service (MBMS); Architecture and Functional Description* (3GPP TS 23.246 version v10.1.0 **Release 10**).

[ETSI TS 126 234]ETSI TS 126 234 v10.1.0 (2011-06); *Transparent End-to-End Packet-Switched Streaming Service (PSS); Protocols and Codecs* (3GPP TS 26.234 v10.1.0 **Release 10**).

[ETSI TS 126 237]ETSI TS 126 237 v10.2.0 (2011-06); *IP Multimedia Subsystem (IMS) Based Packet Switch Streaming (PSS) and Multimedia Broadcast/Multicast Service (MBMS) User Service; Protocols* (3GPP TS 26.237 v10.2.0 **Release 10**).

[ETSI TS 126 247]ETSI TS 126 247 v10.0.0 (2011-06); *Transparent End-to-End Packet-Switched Streaming Service (PSS); Progressive Download and Dynamic Adaptive Streaming over HTTP* (3GP-DASH) (3GPP TS 26.247 v10.0.0 Release 10).

[ETSI ES 282 001]ETSI ES 282 001 V3.4.1 (2009-09), *TISPAN, NGN Functional Architecture.*

[ETSI ES 282 003]ETSI ES 282 003 V3.5.1 (2011-04), *TISPAN, Resource and Admission Control Sub-System (RACS); Functional Architecture.*

[ETSI ES 282 004]ETSI ES 282 004 V3.4.1 (2010-03), *TISPAN, NGN Functional Architecture; Network Attachment Sub-System (NASS).*

[ETSI ES 282 007]ETSI ES 282 007 V2.1.1 (2008-11), *TISPAN; IP Multimedia Subsystem (IMS); Functional Architecture.*

[ETSI TS 182 019]ETSI TS 182 019 V3.1.1 (2011-06), *TISPAN, Content Delivery Network (CDN) Architecture.*

[ETSI TS 182 027]ETSI TS 182 027 V3.5.1 (2011-03), *TISPAN, IPTV Architecture; IPTV Functions Supported by the IMS Subsystem.*

[ETSI TS 182 028]ETSI TS 182 028 V3.5.1 (2011-02), *TISPAN, NGN Integrated IPTV Subsystem Architecture.*

[ETSI TS 183 063]ETSI TS 183 063 V3.5.2 (2011-03), *TISPAN; IMS-Based IPTV Stage 3 Specification.*

[ETSI TS 183 064]ETSI TS 183 064 V3.4.1 (2011-02), *TISPAN; NGN Integrated IPTV Subsystem Stage 3 Specification.*

[IETF ALTO]draft-ietf-alto-protocol-10, *ALTO Protocol,* October 2011.

[IETF RFC 3588]IETF RFC 3588, *Diameter Base Protocol.*

[IETF RFC 3926]IETF RFC 3926, *FLUTE—File Delivery over Unidirectional Transport.*

[IETF RFC 2616]IETF RFC 2616, *Hypertext Transfer Protocol—HTTP/1.1.*

[IETF RFC 3376]IETF RFC 3376, *Internet Group Management Protocol, Version 3.*

[IETF RFC 3810]IETF RFC 3810, *Multicast Listener Discovery Version 2* (MLDv2) for Ipv6.

[IETF PPSP]draft-ietf-ppsp-reqs-05, *P2P Streaming Protocol (PPSP) Requirements,* October 2011.

[IETF RFC 3550]IETF RFC 3550, *RTP: A Transport Protocol for Real-Time Applications.*

[IETF RFC 2326]IETF RFC 2326, *Real Time Streaming Protocol (RTSP).*

[IETF RFC 3261]IETF RFC 3261, *SIP: Session Initiation Protocol.*

[ITU-T Y.1910]Recommendation ITU-T Y.1910 (09/2008), *IPTV Functional Architecture.*

[ITU-T Y.2012]Recommendation ITU-T Y.2012 (04/2010), *Functional Requirements and Architecture of the NGN.*

[ITU-T Y.2014]Recommendation ITU-T Y.2014 (03/2010), *Network Attachment Control Functions in Next Generation Networks.*

[ITU-T Y.2019]Recommendation ITU-T Y.2019 (09/2010), *Content Delivery Functional Architecture in NGN.*

[ITU-T Y.2111]Recommendation ITU-T Y.2111 (11/2011), *Resource and Admission Control Functions in Next Generation Networks.*

[ITU-T Y.dsnarch]Draft Recommendation ITU-T Y.dsnarch, *Architecture of DSN,* TD 161 (WP5/13), ITU-T Study Group 13, Geneva, October 2011.

[ITU-T Y.dsncdf]Draft Recommendation ITU-T Y.dsncdf, *DSN Content Distribution Functions,* TD 164 (WP5/13), ITU-T Study Group 13, Geneva, October 2011.

[OIPF_ARCH2]OIPF *Functional Architecture [V2.1]* – [2011-03-15] http://www.openiptv forum.org/Release_2.html

[OIPF_VOL4]*OIPF Release 2 Specification,* Volume 4—*Protocols [V2.1]*—[2011-06-21] http://www.openiptvforum.org/Release_2.html

[W3C HTML5]*HTML5,* http://www.w3.org/TR/html5/

MEDIA CONTENT DELIVERY AND QUALITY OF EXPERIENCE

Chapter 7

Collaboration between Networks and Applications in the Future Internet

Selim Ellouze, Bertrand Mathieu, Toufik Ahmed, and Nico Schwan

Contents

7.1 Introduction on Future Internet Media Applications 122
7.2 Collaboration Concept in the Future Internet...123
7.3 Existing Collaboration Techniques 126
    7.3.1 The ALTO System 126
        7.3.1.1 Presentation 126
        7.3.1.2 Architecture and Protocol...127
        7.3.1.3 Discovery Protocol...128
        7.3.1.4 ALTO Protocol..129
    7.3.2 The ENVISION System 130
        7.3.2.1 Presentation 130
        7.3.2.2 High-Level Architecture 131
        7.3.2.3 CINA Interface..134
7.4 Conclusion 138
References 140

7.1 Introduction on Future Internet Media Applications

The Internet is witnessing a massive growth for supporting real-time applications such as streaming of audio and video services. It is moving from being a simple monolithic data service network to a pervasive, ubiquitous, and multiservice network in which different stakeholders may dynamically interact for offering value-added services to end users. Nowadays, the domestication of media and technology has allowed the users to take up different roles in addition to being consumers. This concept represents a shift away from models where users are only passive consuming services online to a model where they have taken active roles in many parts of the service lifetime. User Generated Content (UGC) is changing radically the way users consume and interact with digital content, including content sharing, streaming audio and video, augmented reality, massive multiplayer games, virtual 3D worlds, and so on. Besides this, users are taking on different active roles in content distribution such as in Peer-to-Peer (P2P) systems and in social networks. They can create, distribute, and consume digital content. But they can also actively rate, comment and tag content, communicate on its quality, remix part of it, provide subtitle, lyrics, and more. This concept sheds light upon creating new viewing patterns, media experience, social interactions, and community aspects. It allows empowering users to be more creative, and developing new business opportunities to all stakeholders involved in the digital media ecosystem.

The content coming from outside of the operators' network, namely Over the Top (OTT), and content generated by the user is dominating the Internet traffic [1]. More growth is expected in the future due to technological advancements of convenient and portable media-capturing devices, and recent advances in media coding standards (MPEG-4, H.264 AVC, SVC, MVC, DVC, 2D/3D media, etc.). Actually about 70% of Internet traffic is video. In 5–7 years, it is predicted that most of the video content (for handheld device, PC and TV Set-top box) will be delivered from the Internet and in High Definition (HD) quality.

However, the problem of how to provide end-to-end quality of service for these future Internet multimedia applications has not been solved satisfactorily to-date due to the lack of cooperation between the involved stakeholders in the delivery chain.

On the one hand, delivering high-quality services is inevitably calling for the invocation of network services such as multicasting, caching, bandwidth reservation, fallback to lower bitrates by transcoding, quality adaptation, identity authentication, authorization, and geo-location among others. These network services, which are today hidden in the walled gardens, need to be made available to other stakeholders to bridge the gap between them and to support efficient implementation and integration of the different needs and requirements for each one in a cooperative manner. Various initiatives and development approaches have been investigated with the deployment of network agnostic services; however, these approaches struggle with performing with high quality of service as they cannot benefit from accessing the above-mentioned capabilities.

The significant challenge in integrating these network services arises when attempting to blend them into user applications and seamlessly and dynamically invoke various functionalities without violating any privacy rules mandated by their owners, or exposing those services to security threats.

On the other hand, there is a continuous battle between network providers and users' applications generating or consuming multimedia content. The former intend to restrict and control application traffic passing through their networks whereas the latter try to evade from being captured by implementing sophisticated reverse engineering mechanisms. In such network oblivious scenario, the user application may generate content which passes through network links that may be costly or may be unable to support the requested quality. This results in the degradation of the delivered quality of service. This situation is stimulating for the network providers, who are missing the opportunities to earn revenues from the external peering and internal traffic passing through their physical networks but it becomes frustrating when network providers have to deploy more resources to offer their own services. Thus, it is important to bring these entities toward cooperation to cease this no-winner situation by expanding and enhancing the cooperation and interaction between user's application and providers.

7.2 Collaboration Concept in the Future Internet

High-definition, highly interactive networked media applications pose major challenges to network operators. Multisourced content means higher quantities of unpredictable data throughout the network, putting additional pressure at the network edge for unprecedented upload capacity in access networks. If the entire burden of supporting high volumes of HD/3D multimedia streams is pushed to the ISPs with highly concurrent unicast flows this would require operators to upgrade the capacity of their infrastructure by several orders of magnitude. Rather than simply throwing bandwidth at the problem, a new trend is to develop intelligent cross-layer techniques via collaboration between the applications and the network operators. This can be done by optimizing application overlay networks to make best use of the capabilities of the underlying networks and the participant end users; by providing the means by which applications can inform the network operator about their usage; by providing the means by which service providers can access and mobilize specialized network resources to achieve efficient distribution of highly demanding content streams; and by enabling dynamic adaptation of the content to meet the abilities of the underlying networks and to satisfy user preferences.

The P4P/ALTO initiative [2,3] investigated how overlay networks and ISPs can cooperate to optimize traffic being generated by P2P applications and transported over the ISP's infrastructure. The ISP is able to indicate a preference for which peers should exchange data to avoid over-utilization of its network or the unnecessary loading of high-cost resources such as interprovider links. The P2P network benefits by avoiding congestion in parts of the network, resulting in higher average throughput. Several large-scale trials have been performed (in the United States with the Verizon and Comcast network, in South America with Telefonica, in China with China Telecom) and all proved that having such a collaboration reduces the network load in the ISP network, while improving the download time for end users [4,5].

However, the current ALTO specification is limited in terms of exchanged information. The ISP can, for instance, provide an abstracted map illustrating the network topology. But in ALTO, this topology can just be done with costs (as it costs for the ISP) but it might be valuable to extend it with specific metrics that are of interest for the applications; for example, topology based on the latency metric for a delay-sensitive application or topology based on available bandwidth for a traffic-sensitive application, and so on.

Furthermore, the overlay–ISP interaction in ALTO only concerns network information, provided by ISPs, processed by applications. The ISP is blindfolded to the services that their customers subscribe to. To enable the delivery of high-volume content (such as future services like HD or 3D might provide) in the best possible quality, ISPs have to make application-aware network management. It started some years ago, with the introduction of HTTP proxies in order to redirect and deliver the content from local proxies instead of remote Web servers. More recently, a new kind of network equipment appeared which aims at enabling ISPs to limit traffic transiting their networks while still offering services with good QoE. Such equipment includes P2P caches, HTTP caches, or even CDN (Content Delivery Networks) nodes. It can be used at different locations of an ISP's network (aggregation network, peering points, etc.) depending on the ISP objectives and where it wants to reduce the traffic. This hardware is beneficial for the ISPs as well as end users but is still fixed and dedicated to a given role and only concerns the ISP, without interaction with the applications.

Collaborating applications, giving more information about their needs and network usage may give the opportunity to the ISP to allocate network resources more efficiently (e.g., reserve some resources in anticipation or route traffic differently, etc.). For example, the ISP can request to retrieve information from the overlay application: (1) like the application characteristics, such as the encoding bitrate, encoding type, the application traffic requirements in terms of traffic, bandwidth, or delay; (2) about the users of the application, such as the estimated number of end users (swarm popularity), the users location, the list of clients, the list of sources; (3) about the use of specialized nodes in the overlay such as caching nodes, content adaptation nodes, and so on. This could help to better reserve resources or route data.

We can also expect that the collaboration goes further and that ISPs open their network so that applications can dynamically invoke some network services, currently only available for ISPs walled-garden services. Typically, it can provide multicast service, provide bandwidth on demand for some access network links, provide caching on demand, provide storage for media providers, and so on. The list of network services the ISP offers could be retrieved by the application in the first steps of their communication.

With an intelligent collaboration, the ISP could also request from the overlay to start or stop using network services.

For example, in case of huge traffic generated by one application, one possibility is to dynamically set up a multicast tree (at the necessary scale, might be in a local region or wider) when the number of users in the region exceeds a given threshold. In this case, the overlay application might request the ISP to establish a multicast tree from a given source and toward given destination endpoints. This service will allow improving the QoE delivered to end users as the number of clients grows and also to reduce the network load, removing many unicast streams for a multicast delivery. On the contrary, in case a multicast tree is set up by an application but with a limited number of users, ISP could request this application to stop the multicast delivery in case another application with a higher priority requires it.

Caching content is another natural option for saving bandwidth in the network, and delivering content more rapidly to end users by making the content closer to actual user location. The ISP could then deploy some caches in its network. Typically, there are currently two main approaches for network caching: *transparent caching*, where the ISP deploys caching nodes in its network that will cache content and deliver them in a transparent way to end users. In this sense, transparent means that both end users and content providers are not aware of the presence of caching nodes. Opposite to this approach, the *explicit caching* technique is different since, in this case, the caching nodes are clearly known by end users and the latter connect to the caching node to get the content. The ISP can then propose this caching function as a network service to applications.

In yet another collaboration scenario where different end users are interested in the same multimedia content but with different preferences and terminals capabilities (e.g., PC, TV, notebook, PDA, cellular phone, etc.) connected through different heterogeneous networks environments, optimal (network) resource utilization is achieved by transmitting the multimedia content to an intermediate node, that is, an adaptation proxy or gateway, between the provider and the actual end users such that the offered service satisfies a set of usage environment constraints common to all end users. On receipt of the multimedia content, the proxy/gateway, that can be managed by an ISP, adapts and forwards the multimedia content satisfying the individual usage environment constraints of each end user.

The possibility of collaboration between applications and network providers opens a plethora of business and technical possibilities. Applications can make use of these services to drastically improve end-user quality of experience (QoE), to improve resilience and security, or to implement different business models potentially sharing revenue with network providers. For instance, one can offer a function which could manage the delivery of content over different links (in case the device is connected to multiple access networks) and thus share the data delivery over the links according to what the networks can do and the expected QoS. Other known network services that could be offered by ISP are about prioritization of traffic or resource reservation, in case of congestion in the network or to ensure a required QoS. Another one could be about content-aware policy and security issues. Indeed, the last decade has seen a dramatic increase in attacks (capability exploitation, denial-of-service, etc.) that is putting increased pressure on several Internet businesses. Mobile devices are particularly vulnerable since low-bandwidth attacks can deny service easily to any application running on them. Such an interface between the applications and the ISPs can be used to actively stop flows being transmitted to the device or, in an extreme scenario, switch to default-off mode of operation where only authorized senders can communicate with the device. Network resilience is another service that network providers can easily make available to applications. If certain applications (e.g., remote medical surgery) need significantly higher levels of resilience they can request this through the collaboration interface.

We can then see that having such collaboration between applications and ISPs could improve the application QoE while optimizing the network usage. Obviously, this collaboration should be established in order to decide which information to be exchanged, which network services the application can access, for how long, which business model, for how many end users, and so on. Security, authentication, trust are also some issues that should be dealt with for such collaboration.

7.3 Existing Collaboration Techniques

7.3.1 The ALTO System

7.3.1.1 Presentation

Today's P2P applications are mostly unaware of the underlying network topology. Although some of them try to reengineer the underlying network topology by end-to-end measurements of various metrics, such as delay, a detailed and exact knowledge is not obtainable. This leads to disadvantageous situations, for example, scheduling algorithms choose to download content chunks from a source peer far away, when in fact the content chunk would be downloadable from a peer locally close-by. Due to this behavior, ISPs suffer from a vast amount of costly inter-ISP domain traffic generated by these kinds of applications. The protocol developed by the IETF working group "Application Layer Traffic Optimization" (ALTO) allows P2P applications to obtain information regarding the network topology and delivery costs, bridging the gap between overlays and the underlying network infrastructure. Through the ALTO protocol, ISPs are able to convey information helping P2P applications making better than random choices when creating their neighborhood overlay topology. This collaboration of layers benefits both sides. Studies [6] show that by using the ALTO protocol applications provide an improved QoE to its users due to the fact that, for example, the overall download time can be reduced. For ISPs on the other hand, the localization of content downloads leads to less traffic on the costly interdomain links.

Also "Content Delivery Networks" (CDNs) are expected to benefit from the information that is provided through the ALTO protocol. Typically, CDNs consist today of several servers attached to networks of different ISP's. Requests of users for a particular content are redirected to a server which is "closest" to the user, whereas closest typically means the geographical or network distance. By doing this, the content origin server and the backbone links are unburdened, and the user receives a higher QoE due to the lower latency and higher throughput of the content download. Due to the rapidly increasing volume of video and multimedia traffic ISP's strategically invest into their own CDN's, placing their cache nodes at locations that are even closer to the end user. This trend requires that a more detailed knowledge about the underlying network is provisioned to the entity that redirects content requests. Naturally, one promising option to obtain this kind of information is proposed in ALTO, thus using it for CDNs [7], in particular for federated CDNs across network domain borders, is envisaged as well.

The remainder of this section will detail the architecture and protocol specifics of the current ALTO specification.

7.3.1.2 Architecture and Protocol

The IETF ALTO working group is chartered to define two protocols: One for the Discovery of the ALTO Service and another one for the information exchange between the ALTO Service and applications. The discovery process [8] utilizes a combination of deployed mechanisms to allow entities of overlay applications to find the contact point of the ALTO server which is responsible for them. The current ALTO protocol specification [9] is still under development by the working group and has not reached RFC status yet. In its current version 8 [9], the Internet draft employs a RESTful interface over HTTP and uses JSON encoding for message bodies. This allows the ALTO protocol to benefit from the already installed base of HTTP infrastructure and the fact that many overlay applications already have an HTTP client embedded in their software. Same for HTTP, each ALTO transaction consists of one request and one response.

7.3.1.2.1 Architecture

Figure 7.1 illustrates the basic ALTO architecture. The ALTO server gathers information from different optional ISP internal and external information sources and creates its own view of the network domain. More specifically the ALTO server defines network regions and costs between those regions, whereas a network region can be an endpoint, a set of endpoints, an Autonomous System, or an ISP. How costs between those regions are calculated is outside of the scope of the ALTO

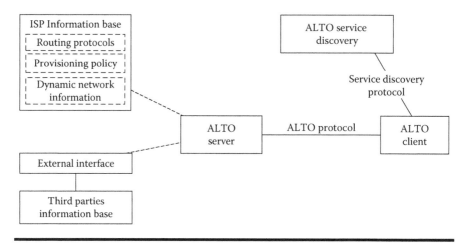

Figure 7.1   Basic ALTO architecture.

specification, typical examples of information that is used include static network configuration databases, routing protocols, provisioning policies or interfaces to external parties, for example, content providers. The ALTO protocol is then used by ALTO clients to query the view of the ALTO server. Thereby the ALTO client can be integrated in different entities, for example, in one scenario it might be a user peer that directly queries the ALTO server. In another scenario it might be more beneficial to have a resource directory, for example, a tracker server in the BitTorrent protocol querying the ALTO server on behalf of a user peer. There are also different scenarios for the CDN use case, but typically it is expected that the entity that routes user requests uses the ALTO server as one information source for its routing decision.

7.3.1.3 Discovery Protocol

In some deployment scenarios a static configuration of the ALTO server's contact point is not possible. For example, some ALTO clients might be multihomed, or mobile. As an ALTO server is usually associated with the access network of the entity that wants to invoke the interface, the contact point needs to be discovered dynamically. In order to cover a broad set of potential network deployments, the ALTO discovery protocol leverages a combination of DNS, DHCP, and manual configuration for the discovery process. In particular, the ALTO server URI is retrieved by a U-NAPTR-based resolution process, which presumes the access network's domain name as input. The domain name is retrieved either by manual user configuration, by DHCP, or by a DNS reverse lookup on the client's IP address. It is expected that the discovery process is used in particular by P2P applications, whereas in CDN use cases the configuration of the contact point is manual and static.

7.3.1.4 ALTO Protocol

The ALTO protocol is divided into services of similar functionality. The Server Information Service provides details on the capabilities of a specific ALTO server, such as supported operations and cost metrics or alternative ALTO servers via the Information Directory. The ALTO Information Service group contains one basic Map Service which provides the current view of the network region to a client based on Network Maps and Cost Maps. Additionally, three services offer additional information: The Map Filtering Service allows the ALTO Server to filter the provided maps according to a specific query of a client. The Endpoint Property Service allows clients to check specific properties of endpoints, such as the Network Location or the connectivity type. Finally, the Endpoint Cost Service allows the ALTO Server to rank endpoints directly. Figure 7.2 illustrates the different available ALTO services.

The ALTO server internally stores an information base which is used to calculate costs for paths between different endpoints. The information base is translated by the ALTO Server into the Network Map and the Cost Map, which an ALTO Client retrieves and uses for path rankings. The Network Map defines the network regions that an ALTO Server considers: It aggregates endpoint addresses, such as IP addresses, together and defines a network location identifier (PID) for each group. The Cost Map then provides pair-wise path ratings among sets of source and destination network locations. The costs can represent different metrics, such as air-miles, hop counts, or generic routing costs, whereas lower values indicate a higher preference of the network operator for traffic to be sent from a source to a destination network location. The advantage of separating network and cost information types into different maps is that both components can be updates independently from each other, for example, in different timescales. While network information is considered to be relatively stable, network conditions and with them the costs may be subject to higher dynamicity. After retrieving both map types an ALTO Client is able to check based on endpoint addresses to which peers it should establish

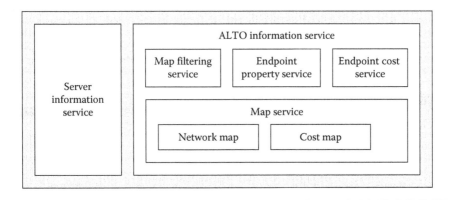

Figure 7.2   ALTO protocol structure.

connections preferably. In cases where the end device should not be burdened with the necessary calculations or where the network operator refrains from publishing exact network maps the ALTO Client may use alternative services in order to get prefiltered maps or rankings based on endpoint addresses.

7.3.2 The ENVISION System

A more elaborated technique providing solutions for collaboration between networks and applications is specified by the ENVISION project [10] under the FP7 European Research Program.

7.3.2.1 Presentation

The proposed technique pushes the collaboration beyond what the ALTO solution allows by defining intelligent cross-layer techniques that offer the possibility not only for exchanging detailed information between networks and applications but also for mobilizing network and user resources to provide better support where needed. This collaboration scheme is intended to offer each side, networks and applications, the opportunity to make use of the information, and the resources available to adapt itself and the content it is conveying to the content distribution context.

This collaboration is a step forward toward the Future Internet as it provides a model based on communication between the applications overlays and the underlying networks which are up to now almost completely agnostic to each other. Figure 7.3 shows an overview of the system based on the concept of collaboration

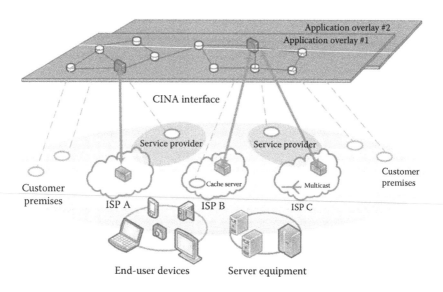

Figure 7.3 Overview of the collaboration between overlays and networks.

between overlays relying on different underlying networks through the Collaboration Interface between Network and Application (CINA) interface. It is required from the application to have an interface with each ISP as they are each responsible for its domain. The role of the application is to consolidate the information gathered from different ISPs to make use of it at an end-to-end level. ISPs, however, are expected to take actions at the ISP level, for example, setting up a multicast tree for the application within its domain.

Some ISPs, on the other hand, could decide to ignore the collaboration scheme, in which case the application will rely on empirical measurements and self-tools in order to cope with the situation in the best possible manner.

To better illustrate the benefits of such collaboration, it is desirable to consider an application use case and analyze the impact of the CINA interface on the system. One example of such an application is the multiviewpoint coverage of sporting events as a bicycle race like the Tour de France. Such an event is tracked by a lot of people in many different countries which could get access to different simultaneous mobile or fixed sources among the professionals, trackside spectators, and even cyclists themselves. Consumers of this content, who may be using various fixed and mobile end devices, can tailor their viewing experience by selecting from many streams according to their preferences, or navigate between streams in real time to zoom or pan around or to follow particular cyclists. Such a context requires the consideration of a lot of parameters in order to achieve a high-quality streaming to the end users and traffic optimization for ISPs. On the one hand, ISPs are unaware of the changing popularity of content sources, its characteristics, the locations of the consumers of that content, or the heterogeneous end-terminal capabilities. From their perspectives, huge amount of traffic is generated by the application between unpredictable locations. On the other, the application, tracking and matching content sources and consumers, strikes to enhance the streaming quality as efficient distribution overlays that can only be built with knowledge of underlying network capabilities. By using the CINA interface to exchange their constraints and resources, the application and the network are prepared to collaborate for the benefit of each other. The application is able to use resources where they are required and are most effective, for example, a network caching server to reduce an interdomain traffic load, or an overlay adaptation node to relieve congestion threat in a subdomain region by decreasing content throughput. Similarly, ISPs could mobilize network services such as multicast to reduce the overall traffic and improve the QoE for the users.

7.3.2.2 High-Level Architecture

Behind the suggested collaboration scheme stands a high-level architecture (Figure 7.4) designed to meet a defined set of functional requirements. As it is a functional architecture, neither specification is defined regarding the location of the functional blocks, whether they are centralized or distributed over specific nodes, nor recommendations are given about how to implement them (grouping

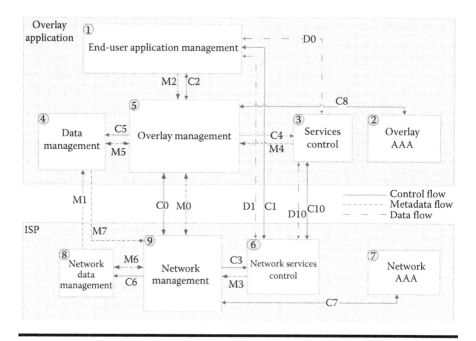

Figure 7.4   High-level architecture.

some blocks together in the same server or distributing them), and so on. The architecture is rather defined to meet the required functional entities, their roles, and the interfaces between them. It was designed to meet the requirements for the expected collaboration between the entities taking part in the content distribution, the end users, the overlay, and the underlying network. The set of fundamental functionalities identified for the collaboration includes the management of (1) the end-user front-end application and terminal, (2) the information data collected at the overlay level, (3) the overlay, (4) the overlay services, and (5) the overlay AAA. At the ISP level, it includes the management of (6) the network, (7) the information data at the network level, (8) the network services, and (9) the network AAA.

Three interface types are considered within this architecture: (1) Control interface for conveying control primitives, (2) Metadata interface for conveying metadata information, and (3) Data interface for conveying the distributed data. Each functional entity's role is explained hereafter.

At both levels two functional blocks are assigned for Authentication, Authorization, and Accounting (AAA) purposes. These aspects are required to deal with situations involving confidential information disclosure or business activities. They enable authentication of users, applications, ISPs, and potentially third-party service providers.

At the network level, the block 8, *"Network Data Management"* processes, stores, and provides information about the network such as the network topology, the offered network services, the content distribution cost, the ISP policies and preferences, and so on. Interface M1 is used to convey the information to *"Data Management"* functional entity.

"Network Services Control" controls the available network services, for example, multicast and caching servers. These services are managed by block 9 via the command interface C3 to mobilize the resources when needed for the data. Monitoring results are reported back using interface M3.

"Network Management" is responsible for the overlay management at the ISP domain (intra-ISP). It receives (1) information about the overlay via M7; (2) requests for resource mobilization via C0; (3) network information via M6; and (4) feedback about services execution and network status via M3. This information is analyzed and by optimization algorithms to achieve a set of targets: (1) define appropriate actions for this distribution context; (2) mobilize network resources if needed; (3) provide feedback and answers to block 5; and (4) update the network information within block 8 via M6.

At the overlay level, *"Data Management"* manages the information provided by the different ISPs and the application. Interface M7 is used for conveying information about the overlay to the ISPs, for example, number and location of users, traffic predictions, and so on.

"Services Control" controls the execution of the services provided at the overlay level, for example, content adaptation, caching, or multiview. These services could be provided by the end users or by third-party entities and are managed by the *"Overlay Management"* via C4.

At the user level, *"End-user Application Management"* represents the end-users front-end. It implements functions for (1) generating and retrieving content, (2) handling data flows, (3) providing information about the user and the terminal, and (4) providing feedback about QoE.

The last functional entity is the *"Overlay management."* It is responsible of managing the end-to-end distribution overlay. It implements the overlay optimization algorithms. This entity is aware of the complete content distribution context, that is (1) content characteristics and processing possibilities, for example, adaptations; (2) end-user context, for example, access network, terminal capabilities; (3) underlying network's capacities and available services, for example, multicast; and (4) available overlay services, for example, multiview, transcoding, caching, and so on. It defines a set of actions to optimize the overall QoE by (1) optimizing the distribution topology, for example, selection of servers or peers, requesting multicast, and so on; (2) performing content adaptation; (3) reacting to networks situations and requests, for example, reducing the data rate for a specified region under congestion threat; and (4) sharing the valuable information with ISPs for better consideration at their level. Obviously some of these actions are subject to security checks performed by block 2 via interface C1.

7.3.2.3 CINA Interface

The CINA interface is the fundamental key for the collaboration between the network and the application. It provides the communication tool used by each side to feed the other with information and to negotiate resources. Similar to the ALTO approach, the CINA interface is a communication channel between a CINA server and a CINA client.

Figure 7.5 shows a possible implementation of CINA server and client. The server should at least implement the Network Data Management and the Network Management. The Network Services Control and the Network AAA functional entity could be externalized or omitted in special cases. At the application side, the CINA client should implement at least the Data Management and part of the Overlay Management. Depending on the application type and constraints, other functional entities could be implemented within the client or externalized.

This interface lays on three aspects: (1) the discovery for locating the CINA servers, (2) the security mechanisms for securing the collaboration, and (3) the communication protocol used for information exchange and resource negotiation. These aspects will be detailed in the following subsections.

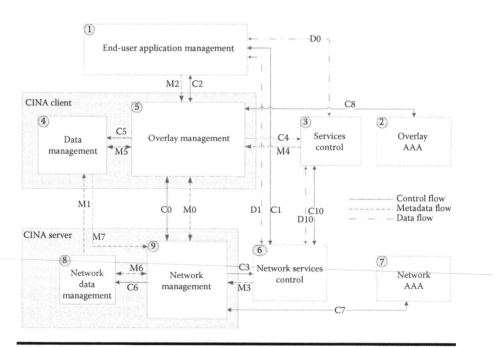

Figure 7.5 Overview of the CINA interface.

7.3.2.3.1 Discovery Mechanism (NS 1/2-1 page)

Similar to ALTO a CINA client needs to first discover the right CINA server of the ISP, typically associated with the access network where the client is registered. The requirements for this discovery process are equivalent to those of ALTO, thus the technique in ALTO-specified discovery procedure can be applied for CINA as well. The CINA client also uses a combination of manual configuration, DHCP, and reverse-DNS in order to retrieve the domain name of the access network where it is registered. Then in a second step this domain name is resolved by a DNS U-NAPTR query to the URI of the CINA server. Therefore, the ISP registers a CINA-specific U-NAPTR service entry in its DNS system. The CINA client then uses this URI to contact the CINA server.

7.3.2.3.2 Security Mechanisms

The collaboration scheme between networks and applications is based on disclosure of information which could be confidential at certain level and intended to be used by a very identified and trusted side. Moreover, network resources mobilized and used for the applications are a critical point that requires a secure environment. These issues are addressed by providing security andAAA mechanisms to the interface.

The requirements identified for the security aspect include: (1) authentication of the CINA server and client mutually, (2) authorization to determine each side access privileges to information and resources, (3) secure communication to prevent interception or modification of the exchanged data, and (4) data usage to ensure that the data securely exchanged remain confidential and used intentionally for the collaboration process.

7.3.2.3.3 CINA Communication Protocol

The CINA protocol specifies the methods a client or a server could use to retrieve information or invoke services. It is based on the protocol that the IETF ALTO WG provides. The ALTO protocol defines a set of basic methods that can be used to retrieve different kinds of information from a network operator although it is currently limited to static information, such as routing costs. Its design is flexible and allows the definition of extensions. The CINA protocol specification is thus aiming at extending the ALTO protocol with more methods to integrate the CINA information and services.

The elemental methods provided by the ALTO protocol consist of:

■ *Network map*: This structure provides a description of the network topology by grouping endpoints of the network into a provider-defined network location identifier called a PID. Endpoints could denote an IP address, a subnet, a set of subnets, a metropolitan area, an autonomous system, or a set of autonomous systems.

■ *Cost map*: This structure provides information about the cost of data delivery between the PIDs defined within the network map. A basic cost type of routing cost is defined within the ALTO protocol.

■ *Filtering map*: This method allows the client to request a restrained set of information from the network map or the cost map.

■ *Endpoint service*: It consists of two methods allowing the client to request information about an endpoint or the delivery cost to that endpoint.

In addition to ALTO, the CINA protocol provides more services and attributes for the overlay to network methods including:

■ *Real metrics for the cost map*: Instead of using an abstract range of values for describing the delivery cost between PIDs, the CINA methods support real metrics such as available bandwidth or latency to qualify the delivery cost. These metrics allow the application to take into consideration not only the delivery cost but also the network capacity.

■ *More properties for endpoints*: New attributes such as the real uplink capacity are available for qualifying an endpoint using the endpoint properties service method.

■ *Network service map*: This method allows the CINA client to retrieve information about the available network services and their location or execution domain. As an example, it acknowledges the client about the set of PIDs where the multicast service is available for request.

■ *Network service instantiation service*: These methods give the client a limited control on the network services by allowing the client to request an instantiation or shutdown of network services when available within the network service map.

Besides the aforementioned services that allow an application overlay to retrieve network information and to instantiate network services, the CINA protocol additionally defines a set of methods for information exchange from the overlay to the network, including:

■ *Footprint map*: This structure provides information about the application footprint on the network. Using the PIDs defined within the network map, the client provides a macroscopic description of the application node distribution within the network topology, for example, number of users, sessions, types, and locations among the PIDs.

■ *Constraint map*: This structure provides information about the constraints exerted by the application on the network. For instance, the client could provide information or future predictions about the traffic throughput between the network PIDs.

■ *Network service instantiation service*: Same as for the overlay to network method, it is used by the server to request an instantiation or a shutdown of network services.

7.3.2.3.4 Collaboration Scenario Illustration

A scenario based on network multicast instantiation for the application data delivery is described in this section to depict the role of the CINA interface in the collaboration scheme. It demonstrates how an example of one network service, multicast, can be dynamically requested by the overlay application. Figure 7.6 schematizes this use-case, with one overlay P2P application, running on top of 4 Autonomous Systems (AS), managed by 3 ISPs. The orange ISP, AS3, offers the multicast facility to the overlay applications.

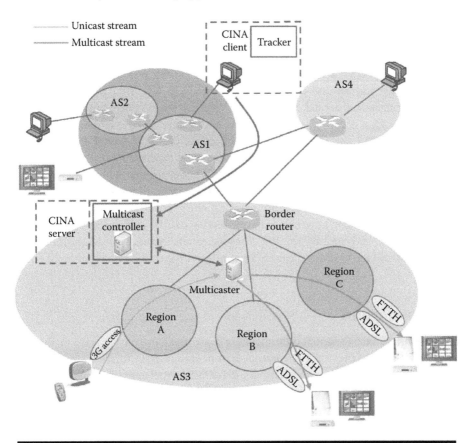

Figure 7.6   Multicast testbed.

Assuming that at a given moment, the application attracts a lot of end users and has a great success, it detects that there are a lot of receivers in the orange ISP domain (IS3) (and for instance that the available uplink bandwidth is insufficient for the target quality). Thus, using the multicast service offered by the orange ISP would enhance the users QoE and reduce the burden on the network.

The first step is the discovery of the ISP CINA server by the CINA client (e.g., tracker of the P2P system). Then the client starts a secure session with the server by executing security procedures, that is, authentication, data encryption and signing on, and so on. The following steps are depicted by the call flow presented in Figure 7.5.

The client requests the network map and the cost map (with appropriate cost attribute) (1). The server replies with the requested maps (2). The tracker detects that the number of clients is growing and exceeds the multicast switching threshold. It requests the network services map (3). The server replies with the map (4) acknowledging the client that multicast is available in its domain. The tracker decides to request multicast using the network services instantiation service (5). The multicast controller within the server checks the availability of the requested resources and then initiates a multicast procedure toward the multicaster (6). The latter sets up the multicast tree, reserves a group address, and opens a port within a unicast address then transmits back these parameters to the server (7). The server confirms the multicast instantiation by sending the multicast address and the multicaster unicast address and port number to the client (8). After reception of the grant confirmation, the tracker chooses by internal means one or more nodes to fulfill the function of stream sources for the multicaster. Then it requests from the source(s) to upload the content to the multicaster (9). The elected source starts uploading the content (10) and reports back the good execution to the tracker (11). At this stage, the tracker requests from the P2P nodes located within the Orange ISP domain to switch to multicast providing them with the required parameters (12). The nodes report back the good reception of the multicast stream (13).

This example shows how one network service can be dynamically requested, configured, and activated by the ISP in order to collaborate with applications in order to improve the QoE for the users and optimize the traffic delivery (Figure 7.7).

7.4 Conclusion

Over the last 10 years, research on audiovisual content enhancement has led to important results bringing HD and 3D contents in front of the scene with some casualties in terms of data size and throughput. And the craze for Internet applications and services brought these contents to everyone's hands putting the networks and application on a constant challenge for delivering the data and ensuring the best QoE for the users. This chapter presented a concept of collaboration between the networks and applications, which we believe constitutes an important aspect for the future Internet. This collaboration allows the different parties and stakeholders, such as users, applications, third

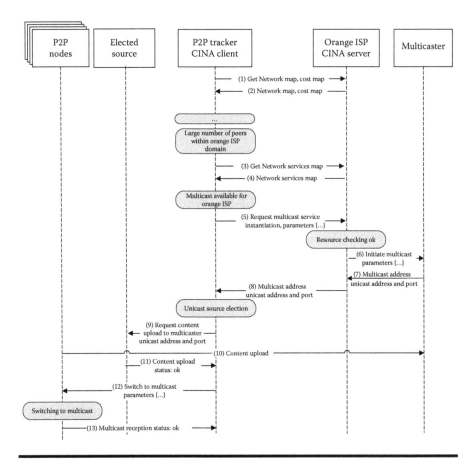

Figure 7.7 Dynamic multicast triggering call flow.

parties, and ISPs, to take active part in the service, to communicate and exchange resources and information. The information provided by each side helps the other one making better consideration of the service constraints. Within an advanced collaboration scheme, these parties could be keen on providing resources such as third-parties storage, network multicast, and content adaptation.

One existing technique of this concept is the ALTO framework. This collaboration is based on network guidance for the overlay applications. It is being considered for P2P as well as CDN applications. This solution is, however, a unidirectional collaboration in the sense that all the optimizations are taken at the application level with indirect gain at the network level. These limitations are addressed by the second presented solution, based on the CINA interface. Although, the CINA protocol is specified taking into account the ALTO protocol framework, it includes more advanced methods and structures for providing important features taking the collaboration between applications and network to the next level. Both approaches,

ALTO and CINA, are still under specifications and developments for addressing open questions about their integration within different contexts. However, this concept of collaboration could be a key feature for designing a Future Internet promoting existing technologies and putting aside any break without neglecting QoE and traffic load issues.

References

1. Cisco systems. Cisco visual networking index: Forecast and methodology, 2010–2015. *Cisco VNI*. [Online] 2011. http://www.cisco.com/en/US/solutions/collateral/ns341/ns525/ns537/ns705/ns827/white_paper_c11-481360.pdf. FLGD 09748 06/11
2. Xie, H. et al. *P4P: Explicit Communications for Cooperative Control Between P2P and NetworkProviders.* [Online] http://www.dcia.info/documents/P4P_Overview.pdf
3. Seedorf, J. and Burger, E. Application-Layer Traffic Optimization (ALTO) problem statement. *IETF.* [Online] 2009. http://tools.ietf.org/html/rfc5693. RFC5693
4. Pando Networks. Large scale ALTO/P4P field trial. *Pando Networks.* [Online] 2010. http://www.pandonetworks.com/Large_Scale_ALTO_P4P_Field_Trial_Affirms_Performance_Improvements_for_Broadband_Networks
5. NAPA-WINE Network Aware P2P-TV-Application over Wise Networks. *Deliverable 2.3 Final report on Task WP2.1.* s.l.: FP7-ICT-2007-1, 2010. Grant Agreement no.: 214412.
6. Seedorf, J. et al. Quantifying operational cost-savings through ALTO-guidance for P2P live streaming. *Incentives, Overlays, and Economic Traffic Control.* Computer Science, 2010, Volume 6236/2010, 14–26, 10.1007/978-3-642-15485-0_3
7. Penno, R. et al. ALTO and content delivery networks. *IETF.* [Online] 2011. http://datatracker.ietf.org/doc/draft-penno-alto-cdn/. draft-penno-alto-cdn-03
8. Kiesel, S. et al. ALTO server discovery. *IETF.* [Online] 2011. http://tools.ietf.org/html/draft-ietf-alto-server-discovery-00. draft-ietf-alto-server-discovery-00
9. Alimi, R., Penno, R. and Yang, Y. ALTO protocol. *IETF.* [Online] 2011. http://tools.ietf.org/html/draft-ietf-alto-protocol-08. draft-ietf-alto-protocol-08
10. UCL, Orange Labs, Alcatel-Lucent, LaBRI, Telefonica, LiveU. Project overview. *Envision-Project.* [Online] 2011. http://www.envision-project.org/overview/index.html

Chapter 8

Information-Centric Networking: Current Research Activities and Challenges

Bertrand Mathieu, Patrick Truong, Jean-François Peltier, Wei You, and Gwendal Simon

Contents

8.1 Introduction 142
8.2 Information-Centric Networking Paradigm..143
    8.2.1 Motivation for a Network of Information 143
    8.2.2 Concepts of Information-Centric Networking..............................144
    8.2.3 Naming Information Objects 146
        8.2.3.1 Properties of Naming..147
        8.2.3.2 Metadata 148
    8.2.4 Addressing and Forwarding for Content Delivery 149
    8.2.5 Transport Layer for Information Objects.....................................150
8.3 Overview of Research Activities on ICN 151
    8.3.1 CCN...151
    8.3.2 DONA 153
    8.3.3 PSIRP...154
    8.3.4 NETINF 156
    8.3.5 Plexus 157

8.4 Challenges..158
    8.4.1 Security...158
    8.4.2 QoS Considerations..159
    8.4.3 Scalability and Reliability 160
    8.4.4 Network Management and Better Manageability........................160
8.5 Conclusion 161
References 161

8.1 Introduction

The Internet usage over the last decade has shifted away from browsing to content dissemination. Host-to-host communications tend to disappear to make way for one-to-many or many-to-many distribution and retrieval of content objects with an increasing request for a better support of mobility. Users only want to know what content is available and how to get it rapidly anywhere, on demand, on any device, and on the go. They do not care about the location of the content.

While imaginative solutions through incremental changes to the network stack or overlay networks have been successfully proposed to address these new usage patterns, it is widely admitted that these solutions also have limitations and drawbacks, so that they do not provide enough real benefits in terms of scalability, security, mobility, and manageability to encourage the creation or deployment of more ambitious and innovative services. They are, for example, confronted to the problem of intercompatibilities due to contentions or tussles between Internet stakeholders that generally have adverse interests and want to promote their solution. In addition, continuously adding overlays or patches to upgrade Internet often results in increased complexity that favors vulnerabilities, limits manageability, and leads to lack of flexibility. All the current workarounds do not actually solve at the root of the problem: they are still host-centric, based on the endpoint addressing and forwarding principles over which Internet was built historically, while the large majority of today's network communications are content-centric, that is, concern production and consumption of pieces of data. Figure 8.1 depicts this concept, which is known as "from hourglass to lovehandles" problem.

Recognizing that the host-oriented paradigm does not cope with the emergence of voracious content-consuming applications, several projects and initiatives have been started over the last few years to envision a clean slate foundation for the future Internet promoting information at the center of the design considerations. In this chapter, we propose an overview of this rising Information-Centric Networking (ICN) paradigm. In Section 8.2, we present the ICN model with the naming, addressing, and forwarding issues. Section 8.3 highlights the most relevant results of the different research activities on the topic in the literature. In Section 8.4, we outline challenges and requirements that need to be addressed for the development of the ICN paradigm. Section 8.5 is devoted to concluding remarks.

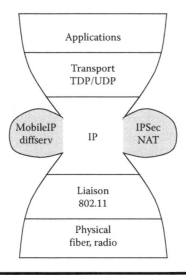

Figure 8.1   Lovehandles problem in IP.

8.2 Information-Centric Networking Paradigm

8.2.1 Motivation for a Network of Information

Internet was originally designed to be an end-to-end connectivity substrate for the delivery of data. All the subsequent enhancements made for improving its architecture revolved around the conversation model that consists of communications between machines using the IP protocol. The current Internet architecture is now a constantly and rapidly evolving interconnection of thousand networks that act as simple carriers providing basic packet delivery services without guarantees, meaning that they make their best effort to try to deliver to receivers anything that senders want to send while only using IP addresses to identify end-hosts for data forwarding and unawarely considering what is being delivered.

Unfortunately, this endpoint-centric communication model does not cope with the overwhelming usage in current Internet anymore. The vast majority of today's traffic is indeed driven by information production and retrieval, ranging from simple RSS feed aggregators to advanced multimedia streaming services, including user-generated or dynamic Web content. The relationships between network entities are not only restricted to the view of network topology, but also represent social or content-aware connections between users that can additionally share common interests (e.g., newsgroups, online photo, and video-sharing service such as Flickr or YouTube, social networks Facebook or Twitter, P2P-based file sharing). The resulting graph modeling today's Internet communications is then very complex. Nodes are mainly pieces of content, rather than endpoints (locators) addressed by the underlying IP protocol, and are connected by one or more types of interdependency based on the notion of intention, interest, or policy-based membership. While still overlaying on top of the host-to-host conversation model, all the current workarounds for Internet support of emerging information-centric applications increase the complexity and do not efficiently map all the relevant ties between the nodes in terms of:

■ *Security:* Today's IP-based network security requires trust on end-hosts and on connections over which content transfer occurs. As a result, the main flaw in IP addressing is that the network accepts anything from senders regardless of information contained in packets, provided that the senders appear legitimate. This situation leads to unsolicited and malicious messages sent to receivers.

■ *Mobility and multihoming:* The dual role of IP addresses as both network layer locators and transport layer identifiers limits the flexibility of the Internet architecture for a more efficient support of mobility and multihoming. In particular, transport protocols are bound to IP addresses to identify communication sessions, which are interrupted when an address changes.

■ *Multicast delivering:* IP Multicast protocol was designed as an after-thought add-on around the original point-to-point communication model to offer the ability to send information to a group of receivers. However, IP Multicast has never taken off outside of LAN environments because of its scalability shortcomings, so that complex overlay solutions are rather preferred to deploy multicast services at large scales.

■ *Scalability and QoS guarantees:* With the rapid proliferation of content distribution services, costly solutions such as overlay networks (e.g., CDN or P2P) have been proposed to alleviate the huge demand of bandwidth and to improve user experience by pushing and caching content to the network's edge, but performance bottlenecks still persist in the last mile and the inability for network operators to control traffic traversing their networks often results in conflicts of business interests or an inefficient network resource optimization.

To overcome these issues from the incompatibility between the usage and communication models on Internet, the Information-Centric Networking, also called content-centric or data-centric networking paradigm has been proposed over few years to target a clean slate architecture redesign by placing content in the foreground, at the heart of network transactions. In the remainder of this chapter, we will use the terms information or data as well as the term content interchangeably.

8.2.2 Concepts of Information-Centric Networking

The ICN paradigm consists of communications that revolve around the production and consumption of information matching user interest. The principal concern of the network is to expose, find, and deliver information rather than the reachability of end-hosts and the maintenance of conversations between them. As a global view, the paradigm can be divided into two functional parts: information dissemination or exposure, and information retrieval. On a rather low level from the ICN perspective, a network is a set of interconnected pieces of information, also called content, information, or data objects, which are addressed by names for routing and managed by applications or services at the higher middleware level. The naming scheme for identifying content objects in ICN is intended to replace the current IP naming scheme, which mixes host locations and content identifiers (like an URL for instance). In particular, the name of a content object is globally unique and independent of its location (i.e., the host holding the data). Content objects is an abstract notion and can be of any type, including, for example, Web applications (e.g., a piece of mail generated by online webmail services), static or user-generated content (photos, videos, documents, etc.), real-time media streams such as VoIP, VoD, Web TV, or online videos and music (a stream being considered as series of chunks of data), more complex interactive multimedia communications, or even devices (e.g., routers, data servers, etc.) for network management. Objects are sometimes organized into clusters to define social relationships or some ontology between them (as illustrated by the notion of scope in PSIRP); they can also be mutable, combined, or aggregated to form new objects.

In ICN, senders do not send content directly to receivers, and any data object delivery is controlled by receivers (cf. Figure 8.2). A sender (or content provider) that has objects to distribute does not actually transmit them in the network, but it rather sends advertisement messages to inform the network that it has content to diffuse, without knowledge of receivers that may be interested in it. A receiver

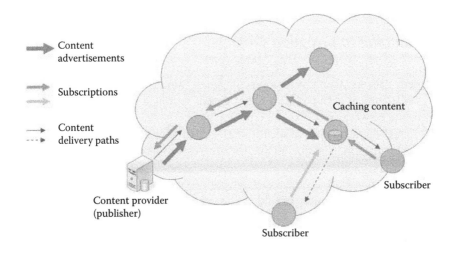

Figure 8.2   Content dissemination and retrieval in ICN (with caching capability).

or consumer declares its interest for some content, without knowledge of potential senders that may have queried content. Only when receiver's intention matches a published information object, the network initiates a delivery path from the sender to the receiver so that content retrieval can now start for the receiver. The match of interest rather than the findability of endpoint that provides content thus dictates the establishment of a communication in ICN. The information-centric network connects producers to consumers, disregarding the underlying hosts involved in communications. The focus is only on content, not on the hosts storing content. Only queried content is delivered to receivers that have asked for it beforehand.

Another key principle that comes with the interest-oriented networking in the ICN paradigm is the use of dynamic content caching to enable fast, reliable, and scalable content delivery with maximized bandwidth to avoid congestion. A router (in the delivery path from the sender to the receiver) can, for instance, cache content objects that traverse it so that subsequent queries for the same objects can be satisfied rapidly by the router. This means that routing in ICN consists of finding and delivering copies of data objects to consenting receivers from the most efficient location in the network.

When using caching for content retrieval, it is important to give users every guarantee that cached content did indeed come from the original source. ICN affixes trust directly on content itself, rather than the end-to-end connection carrying it. This allows users to assert immediately the safeness of content and therefore evading host-related threats and vulnerabilities.

The following sections detail the above-mentioned concepts of ICN and show how they can be achieved through the information object model that involves content dissemination and retrieval.

8.2.3 Naming Information Objects

Retrieving content in ICN can be decoupled into two phases: content discovery and content delivery. Content discovery includes naming and addressing content objects while content delivery defines the rules of routing and forwarding content objects in the network.

Naming content consists of identifying each content object living in the network by a globally unique name, so that the users can address objects by their name to determine the location host or the nearest cache holding a copy of the queried object before the network creates a delivery path from this host to the receiver asking for content. The naming to addressing mapping is a one-to-many relation because of the use of content caching and replication in ICN.

In this section, we first focus on the naming scheme that allows defining the conceptual principles of ICN. We delay our discussion about addressing and forwarding for content delivery in the next section.

8.2.3.1 Properties of Naming

As everything is information in ICN, an emphasis has been put on naming to define the information object model. The objective of naming is not only to uniquely identify content objects in the network, but also to include important properties such as pertinence, usability, scalability, and security.

- *Uniqueness*: Information objects have to be named in a globally unique way in the network. This uniqueness is necessary to route content by their name.
- *Persistence and location-independence*: The name is invariant and independent of the location of the host that stores data. Content can then be replicated or moved from one hosting location to another in the network topology without service disruption, provided that the provider continues to serve it.
- *Usability and scalability*: As a network of information, we can expect to have a huge number of data objects in the network, which are interconnected and can have multiple interdependencies between them. Moreover, objects are not necessarily static and can be mutable, for example, being fragmented in small segments or evolving from one version to another at some time in the future such as a weather RSS feed. In addition to be scalable, the naming scheme should then be usable for dynamic objects and should also allow deletion of objects (e.g., can embed a TTL (Time-To-Live) parameter, which allows to invalidate the object at the expiration of the TTL).
- *Security*: ICN embeds security directly in content rather than assuming trust on users or securing the communication channels that deliver it. Content-based security means that there is a cryptographic binding between content and its name to ensure that information objects are self-certifiable in the sense that when a user receives its queried object, it can rapidly verify that the object was truly originated by the genuine provider. This binding generally consists of a hash of the content provider's private key in the name of the content object. This allows the authentication of the content provider by creating a signature of the data object using the provider's private key, which can be verified with its self-certificated public keysent as metadata to receivers along with the data object. Confidentiality and integrity of content are also guaranteed by a public key encryption. An important challenge for self-certification of content is to handle revocation of objects when the public key is compromised or data are updated.

While the name structure is often made up of several parts to reflect these properties, there are actually two types of namespace in the ICN literature: hierarchical and flat. Hierarchical names, as proposed by Van Jacobson, make it possible to use the concepts of IP routing lookups based on longest prefix matching and to organize content in a way similar to DNS, that is, aggregating content at different levels over trusted domains to improve security and the authenticity of the provenance of content which is useful for Digital Rights Management (e.g., parc.com/videos/widgetA. mpg/). However, the binding between hierarchical names and administrative domains compromises the persistence of content objects. In this case, a flat namespace may be preferred.

The presented properties of naming allow us to assert the following principles for ICN:

■ Resiliency to service disruption and network failure by replicating data in various points in the network.
■ Increased performance by enabling caching for content retrieval throughout the network (the nearest copy of a data object is returned to users).
■ Native support of multicast, mobility and multihoming, rather than add-on solutions. This is achieved with the decoupling between naming and location of data.

8.2.3.2 Metadata

In addition to naming, content objects also have attributes represented by means of metadata. The concept of metadata is important in ICN to provide information or description about a content object or its relations with other objects. For example, metadata associated to a photo can give indications about the image resolution, the author, the date, or any other data inserted by software.

Metadata attributes are used for several purposes in ICN. The semantic nature of these attributes is firstly profitable for applications to manage content objects and to understand how they can be used. For example, in the Van Jacobson's implementation of VoIP using the content-centric parading [1], metadata contain descriptive information about users involved in the conversation. As object names are numerical identifiers, it is usually necessary in ICN to let users to make keyword- or description-based searches for content. This means that a resolution often exists at the application level to map human understandable attributes into object names. Most search engines use metadata associated with objects to implement this resolution in a distributed way.

Beyond semantic meanings, metadata can provide cryptographic inputs to perform more complex security checks on content objects. A simple example consists of including in data objects metadata carrying the content provider's public key combined with a digital signature so that receivers can assert that content did exactly come from the provider. The network can also rely on metadata to perform QoS guarantees on information objects, especially coming from real-time streaming multimedia applications (e.g., prioritizing content objects from a VoIP session). In particular, we can define network metadata to perform some network access control or to collect statistics on traffic usage to monitor network health. Generally speaking, metadata provide an efficient tool for ICN to supervise the network.

There are different ways for implementing metadata in ICN. As metadata are by definition nothing other than data about data, any content object can be metadata for other content objects, and in other words, metadata are simply content objects as any other ones in the network [2]. This design consideration involves an additional class of identifiers to use in conjunction with the name of content so that we can know which metadata items are related to a content object. However, for simplicity, metadata are often considered as labels at the same level as the name.

8.2.4 Addressing and Forwarding for Content Delivery

Addressing defines the reachability of information objects in the network by mapping names to hosting locations. In ICN literature, it is also referred to as name resolution to make an analogy with the current DNS resolution that resolves host names to their corresponding IP.

Name resolution is equivalent to the network layer of today's Internet and includes content location (routing queries) and forwarding for content dissemination and content delivery. As a network of information revolves around content producers and consumers, the publish/subscribe communication paradigm [3] which decouples the sender from the receiver appears as the most relevant networking concept from which ICN takes its inspiration. Most of the information-centric architectures proposed in the research community fairly reuse the principles of the pub/sub-networking communication to implement content forwarding, which is thus based on two functional steps we can mention as REGISTER (or PUBLISH) and FIND (or SUBSCRIBE).

It is not clear if the ICN paradigm can (or should) replace or not the IP layer completely (clean slate approach for the future Internet). ICN can be actually overlaid over any forwarding layer, including IP itself. While the debate is ongoing, current technologies proposed to implement routing in ICN are overlaid over IP facilities and for the time being, can be classified into two categories:

- *One-step name resolution*: name-based routing [4], FIB-based forwarding [5].
- *Two-step name resolution*: Name Resolver Service for translating an object ID into one or multiple source locators [6], Rendezvous-based communication [2] or probabilistic routing (e.g., using Bloom Filters) [7].

In a one-step name resolution, the request message for an information object is directly forwarded from the requester to the source or any cache which can serve the query. Instead of exchanging route information based on IP prefix advertisements as does current BGP routing, routers in the one-step name resolution advertise names of information objects, so that each router can determine efficient routes to objects. Forwarding then consists in finding the better match between the name of the requested information object and the entries of the routing tables.

The second two-step approach, based on the publish/subscribe model, relies on an intermediate mapping proxy to resolve object names into network identifiers (or locators) which are used for routing queries to content sources. Users send a subscription message to the proxy with the name of the desired information object. The proxy is then responsible for getting the requested object back to users. Routing protocols are generally based on Distributed Hash Tables (DHTs) to distribute this name to locator mapping function over the global network.

Examples of these two different routing approaches will be detailed when describing the architectures from research projects related to ICN.

8.2.5 Transport Layer for Information Objects

In the current Internet architecture, transport layer functions such as error detection, lost data retransmission, bandwidth management, flow control, or congestion avoidance are implemented at the endhost level as end-to-end communication processes. This is inconsistent with the ICN paradigm in which the role of endhosts is very different compared to traditional IP networks since communication sessions are only information-centric, disregarding the involved endpoints. As a consequence, if naming and addressing content as discussed in the previous sections allow us to implement the functions of the network and lower layers required for content-centric networking, the transport layer also needs to be considered to completely remove the dependence on endpoints. Although some proposals exist in the literature to translate transport functions in ICN, discussions are still opened. For example, senders and receivers are decoupled in ICN, and because of caching, a requester can receive its stream of queried content from multiple sources in an unpredictable way. The challenge is then how to perform transport control per data source under uncertainty that there is no way to know these sources in advance. One solution could be to let the receiver to control congestion avoidance, which is estimated from the source feedback, as described in ref. [8].

The IP layer implements datagram fragmentation, so that fragmented packets can pass through a link with a maximum transmission unit (MTU) smaller than the original datagram size. In ICN, the equivalent notion is chunking which means that a source can serve content in a series of chunks. A chunk is typically the smallest identifiable piece of a content object, but chunking is more than fragmentation in the sense that a chunk can be split up into smaller fragments for the transport over the network. One of the important factors determining the way that an information object is delivered to a receiver is the presence of chunks from several locations (e.g., caches) in the network and the reassembly of these chunks at the receiver. This leads to additional issues that need to be resolved in the ICN transport layer (e.g., receiver-driven synchronization and flow control between different chunk delivering sources, cache policies for chunks, security checks on chunks, should a chunk be identifiable, addressable, and/or authentifiable? etc.).

8.3 Overview of Research Activities on ICN

This section describes the most relevant results of different research activities on the topic in the literature.

8.3.1 CCN

CCN (Content-Centric Networking) [1,5], designed at Palo Alto Research Center (PARC), was one of the pioneers to promote the ICN paradigm.

In CCN, object names are hierarchically organized in a lexicographic ordered tree. Leaves correspond to content of interest, and each internal node represents the common name prefix shared by a collection of content objects. While using this naming tree, CCN can support dynamically managed contents by authorizing users to make requests based on a name prefix for one content that may not have not been created yet, allowing publishers to generate that content on demand. Objects can also be queried without knowing their full name. Suppose that a video, identified by the name prefix /parc.com/videos/WidgetA.mpg is split into small data chunks which are separately named by adding the suffixes $<$version number, chunk number$>$. These chunks, taken as a whole, form the video file. Users usually do not know the full names of different video segments and thereby rely on the name prefix to trigger the video download.

CCN communications are based on two packet types, Interest and Data, identified by the full or relative name of queried content. A consumer asks for one content by sending an Interest in that content. A CCN node hearing this Interest forwards it to its neighbors unless it owns the queried content and can immediately serve the consumer with a Data message. The latter case means that the name in the Interest is a prefix of the content name in the Data packet.

CCN forwarding is actually similar to the IP forwarding plane for fast lookup of content names in the Interest packets. Figure 8.3 describes the functional parts of a CCN node: the FIB to find the appropriate interface(s) to which arriving Interest packets should be forwarded to reach the providers of queried content, a Content Store that is the LRU buffer memory for content caching, and a Pending Interest Table (PIT) to keep track of the inbound interfaces of received Interest packets so that a Data packet sent back as a response to an Interest registered in the PIT table will be delivered to the right interface(s).

The FIB of a CCN node is populated with name prefix announcements encoded as type – length – value (TLV) elements within IP routing protocols. As messages of unrecognized types in the TLV scheme are currently ignored by these protocols, CCN nodes can be deployed with existing IGP or EGP routers.

PARC has started the development of a prototype, called CCNx [9], available as an open-source software. The PARC's idea is to develop a community around CCN to attract many people to work on this demonstrator, to make it evolve and being thus as a de facto standard for ICN solutions.

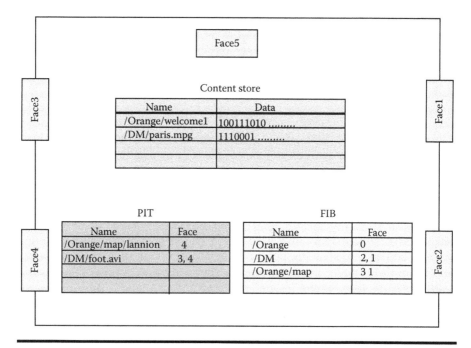

Figure 8.3   The CCN node.

Named Data Networking (NDN) is a new project under NSF Future Internet Architecture (FIA) program based on CCN paradigm [10]. PARC also contributes to this project. The research activities for this project are related to routing and forwarding improvements for CCN. The first goal is to extend the existing routing protocols like BGP or OSPF with the compatibility of named content prefixes. Then, two approaches will be proposed to achieve routing scalability.

- Propose a provider assigned name, which looks like the provider assigned IP address, to achieve the aggregation.
- The user or application may choose a name which is easy to remember and NDN allows the user-selected name, so a mapping service is still necessary for mapping the user-selected name to the provider-assigned name, like DNS domain name resolution.

Since the NDN uses CCN-like longest prefix matching for forwarding based on information stored in the tables FIB and PIT, this computationally intensive task, similar to current IP Lookup, is likely to be the performance bottleneck along the forwarding path due to a potentially huge number of objects in the network. So how to keep an effective forwarding with acceptable performance is an important challenge for the NDN network architecture. The goal consists of finding a trade-off between high-speed longest name prefix match ability and an efficient content storing/deleting/replacing performance in FIB and PIT.

8.3.2 DONA

The DONA (Data-Oriented Network Architecture) [4] proposes a flat naming scheme and a name resolution using a distributed set of network entities, called Resolution Handlers, with caching capabilities to route requests toward the nearest copies of data.

Each information object in DONA belongs to a principal, uniquely identified by a public–private key pair. Names are then defined as $P:L$, where P is a hash value of the principal's public key and L is a label attributed by the principal to ensure that names are globally unique in the network. The principal is responsible for organizing the structure of labels when naming the objects that it manages.

Content delivery in DONA relies on an overlay network of Resolution Handlers (RHs) using a route-by-name resolution through two primitives: FIND($P:L$) and REGISTER($P:L$). Each RH maintains a registration table containing entries of the form $<$($P:L$) or ($P:$*), next-hop RH, distance to a copy$>$, where $P:$* means all data associated with the principal P. When a user asks for content $P:L$, it sends a FIND($P:L$) message to its local RH. If the registration table contains an entry for ($P:L$), the message is forwarded to the corresponding next-hop RH; but if there exist several entries for ($P:L$), the selection depends on local policies or the nearest copy of data, and if entries for both ($P:L$) and ($P:$*) exist, the longest matching label ($P:L$) will be used. Now, if any entry exists for ($P:L$) (or ($P:$*)), the FIND message is sent to the local RH's parent.

When a host is authorized by a principal P to serve a content object named $P:L$, it sends a REGISTER($P:L$) to its local RH (or ($P:$*) if it is authorized to serve all principal's data). If any entry exists for ($P:L$) in the registration table or if the new REGISTER comes from a copy closer than an existing entry, the local RH creates (or updates) the entry for ($P:L$) and forwards the register message to its parents and only to its peers if local policies match. Otherwise, the local RH discards the REGISTER. Before forwarding the REGISTER message to the next-hop RH, the local RH adds to the message header: its signature to protect the authenticity of the message and the distance to the previous-hop RH to keep track of the total distance to the copy of data.

Content forwarding in DONA is based on domain-level label switching. Within each domain, hosts are addressed with a label that is only unique to this domain. When a node sends a FIND message for some content, it appends its domain-specific label to the message header. When forwarding the FIND message from one RH to another RH, labels are pushed onto the stack within the header, so that reversing these labels allows the hosting entity to send queried data back to the user.

To implement caching, an RH has to replace the source address (or label) of an incoming FIND message with its address before forwarding it to the next-hop, so that the data object sent back as a response to the FIND request will be delivered to the RH that can then store data in its cache. Whenever cache is activated upon reception of a FIND, if the RH has the queried data in its cache it can serve the client directly. TTL field or metadata can also be associated with FIND messages to indicate how long a data object should be valid and to request updates when the timer expires.

8.3.3 PSIRP

The PSIRP (Publish-Subscribe Internet Routing Paradigm) [2,11,12] is a European FP7 research project, started in January 2008 and ended in June 2010, promoting an information-centric publish-subscribe networking paradigm for the Future Internet.

The PSIRP architecture (Figure 8.4) relies on a rendezvous system consisting of a set of interconnected physical devices, called Rendezvous Points (RPs), providing rendezvous functionalities for subscription and publication matching.

At the highest level, each data object is related to an application and can be optionally named with an application identifier that is not required to be unique or

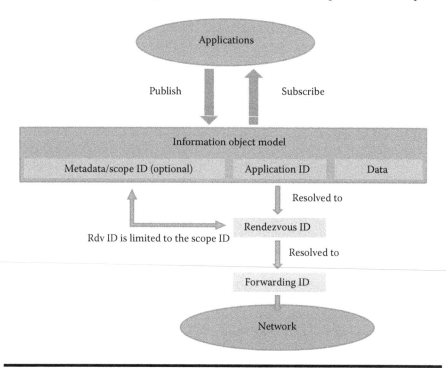

Figure 8.4 Components of the PSIRP architecture.

universal in the network. But when communicating over the network, applications need to resolve application identifiers into a unique and persistent identification. PSIRP relies on a special class of identifiers, called rendezvous identifiers, to name objects in a distributed way, using a flat naming structure. These identifiers could also include some security properties such as data integrity, self-certification, owner authentication or access control, using public-key cryptography whose corresponding private key is only known to the publisher and the public key is widely distributed to receivers.

In addition to have a globally unique rendezvous identifier, PSIRP data objects are also defined within a scope, so that the access to a particular set of data objects can be restricted to only a particular set of authorized users (publishers or subscribers). Scoping is also used for selecting the responsible RPs for Rendezvous identifiers.

PSIRP also makes use of metadata to provide additional semantic information on data objects; application-level metadata such as type of the document, size, author, access rights, caching, and so on; but also network-level metadata, such as network access control, flow control, error or congestion notification, and so on, are possible. For receivers, metadata can be used to describe what they want and to indicate their preferences in how to receive data that they request.

PSIRP forwarding uses a domain-level label switching based on forwarding identifiers, also called labels. A domain allocates a forwarding label for each rendezvous identifier that traverses the domain, and can aggregate several rendezvous identifiers under the same label for scalability.

Typically, if some content is named with a rendezvous identifier RId, the publisher will send a publication related to RId, along with metadata and one or more scope identifiers Sid to limit the reachability of information. When receiving the publication, a rendezvous point forwards it to its neighbors until we reach the rendezvous node where the scope identifier Sid of the publication is registered, and forwarding instructions are updated in the routers on the way. Once receivers subscribe to the rendezvous identifier RId, subscriptions will be relayed from the local rendezvous point to the rendezvous point where the publisher registers. As a subscription progresses to this rendezvous point, labels are appended to the subscription header, so that reversing these labels in the stack allows the publisher to send data back to the receiver. A forwarding tree is thus constructed for this rendezvous identifier; it represents the delivery paths over which content can be conveyed to receivers.

Whenever an RP receives multiple subscriptions with the same Rid, it only forwards a single subscription to the next-hop RP, so that there is also one single returned data flow from the publisher to this RP. This allows support of multicast forwarding trees.

Whenever an RP receives published data, it can store it in its cache to serve subsequent subscribers rapidly with low latency. Caching makes it possible to guarantee high availability of content and services.

The PSIRP has ended, but the work will continue in PURSUIT, which is also an FP7 European project [13,14]. This new project proposes to refine the PSIRP architecture, for both wireless and wireline networks. It is expected to handle further studies on important aspects such as caching mechanisms for better resource utilization and management, transport issues and enhancing mobility with network topologies based on rendezvous points. Another main objective of PURSUIT is to provide various prototypes and development APIs for specifying a reference implementation of information-oriented protocols proposed in PSIRP.

8.3.4 NETINF

NetInf (Network of Information) [6,15] is a network architecture proposed by FP7 Project 4WARD.

NetInf makes a clear distinction between Information Objects (IO) which represent a piece of content or information, identified by a globally unique identifier, and bit-level Objects (BO) which are the basic data object itself. The IO is composed of three fields: the identifier of the content, a set of metadata, and the BO. The identifier contains the type of content and the hash of the owner's (or publisher) public key, hence providing authentication. The Metadata field provides semantic information about the IO, includes the security attributes, such public keys, content hashes and certificates, and can be used by the search service.

Different versions of the same IO may have different authors. Authors are authorized to sign and modify IOs, or delegate modification and signing to other authors. Publishers are not authorized to modify or sign IOs but are authorized by the owners to distribute the IOs. In order to register/unregister an IO, an owner or publisher provides a signature of the IO and the registration time.

The NetInf Name Resolution System (NRS) takes as input either an identifier or a set of attributes describing some properties of the searched object and returns a set of binding records for IOs that matches the input. The IOs include a reference that directly or indirectly can be used to retrieve the BO. In the second case, it means that a two-step resolution is possible, where the application or user may choose in the list of returned IOs, which one to select for requesting the corresponding BOs (contents). The selection of IOs can be based on different criteria for retrieval (cost, download speed, definition, quality . . .).

In NetInf, name persistency is ensured as the ID is independent of the location, only the Name Resolution System's entry will be updated. NetInf also allows persistency regarding evolution of content through versioning which is useful for dynamic objects such as streams.

NetInf defines name resolution zones to separate level of trust for different operators and/or customers. Each zone is responsible for persistently storing a BO with corresponding identifying IO. It is also responsible for caching strategy (prepopulation . . .).

The main routing mechanism used in NetInf is Multiple DHT (MDHT) which is basically a way to implement recursive lookups. When a client asks for an object, a first DHT lookup is made at the first level (e.g., its access networks zone). If it is not found, another DHT lookup is issued at an upper level (e.g., POP zone). If it is still not found, another DHT lookup is made at an upper level (e.g., domain level), and so on. When the DHT lookup is successful at a given level, the result is returned to the client. It is to be noted that despite the hop-by-hop routing and local resolution, this provides the top DHT level has to contain bindings for all data registered in a domain, with possible scalability and possibly performance issues.

The SAIL project [16] is a follow-up project for 4WARD and continues work in the ICN activities. Within the SAIL project, the NetInf solution is under evolution to take into consideration some issues, such as the scalability of the solution, the way to resolve names, and so on.

In SAIL, an objective is also to develop a prototype for the NetInf solution, called OpenNetInf [17], a published and open-source software. Like CCNx, the idea is to attract people working for this prototype.

8.3.5 *Plexus*

The Plexus [18] solution could not be really considered as an ICN solution, because it is not designed with this objective, but it is worth mentioning since it can help in designing alternative solutions to current ICN proposals, mainly for content naming and retrieving in the network.

Plexus is based on previous work called DPMS [7], and is designed with the primary intention to look for contents in a P2P system without exact matching semantic but rather searching a pattern in the content names. Indeed they argue their work telling that current P2P networks need to know exactly the names of the searched contents (e.g., searching "Lord of the Rings" or "Lord Rings" will lead to different identifiers after the hashing function and thus no mean to get it if we have the wrong name as the input parameter). In their approach, they propose to split the names into trigrams and use a Bloom Filter to set bits to indicate whether this trigram is present or not in the name of the content. Thus, the user can find the content even if the searched name is not exactly the one mentioned when inserting the content in the network. Figure 8.5 presents this concept using the Bloom Filter.

It is well known that with Bloom filters there may be false positive (well-known formula), but it is lower to except to have good results in most of the times.

This Plexus system could be seen as a flat system, such as other DHT systems and could lead to inefficient search. However, in the design of DPMS, the authors define kinds of trees: propagation tree, where patterns generated by one peer are propagated in the network and the aggregation tree, where nodes aggregate patterns coming from different nodes below. This second tree avoids a large volume of advertisement data but due to aggregation, information content reduces as we go up in the tree. It is then a trade-off.

We can see that this system could be adapted to fit with ICN requirements where content could be advertised via the trigrams and bloom filters to

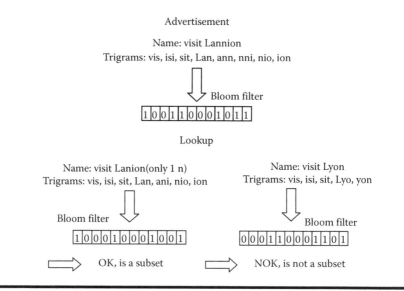

Figure 8.5 Bloom filter in Plexus.

intermediate nodes that could store advertised patterns and thus be able to route to the appropriate sources when an end user is looking for a given content, using the same mechanism.

8.4 Challenges

The aforementioned ICN solutions are still in their early stage and do not address efficiently all the features the future Internet should provide. In particular, some important aspects such as security, QoS considerations, scalability and reliability, and network management need to be addressed in better detail by an information-centric Internet working architecture. We are conscious that there are many other issues, not discussed here, that also need attention, but we think that the following challenges are essential for a first approach of the ICN model.

8.4.1 Security

Security is an essential part of the ICN paradigm. Instead of securing forwarding paths, the ICN model enforces security directly on content objects by placing provision for protection of the authenticity of content objects or against unauthorized access and privacy issues. The use of self-certifying naming based on public-key cryptography implies some issues such as key compromise, revocation, and management that need to be resolved, sometimes by external public key infrastructures and in other cases by the naming scheme itself. For example, CCN advocates for a distribution of keys as a specific type of content objects using the SDSI/SPKI model, and in DONA, key revocations for a principal P can be handled by publications of the form $P{:}L$ for some reserved name L. While ICN provides content-level security based on cryptographic binding between object naming and corresponding content, there remain many other security concerns to be addressed, and most of them are related to denial of service and fuzzy pub/sub message flooding or disturbance. As a possible solution, it may be interesting to explore how far probabilistic data streaming algorithms can be used in ICN to enforce rate limiting or signature analysis on content objects or messages exchanged between users. Policy-based routing with access control carried by network metadata or directly as labels within naming may also contribute to guarantee better security in ICN. Van Jacobson et al. mention, for instance, the notion of content firewall to filter objects based on specific range of the namespace in addition to digital signatures of Interest requests.

8.4.2 QoS Considerations

Quality of Service (QoS) refers to the ability to provide the optimal use of shared resources by scheduling these resources among different classes of traffic carried by the network. The primary goal is to apply different priority to different flows to increase the utility of the network and to guarantee the required level of performance. From the ICN viewpoint, the notion of end-to-end flow does not exist and any communication is an exchange of named content between producers and consumers. The fundamental concern in ICN is then how to prioritize information objects and how to define criteria or mechanisms that allow the network to provide QoS differentiation between objects. In other words, we need to formalize the unifying thread between information objects to define what a flow means in the ICN paradigm, or maybe we should define a novel notion of Quality of Information for a clean slate thinking. The challenge is especially difficult since contents are pervasive, replicated, cached, distributed, and may be accessed or originated from many sources (possibly from different access networks, having different capabilities) via many different paths, traversing ICN routers and non-ICN routers as well. None of the ICN solutions in the literature clearly addresses the QoS issue or proposes a study about resource management. Some just argue that the quality will be better because the content will be retrieved from closer sources; others say that ICN will improve QoS but without detailing why and how. A satisfactory solution to ensure QoS in ICN, depending on the context that binds content objects, might be based on defining a new identification scheme that could allow to embed QoS (and possibly context) information in content identifiers and allow its routing in the ICN. It could also be integrated with routing/queuing/caching mechanisms that enable the delivery with respect to the required QoS. Dynamic routing solutions based on the naming of the content as well as using network topology or other concepts such as network metadata may also be used to help in the distribution of information with the required QoS.

8.4.3 Scalability and Reliability

The main challenge ICN will have to face is to distribute billions of objects to billions of interconnected devices. A related problem is how to synchronize between content to avoid naming conflict or how to guarantee that names are globally unique. ICN must propose an efficient procedure for name allocation that should be distributed and self-manageable to cope with the huge number of objects in the network and to favor dynamically generated content. This reliability feature for naming has, however, a price regarding scalability. Due to the large spatial distribution of objects in the network, scalability is actually an important challenge for content storage and cache policy [8] as well as for naming, whether it is a flat naming as in DONA, a recursive naming as in CCN or a Multiple DHT as in NetInf where at the upper level every IO must be referenced and a total number of objects to be referenced reaching a factor of 10 over current addressing space. Scalability of routing (or name resolution capacity) in the ICN paradigm is also a concern. For example, in Van Jacobson's CCN, unbounded namespace could result in order of magnitude more prefixes in the FIB table. A CCN name is actually designed as a URL-like chain of characters, so how to name an object is quite an open issue, and how to efficiently build the routing and forwarding tables (PIT & FIB) for fast and efficient lookups needs to be analyzed to ensure system scalability. In particular, we need to find out if existing IP lookup techniques (such as hash-based design, using TCAM, trie-based schemes, etc.) could be brought to the CCN network. All these aspects should be carried out in a further research work. Proposed routing solutions based on DHT like NetInf or PSIRP may also suffer from the burden of resource discovery overheads. In actual fact, an important open question is the potential issues of backward compatibility with the current Internet architecture: How the ICN forwarding layer would behave and scale over the IP layer since it seems hard to do without IP?

8.4.4 Network Management and Better Manageability

Network management is lacking in most of the exiting ICN solutions and is only mentioned in PSIRP but without further description. As the term content object is generic in ICN, it does not necessarily refer to application-level data, but can also name network entities (host, link, domain, etc.) or more generally anything material or mental that may be perceived by the senses. Based on this extended definition, PSIRP argues that we can reuse the ICN model for network monitoring and management. It seems promising to explore deeply into this direction as ICN can be used to capture the knowledge about context information and may thus be more efficient to design networks with ease of configuration for management, leading to self-configured and self-optimized networking. Translating network protocols for collecting IP traffic information such as NetFlow or SNMP into the specific ICN context can also be of interest, for example, for content-based network billing. This appears particularly problematic since identity is separated from locator of content providers or consumers.

8.5 Conclusion

In this chapter, we have introduced the new promising Information Centric Networking paradigm that aims at overcoming limitations of the current Internet. We then presented and compared the main ICN solutions that currently exist, some having demonstrators proving the feasibility of the solution. Even if, for pragmatic issues and more short-term deployment, those prototypes are currently running over Internet, the defined ICN solutions have been designed to be run without Internet, but by replacing it (as the Internet did with the telephony networks). We then may imagine that future Internet will be based on ICN concept in some years; that is the reason why network operators as well as network equipment providers carefully investigate this paradigm, directly and via collaborative projects with universities. However, even if the existing solutions are promising, several issues and challenges we have identified at the end of this chapter still remain to be addressed accurately before a real and successful deployment can take place.

References

1. V. Jacobson, D. K. Smetters, N. H. Briggs, M. F. Plass, P. Stewart, J. D. Thornton, and R. L. Braynard, VoCCN: Voice-over content-centric networks. In *Proceedings of ACM ReArch'09*, Dec. 2009, Rome, Italy.
2. Conceptual Architecture of PSIRP Including Subcomponent Descriptions, Public Deliverable (D2.2) of the PSIRP project, http://www.psirp.org/files/Deliverables/FP7-INFSO-ICT-216173-PSIRP-D2.2 Conceptual Architecture v1.1.pdf
3. P. Eugster, P. Felber, R. Guerraoui, and A.-M. Kermarrec, The many faces of publish/subscribe, *ACM Computing Surveys*, 35(2):114–131, 2003.
4. T. Koponen, M. Chawla, B.-G. Chun, A. Ermolinskiy, K. H. Kim, S. Shenker, and I. Stoica, A data-oriented (and beyond) network architecture. In *Proceedings of SIGCOMM'07*, Aug. 2007, Kyoto, Japan.
5. V. Jacobson, D. K. Smetters, J. D Thornton, M. Plass, N. Briggs, and R. L. Braynard, Networking named content. In *Proceedings of ACM CoNEXT 2009*, Dec. 2009, Rome, Italy.
6. M. D. Ambrosio, M. Marchisio, V. Vercellone et al., Deliverable D6.2 of the 4WARD project, *Second NetInf Architecture Description*, Jan.2010, http://www.4ward-project.eu/index.php?s=filen downloadn&id=70
7. R. Ahmed and R. Boutaba, Distributed pattern matching: A Key to flexible and efficient P2P search. *IEEE Journal on Selected Areasin Communications (JSAC) issue on Peer-to-Peer Communications and Applications*, 25(1), 73–83, 2007.
8. S. Arianfar, P. Nikander, and J. Ott, On content-centric router design and implications. In *Proceedings of ReArch*, 2010, Philadelphia, USA.
9. http://www.ccnx.org/

10. PARC Technical Report NDN-0001, *Named Data Networking (NDN) Project*, October 2010.

11. PSIRP project, Publish–Subscribe Internet Routing Paradigm, http://www.psirp.org/

12. P. Jokela, A. Zahemszky, C. Esteve, S. Arianfar, and P. Nikander, LIPSIN: Line speed publish/subscribe inter-networking. In *Proceedings of SIGCOMM '09*, Aug. 2009.

13. N. Fotiou, P. Nikander, D. Trossen, and G. C. Polyzos, Developing information networking further: From PSIRP to PURSUIT, *International ICST Conference on Broadband Communications, Networks, and Systems (BROADNETS)*, October 2010.

14. www.fp7-pursuit.eu/

15. Netinf Website, Network of Information, http://www.netinf.org

16. http://www.sail-project.eu/

17. http://code.google.com/p/opennetinf/

18. R. Ahmed and R. Boutaba, Plexus: A scalable peer-to-peer protocol enabling efficient subset search. *IEEE/ACM Transactions on Networking*, 17(1), 130–143, February 2009.

Chapter 9

Toward Information-Centric Networking: Research, Standardization, Business, and Migration Challenges

Wei Koong Chai, Michael Georgiades, and Spiros Spirou

Contents

9.1 Introduction 164
9.2 Open Research Issues 165
    9.2.1 Content Resolution Approaches 165
        9.2.1.1 Lookup-Based Resolution Approach 165
        9.2.1.2 Hop-by-Hop Discovery Approach 166
    9.2.2 Caching in ICN 166
    9.2.3 Information Object Security 170
    9.2.4 Toward Mobility Support in ICN 171
    9.2.5 Anycast and Multicast in ICN 171
9.3 Business Incentives and Migration Challenges 172
    9.3.1 Content Owners 173
    9.3.2 Content Providers 174

9.3.3 CDN Providers 174
9.3.4 Network Operators 174
9.3.5 Application Developers 176
9.3.6 Content Consumers 176
9.3.7 Data–Money Flow Examples 177
9.3.8 Energy-Saving Incentives 179
9.3.9 Migration Challenges 179
9.4 Research and Standardization Activities 181
9.4.1 Research Activities 181
9.4.2 Standardization Activities 183
9.5 Summary and Conclusions 184
References 184

9.1 Introduction

In the previous chapter, various Information-Centric Networking (ICN) architecture proposals have been reviewed, highlighting the different approaches to achieve an Information-Centric Internet. This chapter continues the discussion on ICN, focusing on the different issues for bringing the concept into reality. These issues include the impact of ICN on the deployed Internet technologies, the business considerations of ICN, research and standardization activities and finally the challenges in smoothly deploying ICN over the current Internet.

Since ICN advocates redesign of some fundamental principles of the current Internet, its deployment will certainly affect the current networking protocols and technologies. Several open research areas where ICN will have strong implications are explained. This includes the overarching approach to how content* requests are resolved at the top-level and those issues concerning network operations, such as support for several sophisticated components of content distribution over the Internet including content caching, security, mobility, anycast, and multicast.

ICN envisions an open Internet marketplace, in contrast with the current view of the Internet as a bit-pipe. ICN is believed to have the effect of lowering market entry barriers (especially for new and small businesses) while allowing flexible business models. The implications of ICN on the future Internet business are discussed, highlighting the impact and incentives of different stakeholders. Already there are various research projects and standardization activities in progress around the world in the context of ICN. A specific section is dedicated on summarizing these efforts and provides readers a view of the momentum the ICN concept is carrying. The chapter finally elucidates on some of the main challenges on the migration of the current Internet toward ICN.

---

* "Information" and "Content" are used interchangeably in this chapter.

9.2 Open Research Issues

The focus of this section is on ICN open research issues at the networking level, that is, on ICN building blocks that are more related to the networking technologies deployed in the Internet today. As such, this pertains to those functions or capabilities that impact on the data plane where information packets are routed and forwarded under the different architecture paradigms.

9.2.1 Content Resolution Approaches

The first and foremost question on realizing ICN is on how content resolution is done. This includes the naming of content and has vast implications to most networking functionalities. It is important to first have a clear idea on the resolution strategies adopted by the various ICN proposals as the resolution and the delivery operations of information objects are often interrelated or in some cases tightly coupled. There are two main approaches to resolving a content request.

9.2.1.1 Lookup-Based Resolution Approach

The first being the *lookup-based* approach where the requested information object is discovered by querying some database or repository (most probably a distributed one for scalability and feasibility reasons). Examples of ICN architectures adopting such an approach are [1–4]. The approach shares some common features with the domain name system (DNS) [5,6] we have today. In DNS, a distributed hierarchical database, mapping Fully Qualified Domain Names (FQDNs) to IP addresses, is maintained in different DNS servers while analogously in ICN, an infrastructure of servers similar to the DNS servers maintains a distributed database translating content name to location(s) of the requested content. In this approach, the resolution and the delivery operations are decoupled whereby one operation follows another. This further implies that the whole content access will take two steps—first to contact the overlay resolution service to obtain the content location information and second to contact the server hosting the content for the initiation of the content delivery. A branch in this method exploits distributed hash table (DHT) (e.g., [7]) techniques such as Ref. [8], instead of relying on hierarchical tree structure. Conceptually, the resolution methodology is similar with the main difference being that instead of traversing specific branch of a tree in the hierarchy of the distributed database, the resolution follows a "trigger" (or finger) from one node to another which finally leads to the required information regarding the requested content. Note that the route from one trigger to another is not related to the actual content delivery route, in the general case. In other words, the final node discovered to "know" the host of the content may be far from the physical content itself. Figure 9.1 illustrates a generic architecture following this approach. The delivery of

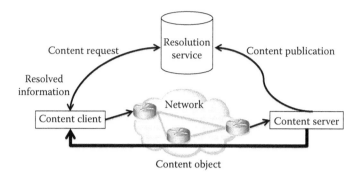

Figure 9.1 **Lookup-based resolution approach relying on a dedicated resolution service external to the network routing and forwarding functions.**

the content object only takes place after content client receives the resolved information from the resolution service.

9.2.1.2 Hop-by-Hop Discovery Approach

The second approach follows a *hop-by-hop discovery* concept where the content request is routed following some hints or information left in the network in order to find the server hosting the required content. Examples of ICN architectures adopting such an approach are [9–11]. This approach requires some sophistication in the content publication operation where information about the content server is inserted at certain places to facilitate the routing of content request toward the content server. When a content request is issued from a user, this request is forwarded following specific rules to find the trail left during the publication operation. The delivery of the content can then follow the "breadcrumb" back to the requesting user. Figure 9.2 conceptually illustrates the operation of this approach. For example, in Ref. [11], the *provider route forwarding rule* is proposed whereby the content publication message is passed along the provider of each domain up until tier-1 domains resulting in a situation where a domain always knows the content housed within and under it (i.e., all its customer domains). By following the same rule, a content request is guaranteed to find the content, unless of course the requested content does not exist.

In the following, we detail the implications and impacts of content delivery within the network, based on the two general approaches toward ICN.

9.2.2 Caching in ICN

One main idea of ICN is to leverage in-network caching technology to enhance content delivery in the Internet. This is made possible with the direct naming of content object and thus, each content object can now be uniquely identified and authenticated.

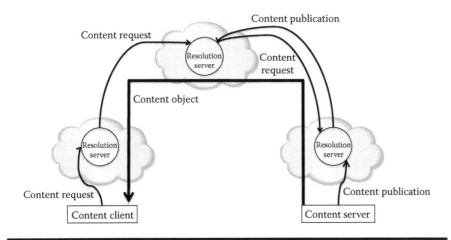

Figure 9.2 **Hop-by-hop discovery approach which resolves a content request following specific rules for guaranteed resolution.**

This, in turn, enables application-independent caching of content pieces in the network that can be reused by other users requesting the same content pieces.

Caching itself, in fact, is not novel. It has been researched in depth in various applications (e.g., microprocessor, RAID disk arrays, Web caching, etc.). In general, caching a less popular item (or conversely, evicting a more popular item) will always decrease the caching gain (i.e., decreasing cache hit probability). However, in practice, the *real* popularity of an item is usually unknown. In fact, if one knows the exact arrival time of each content in the future, then an optimal caching system can be constructed. This is known as Belady's algorithm which provides the theoretical upper bound of the caching gain.

In the context of content distribution in communication networks, caching comes in two forms:

■ Caching *outside* the network—In this form, the content is cached *externally* from the network elements. Typically, this is realized using overlay approaches where the content is stored in entities dedicated for caching purposes. One example of this is the Web proxy caching (e.g., [12,13]) where popular Web pages are cached in Web proxies planted at strategic locations that expect high demand for that content. Content distribution networks (CDNs), on the other hand, leverage caching via surrogate servers [14], which mirror the original server. CDNs then can balance the content demand among them. However, this is usually done without consideration on the underlying network conditions.

■ Caching *inside* the network (also known as in-network caching)—In this form, the content is cached *within* the network elements (e.g., routers). A study on cooperative content caching between Internet Service Provider (ISP) networks is presented in Ref. [15] where it shows such collaboration can reduce overall content distribution costs. Specifically in ICN and on a finer granularity, Jacobson et al. [9] proposed to exploit in-network caching in every router that the packet traverses. Finally, this approach is also being investigated in the IETF Decoupled Application Data Enroute (DECADE) Working Group [16].

The fundamental rationale of in-network caching is to reduce content access latency by enabling routers to provide content delivery service via their cache storage. Specifically, it provides faster response by storing the content (either as a whole, in chunks or in packets) closer to the requesting users and avoids fetching the same item again all the way from the source, which has higher overhead. For instance, a content request will be able to avoid traversing the full path to the content source if the requested content is cached in the intermediate routers.

Figure 9.3 gives a conceptual illustration of the benefit from leveraging in-network caching technology. In today's Internet, any request for content has to be served by a server, which can become easily overloaded. This is especially the case when the server is serving popular content. The current solution to alleviate the server strain is to maintain surrogate servers at strategic locations where the demand is expected to be high. This workaround, however, still requires high manual maintenance and configurations and yet each request still "hits" a server. With in-network caching, every router along the content delivery path can help lessen the content server load by caching the content packets in their cache buffer. By this, effectively, a router emulates the function of a server. When a content becomes very popular, it will quickly populate the caches along many routers (refer to the Figure 9.3b) and thus, majority of the requests will not be hitting the content server itself but being served at intermediate routers. In other words, in-network caching provides a more natural mechanism to lessening server load. This can be further exemplified in the so-called *flash crowd effect* where some content suddenly become very popular (e.g., viral videos, presidential speeches, etc.).

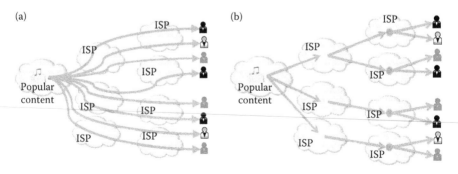

Figure 9.3   In-network caching alleviating demands from the content server: (a) content server must serve each and every request; (b) intermediate caches can also serve content request and thus, not all requests reach the content server itself.

Let us consider a simple numerical example following the proposed caching method in Ref. [9] where a content packet is cached in each and every router it traverses with each router evicting content packets following the least recently used (LRU) rule. We generalize the example for any unit of content rather than just a content packet (i.e., a unit of content can be the whole content object, a chunk of it or a content packet). Further, assume that the arrival of a unit content i, $\lambda = 100$ unit content/s while the arrival of other content, μ is 9900 unit content/s (both following Poisson distribution). Each router has N slot of caching space (homogenous router scenario) whereby each slot can store one unit content. Finally, let n denote the number of hops from content client requesting content i. Based on the model presented in Ref. [17], we show in Figure 9.4 how the mean number of request for content i decreases as the number of hops from content client increases with the simplifying assumption that μ remains constant after each hop. From the figure, we can see that only approximately 13.68% and 65.58% of request propagates upstream after five hops for $n = 10$ and 100, respectively. Clearly, the size of the cache in each intermediate router has a significant impact on the caching gain where we see with 10 times the caching space, the server hit is reduced by approximately a further 50%.

Although the gain in exploiting such in-network caching technique in ICN is clear, it remains a topic for further research on the best strategy to cache content conditioned by the deployability of the solution in the real world.

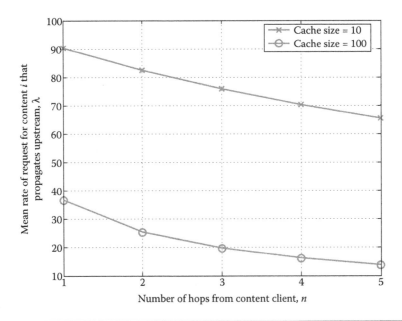

Figure 9.4   Mean rate of request for a specific content lessens after propagating each hop toward content server.

9.2.3 *Information Object Security*

One of the main motivations in advocating ICN is on security, more specifically, the content authenticity and integrity. By binding security features to content names, content can be delivered in an application agnostic manner. Security is one important component that is missing in the original design of the Internet. Over the years, various security mechanisms have been added on top of the Internet. Today, security considerations focus on authenticating the host and securing the transmission infrastructure (e.g., the transmission channel). As such, the content itself is not directly authenticated or secured. A content is usually considered authentic if the host serving the content is able to authenticate itself. Users often care more for the content itself than for the actual content source(s). In ICN, content naming schemes are designed to address this issue by addressing the content directly with certain security features. The basic rationale is that users should be able to verify the content *directly* instead of relying on the so-called *secure* transmission medium or *authentic* content providers vouching for the authenticity of the content.

Well-known security concepts are being applied to content names for content authentication. The most prominent approach is using asymmetric key-based cryptography algorithms. In Ref. [10], the content name contains the hash of the public key of the publisher (known as the "principal"). A drawback of this scheme is that a change in the principal may result in name change (names are not persistent). NetInf [1] avoids this issue by allowing the decoupling of content owner authentication and content authentication itself. The name structure here (called information identifier (ID)) consists of three fields—(1) type, (2) authenticator, A, and (3) identifier tag, L—with an additional metadata structure inclusive of security metadata. The authenticator field binds the ID to a public key of the content owner through a one-way cryptographic hash function which together with the identifier tag must be globally unique. This allows a similar method in Ref. [10] for the authentication of the content owner. The content authentication, on the other hand, relies on *s-certData* which is generated using selected parts of the content and/or metadata of the content. It can be embedded into the L part of the ID for static nonmutable content or into the metadata for dynamic content. The PSIRP project [18,19] follows a similar approach whereby two namespaces are defined—(1) rendezvous identifiers and (2) forwarding identifiers—which can be constructed on either a hash of the content object or the content owner's public key.

In Ref. [9], content packets contain digital signature information for authentication of the binding between the name and their corresponding content. The signature in each content packet includes a standard public key signature to enable verification on the name-to-content binding signed by a specific key. At such fine-grained granularity (i.e., each packet being verifiable individually) the overhead in signature generation alone may be high.

9.2.4 Toward Mobility Support in ICN

The growth of content in the Internet is tremendous. Within this, it is notable that mobile content constitutes a major part of this increase. However, the original design of the Internet has neglected the mobility aspect, designed mostly for static hosts and resources in mind. Mobility capabilities in the current Internet are added after the proliferation of mobile wireless communications clearly underline the need for such support. The most prominent solution to the mobility issue in the Internet is Mobile IP [20] which relies on the concept of redirection whereby when a mobile node exits its *home* domain, the packets destined for this mobile node are first sent to its *home* before being redirected to the domain the mobile node is currently attached to. Various challenges and issues are well known following this approach. Security and inefficiency being two examples. The fundamental problem, however, is the fact that the IP address is used as both locator and identifier, forcing a routing of packets following the location of the host rather than the location of the content.

The Host Identity Protocol (HIP) [21] has been proposed for the support of mobility by separating the host location from the host identity via the introduction of a new namespace with cryptographic capabilities for host identities, with the IP addresses used only for packet routing purposes.

■ Host identity = a public cryptographic key of a public–private key pair

The public key identifies the party that holds the only copy of the private key. Packet transmission is preceded by an identity verification stage using IPSec. Using this approach, a host location change does not break the connection as packets are routed based on the identity. Additionally, thanks to the cryptographic features included in the protocol, a host receiving an unintended packet will not be able to open it since it does not possess the correct private key. Note that this proposal focuses on the host identity separation but does not explicitly consider the content-centric networking issues.

This represents a step forward by decoupling the host identifiers and host locators. In fact, in Ref. [10], HIP is suggested as one of the possibilities for enabling mobility support in ICN. In that proposal, the mobile host follows the *route-by-name* paradigm with the added operation of unregistering from the previous location and reregistering itself to the new location, relying on HIP (or other similar approaches) to maintain the continuity of the content delivery.

9.2.5 Anycast and Multicast in ICN

The host-centric Internet today builds its foundation on a 1-to-1 communication model. As such, the more sophisticated communication models (e.g., 1-to-Many and Many-to-Many communication models) are too complex to realize especially in large-scale and interdomain scenarios. A clear example is on interdomain multicasting (i.e., 1-to-Many) never really gain wide deployment despite various efforts.

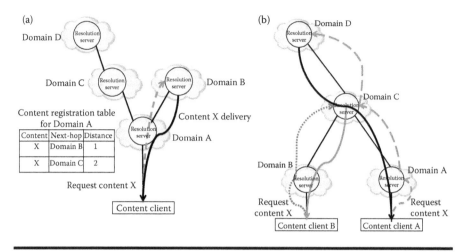

Figure 9.5 ICN enabling more sophisticated communication models: (a) Anycast via name-based primitive; (b) multicast via state-based approach.

In Ref. [10], anycast is realized naturally via the proposed name-based anycast primitive that provides a content discovery mechanism that is based on path-labels rather than global addresses (e.g., IP). A content registration table mapping content name to both the next-hop domain and the distance to the copy (e.g., hop count) is maintained in each domain (see Figure 9.5a). When a request is received, it is routed to the next-hop domain with the shortest distance. In the example figure, domain B is only one domain hop away and thus is nearer compared to the other route via domain C which is two domain hops away. So, the content request is being forwarded to domain B (dashed line).

On the other hand, in Refs. [4,11], interdomain multicast is inherently supported via a state-based content delivery approach with the discovery of content relying on the *provider route forwarding rule*. During the content resolution process, multicast-like content states are installed along the resolution path such that the content is delivered by retracing the path taken by the content request. To illustrate, consider Figure 9.5b where content client A first requested content X. The dashed line indicated the resolution path while the black solid line indicated the content delivery path with states being installed in domain A, C, and D. When content client B requests the same content X, the content is multicasted from domain C without going through domain D (see dotted line for the resolution path and gray solid line for the delivery path).

9.3 Business Incentives and Migration Challenges

This section focuses on the business impact and incentives of the different stakeholders involved in the evolution of their network infrastructure to support the

ICN paradigm. Establishing new business models in the context of ICN is as critical as the development of the technological architectures themselves. The current Internet economic model is quite restrictive as opposed to the flexible and open one targeted by ICN concepts.

Nevertheless, although conceptually, the opening of the Internet marketplace can future-proof the Internet, the traditional stakeholders require clear incentives (especially profits or cost reduction) before embracing ICN ideas. As of now, ISPs are yet to be convinced that ICN can provide sufficient return for covering the new infrastructure deployment cost. Content owners and providers may not trust the ubiquitous caching of their content in the network itself. It is also the nature of an ICN architecture to depart away from how traditional traffic is exchanged across network service providers (NSPs) due to the fact that they are receiver oriented and that the forward and reverse paths are asymmetrical. These intrinsic features should be taken into consideration when forming new business models among the different stakeholders, both the traditional ones and newly emerging actors (e.g., prosumers through user-generated content (UGC)).

This section provides a description of the impact of ICN on the different business actors and associated interactions. It also attempts to identify the incentives of each actor involved to highlight the motivation of moving toward ICNs. In Ref. [22], related deployment issues and incentives of the different actors involved are discussed. The following types of actors are considered here: (1) content owners, (2) content providers, (3) CDN providers, (4) network operators, and (5) content consumers. Advantages and disadvantages are considered and discussed for each actor. Of course, some organizations or companies may cover one or more of the above identities but we divide it categorically for analysis purposes.

Moreover, this section discusses the need for new pricing schemes and summarizes different possible business models between the involved actors, some of which are derived from existing models on CDNs. A note is also made on energy efficiency as a socioeconomic incentive. Finally, migration to ICN is considered.

9.3.1 Content Owners

ICN will trigger the demand for new business models and in this case possible new business deals between content owners, network operators, content providers, and in general all the actors involved. There is ongoing research in projects like the EU ICT project OCEAN [23] which aim to elaborate on business strategies to provide better investment incentives to the different types of players in the value chain. Content owners like, for example, MGM Studios are seen to be the main pushing force behind the deployment of ICN.

ICN will support content owners to distribute their content on a large scale in a fashion similar to how it is done today, that is, ICN should not force drastic changes to how content is published for the access of the public. Content owners may still opt to pay for CDN-like services but the underlying mechanism of content publication (e.g., content naming, resolution) should be transparent to them. This will mean that a Content Owner will save the costs that it will otherwise inherit if it used a service provided by a CDN provider. To achieve this, however, the ICN services must offer the necessary confidentiality, reliability, and quality-of-service assurance to avoid the need of enforcing the content owners to establish SLAs with every network operator covering the regional area or targeted customer group of interest.

9.3.2 Content Providers

The introduction of new ICN business models as with CDN models will have a heavy impact on content providers like YouTube or other content hosting services as the primary aim is to reduce the traffic load and hence external hosting requirements. The aim of both CDNs and ICNs is to support content provisioning from local servers and hence reduce the load on origin servers, reduce bandwidth requirements, which will imply fewer and smaller hosting servers and mirrored servers.

Content providers will need to rethink of their business plans and may have to consider models for supporting CDN and ICN deployment. In the case of CDNs they should focus on out-of-network distributed hosting services to support ISPs and NSPs as well as consider offering CDN services themselves. In the case of the ICNs and in network caching architectures, Hosting Providers may provide, for example, backup services, mirrored services, or better content distribution on a global scale.

9.3.3 CDN Providers

Information-Centric Networks will have a major impact on CDN providers like Akamai as in contrast they can support in-networking caching, as opposed to out-of-network caching, which is how currently CDN providers approach the problem of distributing content Internet-wide.

However, CDN providers can opt to provide service to content owners as well as ISPs and NSPs. In all cases they could provide better distribution of content and possibly support external quality-of-service assurance.

CDN providers can also consider supporting ICN between the involved actors. For example, what is known as a Content Bridge Alliance could be responsible for sharing the revenues between all actors involved in the provisioning of ICN and CDN servers.

9.3.4 Network Operators

The steep increase of high-quality content has increasingly strained network operator's access and connectivity services. Paying users are demanding better quality content (e.g., upgrading from standard definition to high-definition/3D resolution) which require higher bandwidth availability. In ICN, via the in-network caching technology, popular content can be cached locally and thus localizing the relevant traffic, enabling ISPs to save the cost of interdomain traffic.

Moreover access-centric CDN models have been considered where network operators pay access-centric CDNs to serve highly demanded content from caches near the network operator's subscribers. By doing this, network operators will save on bandwidth and capacity by utilizing local storage, and since upstream bandwidth is the highest expense for a network operators, the network operator will minimize expenditure and hence profitability.

Besides access provisioning to support end-to-end Internet connectivity, currently a network operator has to pay its neighboring network operators for connectivity according to the link capacity provided and the traffic volume exchanged in both directions. This traffic is generated by the network operator's users either send or receive content both from within the operator's network or (with legacy architectures) more likely from the rest of the network. Deploying ICN services as part of its network, a network operator will be able to store content on local caches in its network and using these as the source of content when distributing content to its users. This will imply that the use of local caches will reduce the volume of traffic to and from other neighboring operators and hence minimize these costs.

Caching inside the network means upgrading legacy routing equipment like routers, to maintain copies of the desired content. This will impact directly on the network operators who are willing to support ICNs in this fashion.

Of course, upgrading to support ICNs there will be an additional cost incurred to accommodate for the additional building blocks required to support the architecture (as discussed in the previous section). For example, the storage required to support ICN caching is not part of the legacy network equipment that access operators currently have. For these to be deployed network operators will face deployment costs as well as operational costs. If both the deployment and operational costs for caching and data delivery are less than the costs charged by neighboring NSPs for the same traffic delivery, ICN will be worth considering.

Transit network operators, that is, network operators who do not provide access but simply receive and forward traffic will be affected somehow differently by the deployment of ICN. If the primary aim of ICN is the better distribution of content, this will certainly reduce traffic between the different intermediate network operators which is the main source of revenue for them and from which for the majority of the traffic received they could simply have been forwarding a copy of the same content end-to-end, time and time again. As mentioned earlier, network operators will have to pay their neighbors for the traffic volume they send and if ICN reduces this traffic exchange-based profits will also be reduced. The same will apply to any network operator acting as a backbone provider.

There are several projects with aim to propose new ecosystems for better interworking between Internet carriers and attempt to go a step further to how the business models work currently with legacy networks and the Internet. One such project is the EU ICT ETICS project [24] which aspires to propose a new technological perspective with respect to the current availability of static intercarrier services. It aims at accelerating the creation, management, and deployment of new business models and network services interconnection for fixed and mobile technologies. New business model proposals that are derived from projects like ETICS or are based solely on interbusiness interests consider ICN applications as one of their major use case scenarios.

9.3.5 Application Developers

The existing ICN proposals have mostly considered the fundamental application protocol interfaces (APIs) for enabling ICN. These are mostly related to the publication and resolution of information object (e.g., primitives such as publish, register, get, find, consumer, etc.). Beyond these, there is wide space for application developers in creating new applications especially when ICN allows very flexible communication models and is not restricted to the current point-to-point communication paradigm. Moreover, the ICN APIs which provide direct access to the information objects open great opportunity to application developers. Information objects can be used by several applications allowing for common quality provisioning, allow for better quality assurance monitoring and make it easier to differentiate between high- and low-quality media.

ICN architectures can enable new types of applications that were too complex to consider for deployment in traditional networks. As discussed in the previous section anycast and multicast supports are inherent in an ICN. Hence, the support for P2P applications, anycast-based applications and multicast-based applications can be accommodated in a more homogeneous environment. Furthermore, high-quality video streaming, time-control Internet/network applications, distributed server/client applications can be hosted by a more supportive underlying architecture.

9.3.6 Content Consumers

ICN should benefit the content users directly without requiring many adaptations (e.g., update to the client machines or applications). In fact, many of the core features of ICN originate from the users' point of view (e.g., better security, support for mobility, multicast, and anycast). Furthermore, the deployment of ICN should be transparent to the users while allowing them to reap the benefits of the change.

By embracing ICN, content access is expected to be defragmented as opposed to today's Internet where specific content is often restricted to some groups of users (content islands) through different intermediaries (e.g., media aggregator such as YouTube and GoogleVideo, CDNs such as AKAMAI and Limelight). Users can reach content regardless of the hosting platform due to the globally deployed content naming scheme.

The delivery of the content to the users will also be much enhanced. The latency of content delivery can be much improved for popular content through in-network caching (as illustrated in Section 9.2.2). Security issues such as content authenticity can now be verified directly without relying on the host server or the communication infrastructure. Mobile users also benefit from the envisioned ubiquitous ICN support through some mobility mechanisms.

9.3.7 Data–Money Flow Examples

Figures 9.6 through Figure 9.8 attempt to illustrate a possible money flow model for purchasing a video movie online by using the legacy system, a CDN-based system and an ICN-based architecture. Figure 9.6 illustrates how a movie may be purchased online by a customer from a video owner using the legacy system, Figure 9.7 illustrates how a movie may be purchased by a customer from a video owner using CDNs, and Figure 9.8 illustrates how this can be achieved using ICNs. In all diagrams we assume that both the customer and the studio pay a certain tariff to their network operator for Internet service access. In the case of Figure 9.6 and the legacy system the customer pays the studio €12.00. From this the Studio banks €8.00, pays the Media Grid Engine System (MGE) provider €0.10 for external quality assurance monitoring at all different points in the network operator chain, and pays the hosting provider €3.90 for storage and download availability. The hosting provider pays their network operator €3.00 for network access provisioning which in the end will be shared according to traffic volume exchange plans and agreements between the other involved network operators and backbone providers.

Figure 9.7 shows a possible movie purchase use case scenario in the case of CDNs. For this scenario the customer is still charged €12.00 for a movie and the Studio banks €8.00. In this case, however, the Studio provider pays €4.00 to the content provider to take care of the content provisioning and distribution of the movie. The

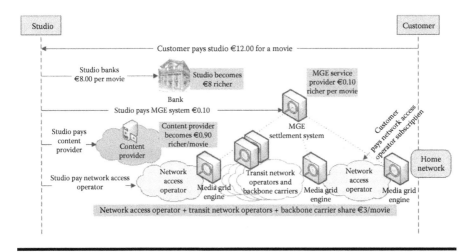

Figure 9.6 A movie purchase use case—the money flow in the legacy model.

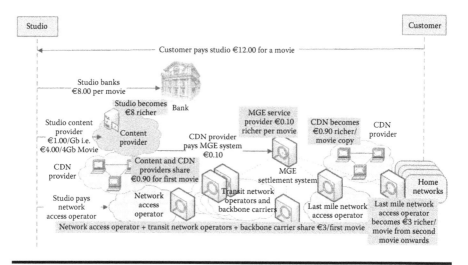

Figure 9.7 A movie purchase use case—the money flow in the CDN-based model.

CDN provider now pays MGE for quality-of-service assurance monitoring, keeps €0.90 and pays the rest to the network operator for their network service provisioning which is shared among them for the first copy of the movie, that is, between all involved network operators and backbone providers. Due to the good distribution of the CDN hosting only the last mile or leaf network operators will be required for any further local downloads and hence only the last mile network operators will gain future revenues for connectivity in transferring future copies of the same content (movie) locally.

Figure 9.8 shows the movie purchase use case scenario but in this case with the network operators following the ICN paradigm. In this case any hosting is taken

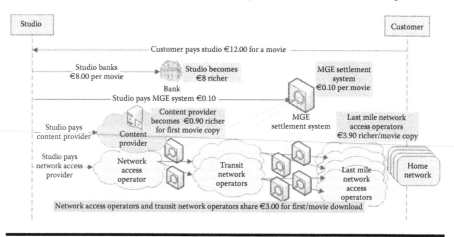

Figure 9.8 A movie purchase use case—the money flow in the ICN model.

care of by the network operator's equipment and their high storage capacity. As with the other scenarios the Customer pays the Studio €12.00 for purchasing a movie, from these the Studio banks €8.00, and pays MGE €0.10 for monitoring the quality of service at different points in the network operator chain. One possible business scenario is that the Studio will then pay the last mile NSP for providing local content provisioning to its customers in a remote location. Other possible business scenarios or general ecosystems can be defined for future scenarios that will be influenced by ICNs. There are currently a number of ongoing research projects attempting to bring to light new ecosystem proposals between all involved business entities [23,24] as the different business interactions between the stakeholders is getting more and more complex making it riskier for them to participate. Figures 9.6 through 9.8 should be treated as possible money flow examples to highlight the different environments, that is, legacy, CDNs and ICNs, and associated money data interactions. Besides research projects, business stakeholders may bring their own attractive business proposals to the table attracting other influenced parties and creating new ecosystems that could evolve around the deployment of ICNs.

9.3.8 Energy-Saving Incentives

Economics and Technologies for Inter-Carrier Services [25] emphasizes on how moving from a host-centric to information-centric networking will open up new possibilities for energy-efficient content dissemination. The analysis shows that content servers are an order of magnitude less energy efficient than networking devices such as routers, gateways, and optical multiplexers. Moreover, as shown in the previous sections anycast models which identify shortest path caches have been proposed and explored for further energy saving.

9.3.9 Migration Challenges

ICN may not be mature yet, but we need to start thinking of ways to deploy it, when that time comes. Much of interesting networking research remained obscure because it lacked a migration path.

The Internet permeates so many parts of our social and business lives that when we want to (or have to) update it we must do it while it is in operation. This has been likened to changing engines on an airplane in mid-flight. The most daring ICN proposals call for nothing less than a change of the Internet engines. Most of the proposals, however, take a less revolutionary stance and suggest a migration path from the current Internet to an ICN Internet. However, "less revolutionary" does not necessarily mean "easy."

Content providers, network operators, and end-users have the most stakes in ICN deployment. These main stakeholders are already using the Internet to publish, distribute, and consume content. If they are to deploy and adopt ICN in their operations they will need clear incentives.

Content providers, of any size, can clearly benefit from ICN, as they will be able to make their content available to every consumer in the world by publishing it only once. Equally important for content providers is the promise of ICN to distribute content quickly and reliably. Content providers today have to pay CDNs, like Akamai, for good quality in content distribution. With ICN aspiring to replace CDNs, content providers can in theory forego that cost. It is conceivable that content providers will pay network operators, instead of CDNs, not just for connectivity but also for ICN deployment. On this point though, the incentives are becoming confusing and deals between providers and operators must be scrutinized before benefits are clarified. But if content providers will only be reaping benefits from ICN without paying to deploy ICN and maintain it, some other stakeholders will have to take over that burden. Wide market reach and premium quality distribution might be good incentives for content providers to support ICN, but they will also need to bill their customers and to control their service. This means that ICN should expose logging and management interfaces, which are somewhat still neglected in current ICN research.

On the other end of the content distribution chain are end users, or more appropriately, content consumers. This is another stakeholder who will mostly benefit from ICN. Consumers will be able to access content uniformly, that is, without having to adapt to the interfaces and tools of content intermediaries, like BitTorrent, YouTube, and so on. They will also have to worry much less (if at all) about the location of content and their own ability to consume that content when they are mobile. Finally, consumption will be faster and more reliable, unlike today's occasional glitches and outages. It is questionable though if content consumers will be willing to pay more for content providers and access network operators to deploy ICN. That burden is again on somebody else's shoulders. On the other hand, end users could donate their computational and storage resources for the deployment of ICN, much like the model for peer-to-peer networking. However, just as in peer-to-peer networks, such an ICN deployment would suffer from the lack of network information and support.

Given the unclear incentives for content providers and end users to finance the deployment of ICN we now turn our attention to the only remaining stakeholder: the network operator. A good incentive for network operators would be the reduction of operating expense. Access network operators spend money on managing their infrastructure and pay higher-tier, "transit" operators for connectivity services, often based on traffic volume. Reducing management cost seems unlikely, as ICN will also require management effort (unless some future proposal emphasizes self-management). Reducing transit connectivity cost with ICN is more realistic, but only for ICN approaches that adopt in-network caching. The effect would be traffic localization with an analogous decrease in transit traffic (and cost). On the flip side of this coin, however, transit network operators would lose revenue. The business model of transit operators is generally to charge more for the more traffic they deliver. Therefore, transit operators would not only resent the deployment of ICN (with in-network caching) at access operators' networks, but would also never deploy ICN in their own networks. In effect, this behavior would create "ICN islands" at the leaves of the Internet and "ICN darkness" everywhere else. As such, end-to-end ICN capabilities would be questionable.

There is clearly a lot of ground to cover for providing incentives to ICN stakeholders. They all need to be convinced that ICN is for the benefit of everybody, before they start deploying it with their own resources.

9.4 Research and Standardization Activities

Several research projects have been set up in the past couple of years to actively pursue technical work on ICN. More recently, standardization of ICN has become a point of discussion. In this section we provide an overview of both of these activities.

9.4.1 Research Activities

Research projects that are exploring ICN are under way both in the United States and in Europe. Cross-pollination among these projects has been possible through a number of events, including the Dagstuhl Seminar on ICN [26], the ICN Session at FIA Budapest [27], and the ACM SIGCOMM Workshop on ICN [28]. We give below a brief description of these research projects:

COMET: COntent Mediator architecture for content-aware nETworks [29]. Telefonica R&D is leading a consortium of 6 partners into an ICN approach involving two planes: the Mediation Plane and the Forwarding Plane. Content publication, name resolution, route discovery, and route selection belong to the Mediation Plane, while route configuration and content distribution are in the Forwarding Plane. COMET assigns hierarchical and globally unique names to content objects and resolves those names to the object's location through a distributed database. The content object information in each node of the distributed database can be coupled to the topology of the node or not. This gives rise to two flavors of the COMET approach that are under investigation for their relative merits. The path from the content server to the content consumer in COMET consists of one or more mediation domains that include forwarding elements. The forwarding elements are content-aware and use this information to prioritize flows.

CONVERGENCE [30]; The Italian Inter-University Consortium for Telecommunications (CNIT) is the coordinator of CONVERENCE, which includes 12 partners in total. The Versatile Digital Item (VDI) is central to CONVEREGENCE's approach and constitutes a generic data container, identified by a unique, host-independent, self-certifying name. The VDI is derived from the MPEG-21 standard and includes content metadata. Providers of VDIs can publish them on the network, while consumers can subscribe to receive VDIs, making use of a middleware that exposes network functionality. The consumer request is routed through the network using the VDI name and content is sent from the serving node making use of source routing.

NDN: Named Data Networking [31]. Ten institutions are participating in NDN, which has sprung from the Content-Centric Networking (CCN) project at the Palo Alto Research Center (PARC). NDN uses hierarchically structured names to identify content. Consumers transmit "Interest" packets for these names on all their interfaces. The Interest packets are picked up and re-transmitted (intelligently, in the steady state) by intermediate nodes, until they reach a node with the matching content. Data packets then flow back along the path of the retransmitted Interest packets and are optionally cached in intermediate nodes for future requests. NDN also proposes a security model that seeks to secure the content itself, as opposed to the endpoints.

CONNECT: Content-Oriented Networking: A New Experience for Content Transfer [32]. This consortium of 6 partners is led by Bell Labs - Alcatel-Lucent Telefonica. CONNECT uses the work of the NDN/CCN project as a departure point. However, CONNECT argues that TCP is not an appropriate transport protocol for ICN and focuses on an enhanced (or new) protocol with fairness and service differentiation characteristics. The scalability of the NDN approach is also a point of investigation for CONNECT, with an emphasis on implementation. Finally, CONNECT looks to devise and evaluate cache management strategies that would be appropriate for in-network caching, as described by NDN.

PURSUIT: Publish Subscribe Internet Technology [33]. A successor to the PSIRP project, PURSUIT is led by Aalto University and involves 8 partners. In PURSUIT, content providers publish content, while content consumers subscribe to content. Content is named with a statistically unique "Rendezvous Identifier" that is associated with a "Scope Identifier" during publication. The in-network Rendezvous functionality matches subscriptions to publication and then multicast forwarding trees are generated to distribute content. While PURSUIT proposes significant changes in the Transport and Network layers, PURSUIT can also function on top of IP.

SAIL: Scalable and Adaptive Internet Solutions [34]. SAIL builds on the work of the earlier 4WARD project and consists of 24 partners, with Ericsson as the leader. A main component of SAIL is the Network of Information (NetInf) architecture. In NetInf, an Information Object has a globally unique hierarchical name, which is used as an index to a record with details about the object. The name and the record are created during publication. A client uses the name to request content and the Name Resolution System (NRS) retrieves the associated record, which points to object replicas. Replicas can reside in provider servers or in in-network caches. The NRS directs the client to the best replica.

9.4.2 Standardization Activities

The multitude of research projects working on ICN constitutes evidence that there are still several open questions in the area. As such, it is difficult to engage in focused standardization of ICN at this stage. However, an ICN Research Group [35] is in the process of chartering within the Internet Research Task Force (IRTF) to pursue pre-standardization work. The ICNRG aims to identify architectural invariants within the various ICN approaches and to pinpoint interfaces toward applications, existing network elements, and between major ICN elements. The group will also foster the creation of an experimentation framework, so that comparison of different approaches becomes possible. A set of documents will be produced, including a problem statement, a survey of existing approaches and an architectural framework. The long-term goal is to provide recommendations to the Internet Engineering Task Force (IETF) for initiating ICN standardization work.

Several active (or concluded) Working Groups (WGs) of the IETF are relevant to ICN aspects. This relevancy is not intentional, but rather circumstantial as ICN tends to cover a large area, ranging from object naming to data routing. We summarize below some of those IETF WGs.

> *urnbis: Uniform Resource Names, Revised* [36]. URNs are location-independent, persistent identifiers for information resources. Names for content objects in ICN should similarly possess location-independency and persistency properties. urnbis is solidifying the definition of the "urn" URI scheme, updating earlier RFCs on URNs with consistent terminology and description, modernizing the foundations of URNs, and revisiting some namespace registrations (such as International Standard Book Number (ISBN)). It is conceivable that ICN registers a namespace with IANA, in which case urnbis would be appropriate in lending support.
>
> *dnsext: DNS Extensions* [37]. Some ICN approaches include distributed and hierarchical databases to store records about content objects and to map object names to object locations. This is similar to DNS and there have been hints and even proposals for extending DNS for use in ICN. Although extending DNS records is not in the dnsext charter, dnsext is the appropriate venue to discuss this possibility.
>
> *cdni: Content Delivery Networks Interconnection* [38]. The cdni WG was recently set up to create protocols for the interconnections of CDNs belonging to different administrative domains. cdni focuses on protocols for request routing, transport metadata, logging, and content control. The interconnected CDNs should support the end-to-end delivery of content from any attached provider to any attached client. This goal is shared by ICN, which on occasion can be viewed as a distributed, public CDN. Moreover, cdni interfaces can have their counterparts, at least in some ICN approaches, so cdni protocol work can prove beneficial for ICN purposes as well.

rtgwg: Routing Area Working Group [39]. rtgwg acts like a "catch-all" WG for routing topics that do not fit in an existing WG or do not justify the creation of a new WG. Routing content from a server to the client, taking into account application requirements and network status, is a major element of some ICN approaches, going as far as to disrupt the existing routing fabric. Discussion on such work within rtgwg can be beneficial to all involved parties.

9.5 Summary and Conclusions

ICN advocates rethinking of the very fundamental design principles of the current host-centric Internet that focuses on point-to-point interconnections. Realization and deployment of ICN will require changes in many aspects of today's network operations. ICN affects basic operations in the Internet, such as signaling, transport, routing, and forwarding functions. Meanwhile, it brings forward natural solutions to many of the issues accrued from the incremental development of the Internet (e.g., mobility, new and emerging application and services) against the ossification of the Internet design itself. The naming of information object opens the avenue for binding security features into the content. Furthermore, the secure binding of name to data enables the exploitation of in-networking caching for enhanced delivery of content. At present, many of these aspects remain as open research issues and are the focus of various research projects and standardization efforts.

As many people have come to rely on the Internet for their social and business needs, any ICN deployment needs to be gradual. More importantly, it needs to make business sense for content providers, network operators, and even content consumers. Current business models might not be compatible with ICN aspirations, so ICN research should also extend to cover socioeconomic issues. Moreover, all business stakeholders should rethink their business plans, as money–data flow interrelationships can change with ICN. The bottom line is that if ICN is too complex or too risky for business, stakeholders may decide not to support it.

Quite a few research projects in the United States and in Europe are investigating ICN, with some good proposals already emerging. A couple of ICN workshops have already been held and projects are increasingly seeking common understanding and collaboration. Even at this early stage, prestandardization activity is under way within the IRTF, targeting IETF standardization when maturity allows it.

References

1. B. Ahlgren et al., *Second Netinf Architecture Description*, 4WARD EU FP7 Project, Deliverable D-6.2 v2.0, April 2010.
2. B. Ahlgren et al., Design considerations for a network of information, in *Proc. ReArch '08: Re-Architecting the Internet*, Madrid, Spain, December 2008.

3. P. Jokela, A. Zahemszky, C. E. Rothenberg, S. Arianfar, and P. Nikander, LIPSIN: Line speed publish/subscribe inter-networking, in *Proc. ACM SIGCOMM '09*, Barcelona, Spain, August 2009.

4. W. K. Chai et al., *Interim Specification of Mechanisms, Protocols and Algorithms for the Content Mediation System*, COMET EU FP7 Project, Deliverable D3.1, January 2011.

5. P. Mockapetris, Domain names—Concepts and facilities, *Internet Engineering Task Force (IETF) Request for Comments (RFC)*, RFC 1034, November 1987. Available: http://www.ietf.org/rfc/rfc1034.txt

6. P. Mockapetris, Domain names—Implementation and specification, *Internet Engineering Task Force (IETF) Request for Comments (RFC)*, RFC 1035, November 1987. Available: http://www.ietf.org/rfc/rfc1035.txt

7. I. Stoica, D. Adkins, S. Zhuang, S. Shenker, and S. Surana, Internet indirection infrastructure, in *Proc. of ACM SIGCOMM '02*, pp. 73–86, Pittsburgh, PA, USA, August 2002.

8. I. Stoica, R. Morris, D. Karger, M. Frans Kaashoek, and H. Balakrishnan, Chord: A scalable peer-to-peer lookup service for Internet applications, *ACM SIGCOMM 2001*, pp. 149–160, San Diego, CA, August 2001.

9. V. Jacobson, D. K. Smetters, J. D. Thornton, M. F. Plass, N. H. Briggs, and R. L. Braynard, Networking named content, in *CoNEXT '09*, pp. 1–12, New York, NY, USA: ACM, 2009.

10. T. Koponen, M. Chawla, B-G. Chun, A. Ermolinskiy, K. H. Kim, S. Shenker, and I. Stoica, A data-oriented (and beyond) network architecture, in *Proc. ACM SIGCOMM '07*, Kyoto, Japan, August 2007.

11. W. K. Chai, N. Wang, I. Psaras, G. Pavlou, C. Wang, G. G. de Blas, F. J. Salguero, et al., CURLING: Content-ubiquitous resolution and delivery infrastructure for next generation services, *IEEE Communications Magazine, Special Issue on Future Media Internet*, pp. 112–120, March 2011.

12. Y. Chen et al., Efficient and adaptive Web replication using content clustering, *IEEE Journal on Selected Areas in Communications*, 21(6), 979–994, August 2003.

13. N. Fujita et al., Coarse-grain replica management strategies for dynamic replication of Web contents, *Computer Networks,* 45, 19–34, May 2004.

14. G. Pallis and A. Vakali, Insight and perspectives for content delivery networks, *Communications of the ACM*, 49(1), 101–106, January 2006.

15. G. Dan, Cache-to-cache: Could ISPs cooperate to decrease peer-to-peer content distribution costs?", *IEEE Transactions on Parallel and Distributed Systems*, 22(9), 1469–1482, September 2011.

16. The IETF Decoupled Application Data Enroute (DECADE) Working Group, https://datatracker.ietf.org/wg/decade/charter/

17. I. Psaras, R. G. Clegg, R. Landa, W. K. Chai, and G. Pavlou, Modelling and evaluation of CCN-caching trees, in *Proc. of IFIP Networking*, Valencia, Spain, May 2011.

18. Publish-Subscribe Internet Routing Paradigm, http://www.psirp.org/

19. S. Tarkoma, M. Ain, and K. Visala, *The Publish/Subscribe Internet Routing Paradigm (PSIRP): Designing the Future Internet Architecture*, IOS Press, 2009.

20. The IETF IP Routing for Wireless/Mobile Hosts (MOBILEIP) Working Group (Concluded), http://datatracker.ietf.org/wg/mobileip/charter/

21. R. Moskowitz and P. Nikander, *Host Identity Protocol Architecture*, RFC 4423, IETF, May 2006.

22. Future Internet, *6th GI/ITG KuVS Workshop* Monday, Hannover, November 2010.

23. Open Content Aware Networks (OCEAN), http://www.ict-ocean.eu/

24. Economics and Technologies for Inter-Carrier Services, https://www.ict-etics.eu/

25. U. Lee, I. Rimac, and V. Hilt, Greening the internet with content-centric networking, *e-Energy '10 Proceedings of the 1st International Conference on Energy-Efficient Computing and Networking,* pp. 179–182, Passue, Germany, April 2010.

26. *Dagstuhl Workshop on Information-Centric Networking*, December 2010, Dagstuhl, Germany, http://www.dagstuhl.de/10492

27. COMET Information-Centric Networking (ICN) Session at Future Internet Assembly (FIA), May 18, 2011, Budapest, Hungary, http://www.comet-project.org/event-icn-fia-2011.html

28. *ACM SIGCOMM Workshop on Information-Centric Networking*, Toronto, Canada, August 2011, http://www.neclab.eu/icn-2011/

29. Content Mediator Architecture for Content-aware Networks (COMET), http://www.comet-project.org/

30. The CONVERGENCE project, http://www.ict-convergence.eu/

31. Named Data Networking (NDN), http://www.named-data.net/

32. Content-Oriented Networking: A New Experience for Content Transfer (CONNECT), http://www.anr-connect.org/

33. Pursuing a Pub/Sub Internet (PURSUIT), http://www.fp7-pursuit.eu/

34. Scalable and Adaptive Internet Solutions (SAIL), http://www.sail-project.eu/

35. Proposed Information-Centric Networking Research Group (ICNRG), IRTF, http://trac.tools.ietf.org/group/irtf/trac/wiki/icnrg

36. Uniform Resource Names Working Group (urnbis), IETF, http://datatracker.ietf.org/wg/urnbis/charter/

37. DNS Extensions Working Group (dnsext), IETF, http://datatracker.ietf.org/wg/dnsext/charter/

38. Content Delivery Networks Interconnection (cdni), IETF, http://datatracker.ietf.org/wg/cdni/charter/

39. Routing Area Working Group (rtgwg), IETF, http://datatracker.ietf.org/wg/rtgwg/charter/

Chapter 10

Content Delivery Network for Efficient Delivery of Internet Traffic

Gilles Bertrand and Emile Stéphan

Contents

10.1 Introduction 188
10.2 CDNs: From the Early Internet to Today's Video-Driven Landscape 189
    10.2.1 History of CDNs 189
    10.2.2 Definition of CDN 190
        10.2.2.1 Peer-to-Peer 190
        10.2.2.2 Transparent Caching 190
        10.2.2.3 Multicast 191
    10.2.3 A Taxonomy of CDNs 191
        10.2.3.1 Over-the-Top CDNs 191
        10.2.3.2 Content Service Providers' CDNs 192
        10.2.3.3 Carriers' CDNs 192
        10.2.3.4 Internet Service Providers' CDNs 192
        10.2.3.5 Comparison on the CDN Categories 192
10.3 CDN Functioning 193
    10.3.1 High-Level Architecture 193
    10.3.2 CDN Operations 194

10.3.2.1 Content Preparation (Outside CDN) 194
10.3.2.2 Content Ingestion/Acquisition.......................................194
10.3.2.3 Deployment/Distribution 194
10.3.2.4 Cache Selection, Redirection, and Delivery 195
10.3.2.5 Other Important CDN Operations 195
10.3.3 Request Redirection...195
10.3.3.1 A Two-Level Process..195
10.3.3.2 Clustering...197
10.3.3.3 CDN Paradox...197
10.3.3.4 Comparison of DNS and Application-Level
Redirection 198
10.3.3.5 Content Ingestion and Request Routing......................199
10.3.4 Cache Behavior...199
10.3.4.1 High-Level Principle...199
10.3.4.2 Memory Management 199
10.3.4.3 Cache Hierarchy... 200
10.3.4.4 Handling of Dynamic URLs... 200
10.3.4.5 Detection of Stale Content 201
10.3.5 Logging 202
10.4 Securing a CDN 202
10.4.1 Cache Poisoning 202
10.4.2 Unauthorized Access to Content...203
10.4.3 Denial of Service Attacks on the CDN or through
the CDN 203
10.4.4 Overview of CDN Security Mechanisms................................. 204
10.5 Trends and Recent Advances on CDNs.. 204
10.5.1 Industrialization of Content Networking................................ 204
10.5.2 CDN Interconnection 205
10.5.3 From Content Storage to Dynamic Object 205
10.5.4 From End-to-End Internet Architecture to Client-to-Content
Paradigm 206
10.6 Conclusion 206
Acronyms..207

10.1 Introduction

The Internet is built over a set of administrative networks interconnected by peering links which exchange datagrams, thanks to the Internet Protocol. The multiplicity of the interconnection guarantees the global availability of the network. The exchange of high-level routing information through BGP guarantees hosts reachability among administrative networks.

The generalization of both Internet connectivity and terminals supporting Web browsers has transformed the Internet into a global entertainment network.* However, the evolution of the underlying Internet protocols and architectures has been slower and delivery performance has become an issue for most content providers (CP). For instance, end-users' video experience is affected by buffering and artifacts.

Content delivery networks (CDNs) provide mechanisms to adapt the Internet content delivery performance to the usages and to fulfill consumers' expectations. They optimize networks' and servers' resources by replicating content in the network. CDNs efficiently complement IP multicast technologies, which optimize the delivery of a single data stream to multiple users: CDNs not only support flow replication (application-level multicast) but also optimize asynchronous content delivery.

This chapter presents CDNs and how they enable operators to support new Internet usages. It provides a brief history of the CDNs and introduces the recent changes in the Internet eco-systems that have led to an increased interest in CDNs. Then, it presents the evolution of CDNs. The chapter describes the internal machinery of a CDN: its architecture, its operations, and so on. In addition, it explains the main security issues introduced by CDNs. It concludes with a section on the trends and recent advances in CDNs.

10.2 CDNs: From the Early Internet to Today's Video-Driven Landscape

10.2.1 History of CDNs

Akamai Technologies, the leading CDN company, was founded in 1998. At that time, the main applications of the Internet were much different from today's video-driven landscape. Therefore, CDNs targeted mostly the acceleration of Web traffic. The fundamental concept of the CDNs was to replicate content (e.g., copies of Web pages and images) close to the end users to improve the Web performance. It was part of the first version of HTTP 1.1† protocol standard, as early as in 1997. HTTP standards indeed enable caching proxies to deliver content on behalf of origin servers, to improve performance, save network resources, and enforce traffic filtering policies. For example, caching proxies have been deployed for more than 15 years at the edges of campus networks. The servers that hold the replicated content are often called surrogates, or cache nodes.

---

* The usage of the Internet for other usages, like VoIP or enterprises services, is increasing dramatically too. Nevertheless within five years, 80% of Internet traffic will be video-based. (*Source*: Cisco http://blogs.cisco.com/sp/making-money-in-internet-video/)
† RFC2068 the initial specification of HTTP 1.1 has been complemented over the years.

Today, as the Internet has become the support of a much wider diversity of services, CDNs address many different types of content. In particular, video traffic is subject to stringent quality-of-service (QoS) constraints and requires massive amount of network resources. Therefore, most large video providers rely on CDNs for delivering their content through streaming, video on demand, and so on. In addition, more and more people use the Internet to buy or sell products and services. As customers are sensible to the loading time of Web-shops, Internet sellers use CDNs to accelerate their websites (dynamic site acceleration (DSA)). Last but not least, CDN providers have developed application-specific features to accelerate enterprise applications such as Web services, in-network databases, cloud services, and so on.

10.2.2 Definition of CDN

In this section, we compare CDN to other content delivery techniques, such as P2P, multicast, and transparent caching.

10.2.2.1 Peer-to-Peer

Peer-to-peer (P2P) follows the same rationale as CDN; that is, it uses more distributed servers than in the typical client–server model. However, P2P requires littler infrastructure in the network than CDNs, because it uses the end-user's devices as content servers, whereas CDNs rely on in-network servers. In addition, CDN typically does not require any modification to the end-user's configuration, whereas P2P relies on the installation of client software by the end user.

Interestingly, some CDNs use P2P technologies to optimize the distribution of content among the cache nodes, and to take advantage of end-user's devices for improving content delivery performance. For instance, Akamai has implemented a client-side cache called NetSession.

10.2.2.2 Transparent Caching

Transparent caching is a method that replicates popular content somewhere between the client and server, and which does not introduce additional signaling with either side.

For instance, enterprises and universities commonly perform transparent caching, thanks to caching proxies used to access the Internet.

Transparent caches can be located anywhere on the path between the client and the server. In particular, browsers use a local cache, local area networks (LANs) proxies embed a cache, there have been some examples of ISPs deploying caches, reverse proxies in server farms often cache static content, and so on.

10.2.2.3 Multicast

Multicast enables carrying a single traffic flow in the network to serve several end users that request the same content simultaneously. For example, multicast is commonly used to deliver TV content over IP networks. However, multicast requires specific features in network equipment (e.g., routers, DSLAMs) and is not adapted to services such as video on demand (VoD) where the end users request the same content at different times. On the contrary, CDN is well adapted to such services. CDNs are also used in some cases to improve the distribution of live TV content. In these cases, the caches nodes replicate unicast streaming flows and provide a kind of application-layer multicast, which is especially useful to alleviate the traffic load on the origin servers, to save network resources, and to deliver live TV on networks that do not support multicast.

10.2.3 A Taxonomy of CDNs

As depicted in Figure 10.1, we classify CDNs in four main categories depending on the entity that operates the CDN.

10.2.3.1 Over-the-Top CDNs

The over-the-top (OTT) CDNs are deployed by companies whose business is essentially centered on CDN, who own and operate datacenters in several locations, but do neither own the network interconnecting these datacenters to Internet, nor the content that they deliver. Examples of OTT CDNs include Akamai, Limelight, CDNetworks, and so on.

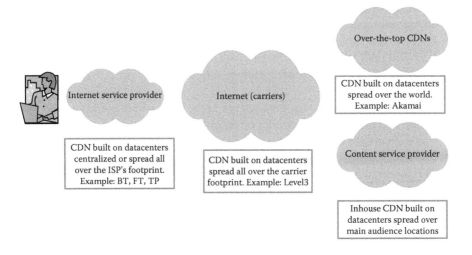

Figure 10.1   A taxonomy of CDNs.

10.2.3.2 Content Service Providers' CDNs

Some content service providers (CSPs, a.k.a. content providers) have deployed their own CDN to support their services specifically. They rely only on their internal CDN or use an external CDN for some content assets. For example, it is clearly visible from traffic captures that the main providers of user-generated video content, namely YouTube and Dailymotion, use an internal CDN to deliver at least part of their traffic.

10.2.3.3 Carriers' CDNs

Carriers, also called transit operators or tier-1s, sell IP connectivity to smaller network operators and CP, to enable them to exchange traffic with the whole Internet. Several carriers have deployed a CDN to complement their IP connectivity services.

10.2.3.4 Internet Service Providers' CDNs

Internet Service Providers (ISPs) have recently begun deploying CDNs in their networks, mainly to gain more control on the quality of experience (QoE) of their subscribers, and to save on network costs.

10.2.3.5 Comparison on the CDN Categories

The CDNs described in the above sections have different strengths and weaknesses, which can be summarized as a footprint versus QoE trade-off. The ISPs' CDNs indeed take advantage of the closest proximity to the end users, and thus, offer the best content delivery performance. However, they typically cover a local footprint, restricted to the ISP's subscribers. In contrast, Carriers' and OTT CDNs cover a global footprint, but cannot deploy cache nodes in the last mile of the network, and thus, suffer from a higher latency. Finally, the CSPs' CDNs target the services of the CSP and their audience specifically.

The location of the cache nodes strongly impacts CDNs' performance, because:

■ CDNs reduce traffic only between the cache node and the origin server
■ CDNs circumvent congestion points (bottlenecks) only between cache node and the origin server

The potential bottlenecks include the access (especially for mobile communications), the backhaul (for instance, ATM networks), and carriers' peering points.

Latency (round trip time) affects content download duration on the Internet, because of the connection and congestion control mechanisms of the transmission control protocol (TCP). Therefore, the round trip time between the end user and the cache node, which depends on the location of the cache node, strongly influences CDN performance.

10.3 CDN Functioning

10.3.1 High-Level Architecture

CDNs rely on a distributed and hierarchical overlay architecture whose main components are the servers which deliver the traffic. These servers are commonly referred to as delivery servers, surrogates, or *cache nodes*. The cache nodes are sometimes organized hierarchically to improve the amount of traffic that is served by the CDN. In this case, we call *parent caches* the cache nodes that deliver content to other cache nodes. The CDN must route the end-users' request for content to the most appropriate cache nodes: this is the role of the *central node*. Finally, the CDN must monitor the cache nodes, manage the content distribution, report about its activities, and so on. We call *back-office node* the component that accomplishes these tasks. Figure 10.2 presents an example CDN architecture which illustrates the four main CDN entities that have been listed above:

1. Central node (a.k.a. request router)
2. Back office node (a.k.a. management node)
3. Parent cache
4. Cache node (a.k.a. Surrogate)

More sophisticated CDN architectures include additional nodes, for instance, specific storage servers for content acquisition, application-specific servers for applications acceleration, and so on.

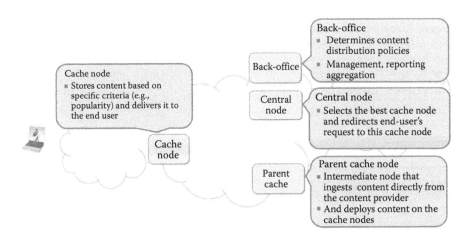

Figure 10.2   An example of CDN architecture.

10.3.2 CDN Operations

10.3.2.1 Content Preparation (Outside CDN)

A piece of content must sometimes be prepared before it can be delivered to the end user. For instance, consider TV content: every video is associated with a summary, encoded for various terminals, and so on. CP typically rely on a content management system (CMS) for the content preparation, and upload ready-for-delivery content assets on the CDN.

10.3.2.2 Content Ingestion/Acquisition

Content ingestion/acquisition is the process by which the first content copy is made available within the CDN. Ingestion can be

- Pulled a.k.a. dynamic content acquisition: The CDN acquires a copy of the content asset only after an end user has requested this content asset. Figure 10.3 describes this process.
- Pushed a.k.a. content prepositioning: The CSP declares new content assets on the CDN, which immediately acquires a copy of these assets. Figure 10.4 describes this process.

Pulled content acquisition is mandatory for large content catalogs such as the ones related to user-generated content, because it is not possible for the CDN to replicate all the content assets. In this case, the CDN dynamically acquires content when it becomes popular.

10.3.2.3 Deployment/Distribution

We name content deployment or content distribution the process through which content is replicated within the cache nodes and parent cache nodes. Similar to content acquisition, content distribution can be

- Pulled: The cache nodes retrieve content assets after an end-user's request
- Pushed: The central back office node manages the deployment of content on the cache nodes

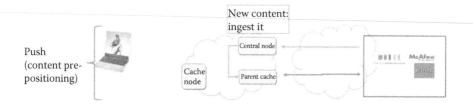

Figure 10.3   Content prepositioning.

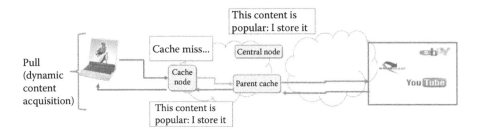

Figure 10.4   Dynamic content acquisition.

10.3.2.4 Cache Selection, Redirection, and Delivery

A large part of a CDN's intelligence reside in its ability to serve end-users' requests from the "best" cache node. This means that the CDN must

- Select a cache, depending on various criteria such the geographical location of the end user and the cache nodes' load and status, the requested file type and so on
- Redirect the end user to this cache
- Deliver content to the end user from the selected cache

These operations are further described in subsections.

10.3.2.5 Other Important CDN Operations

The operations described previously enable a CDN to deliver content. They are not sufficient to fulfill the CSPs' expectations: CSPs need CDNs to provide reporting, monitoring, charging/billing, and interfaces with external enablers (e.g., advertising).

10.3.3 Request Redirection

10.3.3.1 A Two-Level Process

We call redirection the mechanisms that enable the end-user's content requests to arrive to the CDN instead of arriving to the CSP's origin servers. In fact, CDNs require two levels of redirection:

1. CDNs interact with the CSPs to guarantee that the end-user's requests reach the CDN (redirection to the CDN).
2. CDNs internally redirect the end user to the most appropriate cache node (redirection to the cache).

10.3.3.1.1 Redirection to the CDN

The redirection to the CDN can involve various mechanisms. The most common are the following:

- The CSP modifies the links on its portal so as to address content directly on the CDN.
- The CSP sets up an alias at DNS level (CNAME resource record), so that one of his fully qualified domain names (FQDN) is resolved by the CDN. Therefore, all the URLs associated with this domain name directly target a CDN component.

10.3.3.1.2 Redirection to a Cache Node

The redirection to a cache node can involve various mechanisms. The most common ones are the following:

- The CDN resolves the fully qualified domain name (FQDN) of the URL used by the end user to request content and provides the IP address of the selected cache to the end user.
- The CDN uses the redirection message of an application protocol (e.g., HTTP 302 redirect) to ask the end-user's browser to resend the content request to another URL, on the selected cache node.

RFC 3568 (http://tools.ietf.org/html/rfc3568) presents the well-known request routing mechanisms in CDNs.

10.3.3.1.3 Overview

Figure 10.5 describes an example high-level call flow for the redirection of the end user to the most appropriate cache node.

1. The end user browses the CSP's portal and clicks on a URL to request a specific content asset.
2. The CSP has previously written the URL to point toward a content located on the CDN central node instead of on its origin servers.
3. The end-user's terminal sends DNS queries to resolve the FQDN of this link (not represented on the figure). Then, it sends the HTTP GET content request to the CDN central node.
4. The CDN central node selects a cache node and answers the end-user's request with an HTTP 302 redirection message to the URL of the content on a cache node.

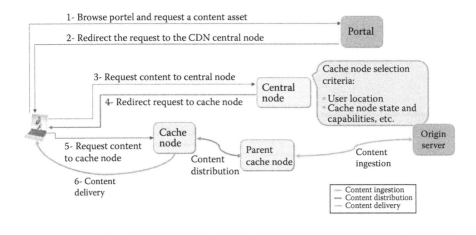

Figure 10.5   An example of the two-level redirection process in CDNs.

5. The end-user's terminal sends DNS queries to resolve the FQDN of the provided link (not represented on the figure). Then, it sends the HTTP GET content request to the cache node.
6. If the cache node has a local copy of the content it delivers it to the end user (HTTP 200 OK). Otherwise, it acquires content, for instance, from a parent cache.

10.3.3.2 Clustering

Caches are often regrouped in clusters; in this case, there is a third redirection level in the process. First, the request is directed to the CDN, then, to a cache cluster, and finally, to a cache node. In this process, the CDN must share the traffic load on the caches of the cluster. At the same time, it must not affect the sessions between the end user and the caches. To fulfill these purposes, a hash if often computed on end-user's IP address to select the same cache for all the requests of a given user. The interested reader might study WCCP (draft-wilson-wrec-wccp-v2) for an example of such mechanism.

10.3.3.3 CDN Paradox

CDNs face a paradox: a large part of their added-value involves *performance and scalability improvements*. However, as Figures 10.6 and 10.7 show, the delivery of the same content asset implies *more complex exchanges* with a CDN than without any CDN. So, the performance and scalability improvements come mostly from the latency improvement between the end user and the server that delivers the content. The conclusion of this paradox is that, for small assets such as gif files, CDNs

Figure 10.6   Overview of content delivery call flow without CDN.

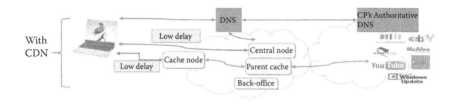

Figure 10.7   Overview of content delivery call flow with CDN.

must be well engineered to improve the QoS because the latency added by the redirections can affect the delivery performance. Nevertheless, content provider benefits of server and bandwidth savings for both small and large files.

CDN is more efficient for larger assets (e.g., VoD) for several reasons: as redirections occur only before the initialization of the data transfer, the delay of the redirections impacts only the arrival of the first packet of the data transfer. The faster TCP rate and the circumvention of peering bottlenecks largely compensate for the extra latency introduced by the redirection operations.

10.3.3.4 Comparison of DNS and Application-Level Redirection

HTTP provides lower redirection performance than DNS because HTTP redirection decisions are not memorized: there is one redirection operation per content request. On the contrary, DNS redirection is highly scalable, thanks to the caching of DNS answers by the user terminal and the local DNS: a single cached redirection at DNS level is valid for many objects downloaded from the same FQDN.

However, HTTP enables more accurate geolocation and supports more sophisticated redirection criteria, because the request router can use both the IP address of the end user and the request details (FDQN and URL) to take the most appropriate redirection decision. For instance, HTTP redirection enables differentiating the request routing operations depending on the requested file extension. In addition and contrary to DNS servers, HTTP redirectors redirecting the end user to a completely new URL, which is useful to rewrite some parameters carried in the URL. In contrast, the authoritative DNS server sees only the FQDN of the request, not the remainder of the URLs and other request details. In addition, the authoritative DNS servers see only the IP of the local DNS, not the one of the end user, which leads to inaccurate geolocation of the end user.

10.3.3.5 Content Ingestion and Request Routing

Content ingestion and request routing operations are closely related: a cache must be able to send a request to the CSP's origin server to acquire content. However, end-users' requests for that content are redirected to the caches. As a consequence, the CDN must provide a mechanism to avoid that the content ingestion requests "loop" due to the request redirection operations. There are well-known solutions to this problem. For example, the cache nodes can use a URL that points directly to the origin server instead of a "CDN-ized" URL. In particular, they can use a specific FQDN for content ingestion, different from the one that is redirected to the CDN.

10.3.4 Cache Behavior

10.3.4.1 High-Level Principle

The basic operation of a cache node (Figure 10.8) is to maintain a table that associates URLs to a popularity measure and potentially to a local copy of the associated file. Upon receiving a content request, the cache checks the URL of the request, and looks in its database if it has a local copy of the answer for this requested URL. We call *cache hit* the situation where the cache has a valid local copy of the requested data and answers the request on behalf of the origin server. We call *cache miss* the situation where the cache does not have a valid local copy and forwards the request to the origin server.

10.3.4.2 Memory Management

In the case of cache miss, the cache node decides if it wants to store the origin server's answer to the request. It typically takes this decision basing on the content popularity: if the content asset is frequently requested, the cache stores the answer, which enables it to directly deliver the content for future requests. Three strategies can be distinguished:

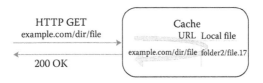

Figure 10.8   High-level principle for cache node's behavior.

■ *Static caching*: The cache node replicates every content asset of the CSPs that have bought the CDN services, without taking content popularity into account.
■ *Dynamic caching*: The CDN uses cache replacement algorithms based on content popularity (hot/cold), to decide which assets the cache node must be delete to create space for caching new assets.
■ *Hybrid*: Some content is primed (always replicated), the other is managed with dynamic caching.

Many dynamic caching algorithms are described in the literature. The two most well-known ones hold self-explanatory names: (1) Least Recently Used (LRU) and (2) Least Frequently Used (LFU).

Most cache nodes use both RAM and solid state drives or hard disk drives to store content. They take decisions on where to store the content depending on its popularity: the hottest content is placed in the fastest memory (RAM); whereas, other content is stored on other supports.

10.3.4.3 Cache Hierarchy

In CDNs that use dynamic caching, cache hierarchy is often use to improve the *hit ratio*, that is the percentage of requests that lead to a cache hit on the CDN. The idea is to replicate the hottest content in multiple locations close to end users, and to store additional (less hot) content on parent caches. Cache hierarchy can be implemented quite easily: the cache nodes deliver content in case of cache hit, and request content to a parent cache in case of cache miss. Therefore, the parent caches see different request popularities:

■ They do not see the requests for the hottest content (cache hit on the edge cache nodes), and thus do not cache it.
■ They see requests for less popular content (cache miss on the edge cache nodes) and cache the hottest content associated to these requests.

10.3.4.4 Handling of Dynamic URLs

As explained in a previous section, cache nodes track the content popularity (request frequency) based on the requested URL. However, many websites use dynamic URLs, which embed one or several parameters instead of just a file name. For example, the following URL includes a content identifier, a user name, and a date: http://example.com/video1&user=gilles&date=09032011.

The problem with dynamic URLs is that different URLs can point to the same file. For example, if the movie "video1" is requested thousand times by thousand different users, the cache node will see thousand different URLs requested a single time. As a result, the cache node will not identify "video1" as a popular content. To solve this problem, cache nodes must be able to rewrite dynamic URLs to extract a content identifier. In the previous example, the cache node should track the popularity of the requests for: http://example.com/video1. Cache nodes require specific rewriting rules (regular expressions) for all the CP that they serve and that use such dynamic URLs.

10.3.4.5 Detection of Stale Content

The content addressed by a URL can change over time or depending on the context of the request (e.g., location and identity of the user). For example, the webpage of your favorite newspaper most probably changes every few hours to include the latest news, and possibly local news for your area. Therefore, CDNs must be able to verify that the content copies that they hold are valid, before sending the content to end users.

HTTP provides specific headers to help the cache determine the validity of cached content. In particular, the "cache-control," "ETag," and "last-modified" headers enable the server to indicate the interdiction to cache a content asset, the content last modification date, and a hash on the content.

Cache nodes and parent caches should always respect the content provider's caching directives in HTTP headers (e.g., no-cache). This is of premium importance for security: imagine if caches stored and delivered to other end users all your banking data every time you administrate your banking accounts on Internet.

When the caches want to validate content, they can send a *conditional* HTTP GET request to the origin server, using, for instance, the "If_Modified_Since" or "If-None-Match" conditions. Figure 10.9 describes a validation based on the date of last modification of the content. In this figure, the cache node sends a message to the origin server to say "if the content has changed since this date (Tue Mar 2011 10:07:04 GMT) send it to me, otherwise send me an HTTP 304 not modified message." As the content version is the same on the origin server as on the cache node the server does not send the content and replies with a (smaller) message that

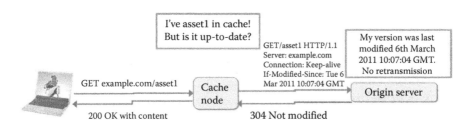

Figure 10.9   An example of content validation based on the if-modified-since header.

confirms that the cached version is up-to-date. This mechanism enables CDNs to check that the version of the cached content is the same as the one on the origin server, and the server to deliver content only when required.

10.3.5 Logging

The core functions of a CDN enable it to deliver contents to end users. They are complemented with side functions, such as logging, which are crucial for a commercial CDN services. Logs on the CDN nodes provide basic information that can be used for reporting, billing, nonreal-time monitoring and for detecting abnormal behavior. Numerous log formats exist, some are proprietary and others come from open-source Web technology (e.g., Apache log format) and standards.

CDN reporting is essential for CSPs; it enables them to check the service performance, to get an insight in the user's behavior, and to measure audience. The information that a CDN must report a CSP depends on many aspects such as the business model of the CSP, the services delivered, the type of content, the CDN service price. The logs provide valuable reporting information but suffer from several limitations. For instance, data transfer information is logged at file transfer end; as sessions can last several minutes for large-file download, the information contained in the logs does not permit accurate real-time reporting.

Log processing requires collecting information on the CDN nodes, adapting and aggregating the invaluable ways. It is a quite sophisticated function, because the number of log entries can be huge and logs contain confidential information related to the user's behavior (privacy concern) and the customers of the CDN (business concern). Such information must be carefully filtered before being presented to the CDN customer. In addition, delivery techniques based on chunking, for example, HTTP adaptive streaming, increase by several orders of magnitude the number of log entries. As an example the delivery of a one-hour-and-a-half movie made of 500 chunks of 10 s and will generate 500 log entries instead of one single entry.

Log processing interacts with several functions (reporting, billing, etc.) of the CDN. For example, the logs can be used to identify out-of-order CDN components. In addition, they provide useful nonreal-time information on the load of the CDN nodes. Finally, the billing function can process them to consolidate the delivery information per content service provider, per delivery service, per geographic area.

10.4 Securing a CDN

10.4.1 Cache Poisoning

End-users trust CDNs and expect them to deliver valid content. However, attackers might try to replace valid content by malicious one, to take advantage of the large delivery resources of the CDN and to spread viruses or other nonlegitimate content, for free. This kind of attack is called "cache poisoning."

CDNs must offer a reasonable level of security to prevent cache poisoning. This involves, for instance:

- Authenticating the origin servers, to be sure that the CDN retrieves content from the legitimate content source.
- Validating the content, to be sure that the content received by the CDN and the content emitted by the origin server are the same.

10.4.2 Unauthorized Access to Content

CSPs trust CDNs and expect them to protect their content from unauthorized access. In particular, video CPs impose other delivery restrictions based on:

- The location of the end users (geoblocking)
- The end-users' identity (e.g., did they pay or not)
- The date
- The number of the time the end user has accessed the content (replay limitations), and so on

They are transparent for CDNs when they are enforced from end-to-end between the CSP and the end user with digital right management (DRM) content protection solutions.

In addition to DRMs, CDNs often use a mechanism called URL signing to check if a content request must be authorized or not. This mechanism relies on tokens (e.g., MD5, SHA-1) and other encrypted information in the URLs used to query content on the caches/surrogates. An example of URLs containing tokens is http://example.com/video1.wmv?expiry=564481213&Client=64.10.212.101& key=5465421.

URL signing requires the exchange of authentication/encryption keys between the CSP and the CDN, so that the CDN can decrypt the tokens in the URLs that the CSP provides to end users. Internally, the CDN must exchange keys between caches nodes and the central node to secure the complete delivery chain, as well as between cache nodes and origin servers to authenticate and validate the content acquisition requests and answers.

10.4.3 Denial of Service Attacks on the CDN or through the CDN

One of the most common attacks in the Internet involves breaking a service by overloading it with a storm of meaningless requests (Denial of Service (DoS) attack). CDNs must be able to detect such attacks and to block malicious requests.

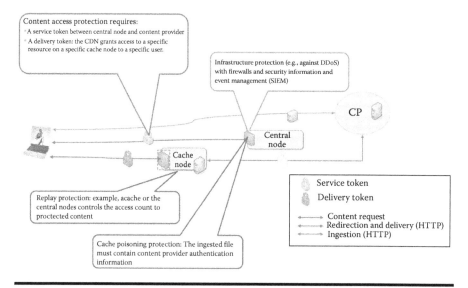

Figure 10.10 Overview of CDN security mechanisms.

In addition to being a target for DoS, CDNs can represent a valuable tool for DoS attackers. Indeed, attackers could try using the cache nodes to send a massive amount of requests to the DoS attack targets. This means that the CDN infrastructure must provide monitoring and alert tools for detecting abnormalities in end-user behavior, as well as firewalls to prevent unauthorized access to the CDN infrastructure.

10.4.4 Overview of CDN Security Mechanisms

Figure 10.10 provides an overview of the previously described security tools that a CDN must provide URL signing, firewalls, monitoring and alert systems, and so on.

10.5 Trends and Recent Advances on CDNs

10.5.1 Industrialization of Content Networking

Content networking faces several challenges, such as the traffic growth, the increase of end-users' QoE expectations, and the CSPs' cost reductions expectations.

Newcomers on the CDN market must compete with existing CDNs to conquer existing or emerging CSPs. Some actors therefore differentiate their services, thanks to specific features, for instance, for right management, content adaptation, bit-rate adaptation, live content support, log processing, request routing.

CSPs distribute their content worldwide and require high QoE experience and robust content delivery solutions to attract and retain their customers. To fulfill these demands, content networking is becoming an industry sector. Interworking between the CDNs is required for more efficiency in content delivery and to avoid deploying expensive redundant infrastructure. We expect CDN internetworking at a massive scale to shape the Future Internet, but this evolution first requires the specification and the deployment of open standard interfaces.

10.5.2 CDN Interconnection

In June 2011, IETF, the international standards development organization (SDO) in charge of producing the Internet specifications, chartered a Working Group (WG) named CDNi (http://tools.ietf.org/wg/cdni/). This WG specifies interfaces so that CDNs can interoperate as an open-content delivery infrastructure.

CDNi tackles the main use cases for CDN interconnection, which are listed below*:

- Footprint extension to satisfy geographic extension, region-to-region interconnection, nomadic usages and geographic policies restrictions
- Traffic offload for overload handling, for dimensioning savings, for resiliency and branding considerations
- CDN capability extensions, for instance, to support other devices and network technologies, to provide interoperability between different vendors' CDN solution, and to improve QoE

The problem statement[†] document identifies the interfaces required to create multi-CDN systems:

- CDNI request-routing interface
- CDNI metadata interface
- CDNI logging interface
- CDNI control interface

10.5.3 From Content Storage to Dynamic Object

CDNs are designed to shift servers' burden to in-network servers, namely, the cache nodes. Initially, CDNs essentially served static files ranging from small gif images in a Web page to large software distribution files. Today, the Internet supports more diverse services, and end-users' interactions are far more complex than 10 years ago. Therefore, CDNs have evolved to support these new usages.

For example, the generalization of e-commerce pushes new requirements on CDNs. In fact, e-commerce revenues are affected by the response time perceived by the customers: customers frustrated by slow interactions abort their transactions

---

* http://tools.ietf.org/html/draft-bertrand-cdni-use-cases
† http://tools.ietf.org/html/draft-jenkins-cdni-problem-statement

with e-commerce site. In addition, e-commerce websites are more personalized and complex (large number of objects), which penalizes the Web-server performances. Consequently, e-commerce vendors try to outsource this burden in the CDNs, for improving the QoE. That is why CDNs start integrating Web technologies such as secure transaction, identity control, dynamic content management. This industrialization cannot be avoided to guarantee end-user experience.

10.5.4 From End-to-End Internet Architecture to Client-to-Content Paradigm

Internet is designed as an end-to-end architecture where hosts located at the edge of the Internet exchange signaling and data are exchanged in IP datagrams. The forwarding of these datagrams over the Internet, from one autonomous system to another, is based on the destination address of the IP header.

The generalization of content delivery infrastructures is changing this routing paradigm. Already 20% of Internet traffic comes from regional caches.* The routing of this traffic toward caches is based on the requested content name (FQDN and URL) and the end-users' address.

A new content networking paradigm may be required in the future for two main reasons:

1. Badly operated CDNs generate a lot of redirections, which sometimes involve end-users terminals and slow down the time-to-service.
2. CDN is designed neither to support end-users private contents nor to support interpersonal communications.

This explains the emergence of concepts such as information centric network (ICN) and content centric network (CCN). These new paradigms follow a clean-slate design approach for the future Internet. They consider a content-asset as a set of information pieces addressed by names and associated to metadata.† The content is disseminated, localized, and delivered according to its name which does not depend on the content localization.

10.6 Conclusion

In this chapter, we have presented the history, the functions, and the trends for CDNs. We have shown that CDNs will be interconnected in the short term and will evolve to distribute legacy live contents.

---

* "Internet Traffic Evolution 2007–2011," http://www.monkey.org/~labovit/papers/gpf_2011.pdf
† "Information-Centric Networking: Current Research Activities and Challenges," *IEEE Communications Magazine*, September 2010.

The Internet framework is currently being reviewed by the IETF to face the huge traffic increase and the changes of usage. It is evolving from an end-to-end client-server framework toward a content-centric approach. CDN is one element of this framework. Its success depends not only on interfacing administratively separated CDNs but also on the interworking with other storage and content distribution techniques like IP multicast, cloud, and transparent caching.

CDNs will face the arrival of new CP with different distribution policies: the end users. Currently, although CDNs increase end-users QoE, they remain transparent for end users. However, end users generate and exchange more and more multimedia contents and other documents. The amount of uploaded data explodes with the generalization of powerful terminals such as smartphones. Content size is increasing due to better encoding quality and longer duration, for instance. Hence, end users will require distribution techniques for storing, backuping, distributing, editing, monetizing, producing, meshing, and sharing privately the content they create.

Acronyms

BGP	Border Gateway Protocol
CCN	Content-Centric Networking
CDN	Content Delivery Network
CMS	Content Management System
CSP	Content Service Provider
DNS	Domain Name Service
DSLAM	Digital Subscriber Line Access Multiplexer
FQDN	Fully Qualified Domain Name
ICN	Information-Centric Networking
IETF	Internet Engineering Task Force
LFU	Least Frequently Used
LRU	Least Recently Used
OTT	Over-The-Top
P2P	Peer-to-Peer
QoE	Quality of Experience
QoS	Quality of Service
RAM	Random Access Memory
SDO	Standards Development Organization
VoD	Video on Demand

Chapter 11

Content Delivery Networks: Market Overview and Technology Innovations

Bertrand Weber

Contents

11.1 Introduction 210
11.2 Concept of CDN 211
11.3 CDN Market Overview 211
11.4 Types of CDN Services Offered 212
    11.4.1 Media Delivery Services 212
    11.4.2 Additional Services 213
        11.4.2.1 Advanced Domain Name System 213
        11.4.2.2 Reporting and Analytics 214
        11.4.2.3 Online Video Platform and Content
                Management Service 214
        11.4.2.4 Multi-CDN 214
    11.4.3 Value-Added Services 214
        11.4.3.1 Website Acceleration (for Static and
                Dynamic Websites) 214
        11.4.3.2 Application Acceleration 214
    11.4.4 CDN Services Evolution 215

11.5 CDN Service Providers 216
    11.5.1 CDN Pure-Play Service Providers...216
        11.5.1.1 Variety of Players 216
        11.5.1.2 Partnerships and Acquisitions......................................217
    11.5.2 Network Operators Entering the CDN Market..........................217
        11.5.2.1 Verizon 217
        11.5.2.2 AT&T 217
        11.5.2.3 Level3...217
11.6 Technology Vendors..218
11.7 Technology Evolution 218
    11.7.1 Akamai Technologies..218
    11.7.2 Limelight Networks...219
11.8 Major Technology Innovations...219
    11.8.1 Adaptive Streaming 219
    11.8.2 Multiscreens Delivery 219
    11.8.3 HTTP-Based Streaming..219
    11.8.4 Quality of Experience Monitoring and Reporting 221
    11.8.5 Peer-To-Peer-Assisted Delivery...221
    11.8.6 Multi-CDN Platforms...221
Acronyms...221
References 222

11.1 Introduction

Content delivery networks (CDNs) optimize and accelerate the delivery of content by caching popular objects on servers distributed on the Internet and located closer to end users. They were first deployed several years ago to overcome some of the Internet's limitations (performance, scalability, network congestion, delay, server overload, etc.) and to improve the performance of websites with static content. Since the early years they were deployed, CDNs have evolved dramatically with the growth of traffic on the Internet. They are now able to accelerate and deliver video traffic, websites, and applications to a wide range of end-user devices.

The worldwide CDN business was worth $1.5 billion in 2010 with leading CDN service providers Akamai Technologies and Limelight Networks representing $1.2 billion. In the last few years, a large number of new players including network operators have entered the market.

The delivery of video and static content remain the most common services on the market today but they have become commoditized. CDN players have recently started to offer services to be able to accelerate the delivery of dynamic websites and applications. Innovative technologies have been implemented by CDN service providers in order to expand their portfolio and differentiate themselves from competitors. It includes the delivery of High Definition (HD) video to a variety of end-user devices as well as Hypertext Transfer Protocol (HTTP)-based adaptive streaming.

11.2 Concept of CDN

CDNs were built more than a decade ago to improve the performance of Internet websites with static content. They have evolved and can now deliver large video files, games, dynamic websites, and applications to different types of devices. CDNs were developed to solve websites performance problems, due to network congestion or origin server overloading, that arise when too many users want to access popular pieces of content (Figure 11.1).

CDNs improve end-user performance by dynamically caching content on edge servers located closer to users instead of using the origin server (Figure 11.2).

11.3 CDN Market Overview

A Tier1 Research report (Davis, 2010) found that the total CDN market was worth $1.4 billion in 2010 with North America representing 63% of the market and Europe 19%.

Akamai Technologies is the dominant CDN service provider with revenues of $1.02 billion in 2010 while Limelight Networks is the second largest provider represented $183 million.

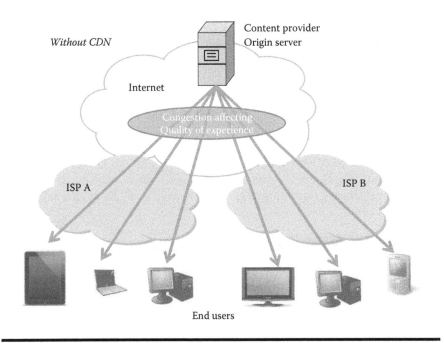

Figure 11.1 Content delivery without a CDN.

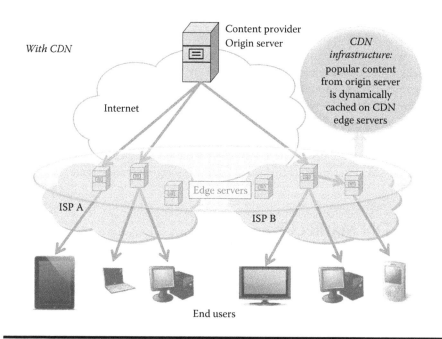

Figure 11.2 Content delivery with a CDN.

This market is becoming more and more competitive. More than 30 new players entered the CDN business during the last few years and drove the pricing as low as $0.05 per GB according to Rayburn (2010).

These new CDN service providers include Highwinds, Edgecast, and Cotendo among others.

In order to differentiate themselves from these new competitors, leading CDN providers had to expand their service portfolio and offer additional services to their customers, for example, encoding, content management, ad insertion, and audience measurement.

According to Tier1 Research, the CDN worldwide market is expected to reach $4.5 billion in 2015. Figure 11.3 from a Yankee Group report (Vorhaus, 2009) shows which companies were front-runners and followers on the CDN market as of 2009.

11.4 Types of CDN Services Offered

11.4.1 Media Delivery Services

Media delivery (especially video) is the most popular service and is now widely offered by CDN service providers. These services are today considered as almost mandatory for a provider portfolio. Even if the pricing has decreased and the margins are now very low, the revenues of CDN service providers for these services are

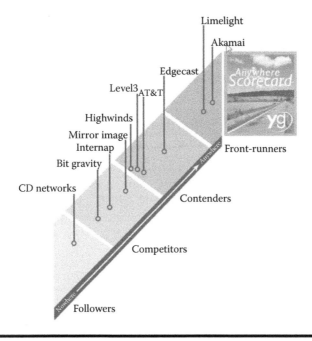

Figure 11.3   Yankee group CDN scorecard for 2009.

still growing because the traffic volume is growing very rapidly and largely compensates the pricing decline.

Media delivery refers to hi-performance content delivery for static media objects including HD. It includes but is not limited to video streaming (e.g., live event), video progressive download, and delivery of video ad, software, game, or large files. Geo-blocking, that is, the ability to limit the access to some contents, and geo-targeting, that is, the possibility to select a version of a specific content according to geographical criteria, are widely supported by most CDN service providers. Major media formats, for example, Adobe Flash, Microsoft Smooth Streaming, Microsoft Windows Media, Apple QuickTime, as well as major media delivery protocols, for example, HTTP, Real Time Messaging Protocol (RTMP), Microsoft Media Services (MMS), Real Time Streaming Protocol (RTSP), or File Transfer Protocol (FTP) are supported too.

11.4.2 Additional Services

11.4.2.1 Advanced Domain Name System

Advanced Domain Name System (DNS) is an efficient, secure, fast, and intelligent DNS resolution based on several criteria, for example, time, geo-location, or load. It can be used for load balancing between different data centers and for selecting the best CDN or group of caches.

11.4.2.2 Reporting and Analytics

Reporting provides statistics on the usage and the users quality of experience in real time. Analytics looks at end-users activity and patterns in order for a CDN customer to better customize its video service or website.

11.4.2.3 Online Video Platform and Content Management Service

CDN media delivery services can be bundled with a complete video solution to manage, publish, measure, and monetize online video content, for example, solutions for encoding, content management, DRM, player development, or ad insertion.

11.4.2.4 Multi-CDN

Some content providers like Netflix or Hulu choose today to rely on several CDNs from different providers at the same time in order to optimize their costs and diversify their delivery sources. Several companies, for example, 3Crowd, Conviva, and MediaMelon, now offer technologies to CDN customers for them to define and enforce rich user redirection rules. It is possible to select in real-time on a per-delivery-session basis the most appropriate CDN based on a combination of criteria, for example, CDN availability, CDN cost, time, end-user location, type of terminal, and type of content.

11.4.3 Value-Added Services

11.4.3.1 Website Acceleration (for Static and Dynamic Websites)

For many companies, for example, in the media, entertainment, e-commerce, banking or travel booking industries, it is very important to be able to accelerate their website and provide the best possible experience to the users. Some CDN service providers like Akamai or Cotendo offer such website acceleration to their customers. Technologies used are different from the ones used for media delivery. Website content is most of the time smaller in size and dynamic and is not directly cacheable. Various content types, for example, small, static, dynamic, or secured objects, are supported for delivery.

11.4.3.2 Application Acceleration

Some CDN service providers offer application acceleration allowing end users to benefit from optimized IP-based and Web-based applications. Different types of applications can be accelerated, including project management software (e.g., Autodesk), expense management systems (e.g., Oracle), contact management systems (e.g., Siebel, Salesforce.com), learning management systems, e-mail applications, ad campaign management tools (e.g., DoubleClick), or inventory management applications.

11.4.4 CDN Services Evolution

Media delivery services have been commoditized and margins are very low. The strong competition between CDN providers for this type of services has constantly driven pricing per Gigabyte delivered down over the years. But CDN revenues for media delivery services will still continue to grow strongly due to the expected growth in traffic, especially video traffic. Figure 11.4 shows the growth of traffic delivered by Limelight Networks over the last five years ("2010 Investor Overview," 2010). The traffic has multiplied 47-fold.

In order to differentiate themselves, CDN providers started to offer some of the just described additional services, for example, encoding, content management, ad insertion, and audience measurement.

But the highest margins are now made with the value-added services, that is, website and application acceleration services. Pricing per GB delivered for these types of services is much higher as fewer CDN providers have developed the expertise and technology to offer these services today. The traffic is of course still lower for value-added services than for media delivery services. Figure 11.5 shows that value-added services accounted for 33% of total Limelight's revenues for the third trimester of 2010 ("2010 Investor Overview," 2010).

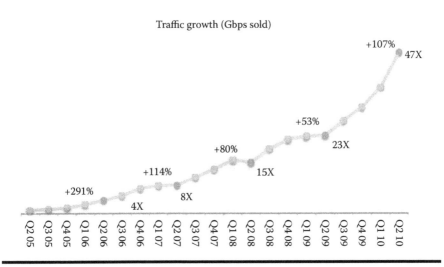

Figure 11.4 Limelight's traffic growth over the last 5 years.

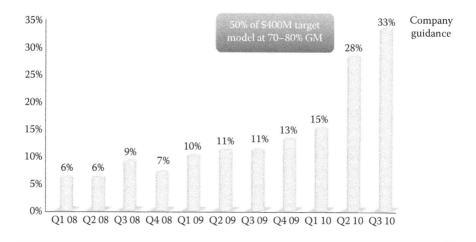

Figure 11.5 **Limelight's value-added services as percentage of revenue growth.**

The total CDN market was worth around $1.5 billion in 2010. About 50% comes from nonvideo services. Website and application acceleration represent about 30% of the total. The margins for these types of services are much higher than for the media delivery services. Also website and application acceleration have lower requirements in terms of network and storage.

11.5 CDN Service Providers

More than 40 service providers compete now on the CDN market. Two categories of providers can be considered: the pure players and the network operators.

11.5.1 CDN Pure-Play Service Providers

11.5.1.1 Variety of Players

A pure-play service provider operates a CDN overlay on top of third-party's network infrastructures. Such providers have for a long time been the only players on the market. They still represent a large portion of the CDN total revenues. The leaders are still Akamai and Limelight, but some other providers lately gained a lot of traction, for example, Edgecast or Cotendo.

During the last few years, competition has been very strong and drove pricing down, especially for media delivery services. According to Streaming Media, even leaders such as Akamai and Limelight have been forced to offer their video services at quite a low price to some of their largest customers, that is, less than 0.06$ per GB for a total yearly traffic of 500 TB or more.

11.5.1.2 Partnerships and Acquisitions

Most of the CDN pure-play service providers have developed their own ecosystems of technology partners to be able to offer additional services which are not directly linked to content delivery but complement their CDN services. It includes but is not limited to content production, creation, encoding, adaptation, management and hosting systems but also digital rights management, ad insertion, analytics, and audience metering platforms.

CDN pure players have also extensively relied on company acquisitions to strengthen their service offerings and market shares, for example, Akamai has purchased Speedera, Nine Systems, RedSwoosh (peer-assisted content delivery), Netli (website and application acceleration), Acerno (targeted advertising) among others.

11.5.2 Network Operators Entering the CDN Market

An increasing number of network operators offer CDN by reselling services from pure-play providers (e.g., Verizon), partnering with a CDN pure player, or building their own infrastructure (e.g., Level3). Their goal entering the CDN market is to be able to generate additional revenues but also to optimize the network traffic on their networks as well as the quality of experience of their broadband subscribers.

11.5.2.1 Verizon

Verizon has been reselling Akamai services in the United States since several years. In addition, Verizon uses a CDN platform deployed within its network and managed by Alcatel-Lucent. This CDN is used to optimize the delivery of content owned by movie studios or TV networks that have contracted with Verizon.

11.5.2.2 AT&T

AT&T started to deploy a CDN infrastructure in 2008 (invested $70 million) and entered the market the same year. AT&T initially decided to develop their own in-house CDN platform by just relying on some external technology enablers provided by different providers, for example, ExtendMedia and Qumu. But in 2010, AT&T started to resell services from Cotendo to be able to offer website and application acceleration to AT&T's customers. And in 2011, they started to use licensed EdgeCast's software to run their CDN.

11.5.2.3 Level3

Level3 acquired Savvis's CDN business in 2006 for $135 million and entered the content delivery market. They expended their CDN to Asia and Australia in 2008. Level3 significantly increased its CDN capacity in 2010, adding 1.65 Tbps and extended its footprint to Canada, Europe, and South America. Level3 was few months later selected as one of Netflix's primary CDN provider and announced 2.9 Tbps of additional CDN capacity.

11.6 Technology Vendors

CDN pure-play service providers like Akamai or Limelight have built their own CDN platform and do not rely on external CDN product providers. However they still need low-cost server hardware from manufacturers like HP or Dell. Akamai has deployed more than 84,000 servers in 72 countries.

Network operators offer CDN services by reselling services from pure-play providers (e.g., Verizon), partnering with CDN pure players (e.g., AT&T with EdgeCast), or building their own infrastructure (e.g., Level3). AT&T stopped its in-house CDN project done with strategic technology partners. AT&T started reselling Cotendo services and using licensed EdgeCast's software to run their CDN. Network operators such as Orange France Telecom, British Telecom, or Telecom Italia started to build their own CDN by relying on product providers like Cisco, Alcatel-Lucent, Juniper, or Verivue.

Cisco introduced its Content Delivery System–Internet Streamer (CDS–IS) product in 2007. It consists of a Content Delivery System Manager (CDSM), one or more Service Engines (SEs) and one Service Router (SR). The SE handles content ingest, distribution, and delivery to user devices. The SR handles user requests and redirects the users to the most appropriate SEs. The CDSM manages and monitors the CDS–IS.

Juniper invested massively during the last few years into CDN technologies. In April 2010, they acquired Ankeena Networks and its Media Flow content delivery and caching technology. In August 2010, Juniper released VXA, a line of CDN appliances leveraging Ankeena's Media Flow. In addition, Juniper offers a complete range of application acceleration solutions, the WXC and WX appliances.

11.7 Technology Evolution

11.7.1 Akamai Technologies

Akamai entered the CDN market in the late 1990s and initially focused on website acceleration. Over time they enhanced their technology and acquired different companies to be able to offer other types of CDN services, that is, media services and application delivery. In 2005, Akamai acquired Speedera Networks, one of their strongest competitors back then for $130 million. In 2006, they purchased Nine Systems for $164 million, an online media management company, to enable Akamai's customers to easily produce, publish, and distribute content on the Akamai CDN. In 2007, Akamai acquired Netli for about $180 million to enhance their application acceleration service. In 2010, Akamai bought Red Swoosh and became the first CDN service provider to leverage peer-assisted delivery to enhance further the scalability of their CDN infrastructure.

11.7.2 Limelight Networks

Limelight was funded in 2001 and is now Akamai's main competitor on the market. Limelight designed its CDN platform right from the beginning with a focus on the ability to deliver massive amounts of video with fewer but larger fiber-based interconnected point-of-presence (PoPs). Limelight also acquired technological companies to enhance its service portfolio, for example, Delve Networks in 2010 for around $10 million, enabling the management, measurement, and monetization of video content and EyeWonder in 2009 for $110 million specialized in interactive advertising. Figure 11.6 shows Limelight's content delivery platform and its different features ("2010 Investor Overview," 2010).

11.8 Major Technology Innovations

In order to enhance the overall performance and value offered to their customers, CDN systems have constantly evolved over the years, been optimized, and implemented new features.

11.8.1 Adaptive Streaming

Adaptive streaming was invented by Move Networks and then generally adopted by the industry, that is, Adobe and Microsoft. Most of the adaptive streaming technologies now available on the market, that is, Adobe Dynamic streaming, Microsoft Smooth Streaming, and Apple HTTP Live Streaming, are now supported by most CDN service providers. Adaptive streaming enables video fast start, mitigates buffering, and allows each user to get the most appropriate bit rate at any given time.

11.8.2 Multiscreens Delivery

CDNs can now offer HD quality on different types of user devices, that is, laptop, connected TV, and mobile, thanks to adaptive streaming among other optimizations.

11.8.3 HTTP-Based Streaming

CDNs have migrated from proprietary streaming protocols, that is, Adobe RTMP and Microsoft RTSP, to HTTP-based streaming enabling simpler and cheaper content delivery architectures.

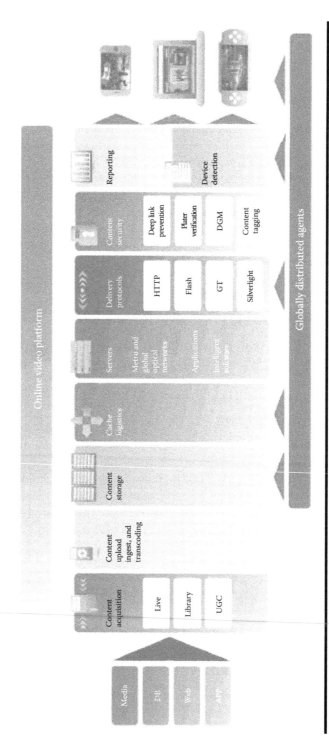

Figure 11.6   Limelight content delivery platform.

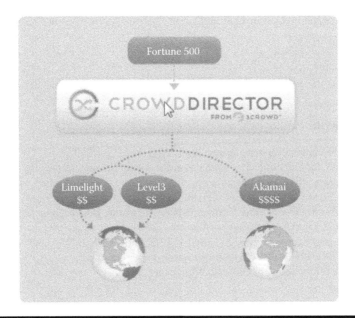

Figure 11.7   3Crowd's multi-CDN platform.

11.8.4 Quality of Experience Monitoring and Reporting

Akamai and Limelight CDNs provide real-time quality of experience monitoring and reporting.

11.8.5 Peer-To-Peer-Assisted Delivery

Peer-assisted delivery like Akamai with its NetSession service allows an optimum CDN scalability and bandwidth efficiency while limiting cost.

11.8.6 Multi-CDN Platforms

CDN customers can now rely on multi-CDN platforms, for example, 3Crowd, MediaMelon, and Cotendo, to manage services from several CDN providers. These platforms can route the user requests to the most appropriate CDN based for instance on cost, performance, or content priority. Figure 11.7 illustrates 3Crowd's multi-CDN technology ("The Multiple CDN Strategy," 2010).

Acronyms

CDN   Content Delivery Network
HD   High Definition

HTTP   Hypertext Transfer Protocol
P2P   Peer-to-Peer
PoP   Point-of-Presence
Tbps   Terabits per second

References

J. Davis, CDN market overview: Spring 2010, *Tier1 Research*, 2010.

D. Rayburn, CDN pricing: The going rate for video delivery, *Streaming Media West, Presentation*, 2010, from http://www.streamingmedia.com/conferences/west2010/presentations/SMWest-2010-CDN-Pricing.pdf.

D. Vorhaus, Anywhere scorecard: Content delivery networks, *Yankee Group*, 2009.

"2010 Investor Overview," *Limelight Networks*, 2010, from http://files.shareholder.com/downloads/ABEA-5OXHL5/1346143878x0x437700/9FDF7AA3-4A23-4B64-8064-7BC0435D4118/Q3_2010_Investor_Presentation.pdf.

"The Multiple CDN Strategy," *3Crowd*, Whitepaper, 2010.

Chapter 12

Quality of Experience in Future Media Networks: Consumption Trends, Demand for New Metrics and Evaluation Methods

Adam Flizikowski, Marek Dąbrowski, Mateusz Majewski, and Krzysztof Samp

Contents

12.1 Introduction 224
12.2 Framing the Environment 226
12.3 Regulations and Standards: Current Duties of Operators and
     Service Providers 227
     12.3.1 Changing QoE Perception in Transforming
          Audiovisual Environment 228
     12.3.2 Anywhere, Anytime TV Services................................... 228
     12.3.3 Interactive TV Services 229
     12.3.4 Social TV Services 230
12.4 QoE of Audiovisual Services 232
     12.4.1 State-of-the-Art QoE Measurement...............................233

12.4.2 QoE Measurements: Practical Perspective of a Provider 236
12.4.3 Conclusions 236
12.5 Redefined Approach to QoE Assessment of IPTV Services 237
12.5.1 Role of the Human in the QoE Holistic View:
Change of Paradigm 240
12.6 A Redefined Approach to QoE Assessment of IPTV Services..................247
12.7 Final Conclusions..250
Acronyms..250
References 251

12.1 Introduction

Multitude of services is emerging on the horizon of media networks; however, one of them is the most promising as it is "all in one." The Internet Protocol Television (IPTV) is a service that does converge multiple accompanying services (voice, presence, trick play, etc.) into one service package, not only promise this. As such, it definitely increases probability of keeping the viewer in the loop (in the form of a TV set) for a longer time. In contrast to linear TV, it brings capabilities of bidirectional communication (user feedback) onboard. This, in turn, means more "entry points" for becoming or staying an active user of the service (e.g., *"someone called me on the TV . . . and then we decided it is worth sharing TV show experience," "I was participating in a quiz on my TV . . . my recommender system suggested me to take distance learning course to catch up with history as I had lots of wrong answers in a quiz,"* etc.). Linear TV as used currently is a synonym of accessing information and way to relax. But emerging IPTV services are more sophisticated and efficient in delivering content—it can be viewed on multitude of devices (PC, TV set, mobile phone), using a variety of network technologies (also during the session), it can adapt to network conditions (e.g., codec, buffering). Even more—it can adapt to user preferences and context (preliminary solutions). Thus, it is essential to understand that the definition of the notion of quality of experience (QoE) should also be adapted to the changing reality. Such a convergent service should be capable of (among all): satisfying viewer's expectations, inducing emotions, giving a pleasure but as we do live in information age also provide *contextual information* (synthesis thereof), assuring any *means for controllability* (parents want to control what their child is watching, for how long), reinforcing *lifelong learning* (through its integration with the underlying platform), and *sensing the context* of a user (*"a movie for fellows from work," "an advertisement targeted to husband and wife before their wedding anniversary," "family comedy,"* etc.). The latter expectation is extremely important as (IP)TV pretends to incorporate more and more elements of personalization (recommendation)—that is rather typical to the personal devices like (smart)phone where a particular person only uses the device (thus the service is being personalized/fit to the person/owner only). Whereas (IP)TV differs in its capabilities to support this aspect as it is a *shared device* and as such should consider (balance) the issue of "whose experience" it is assessing right now (father's, mother's, child's, family, colleagues', etc.)—the notion of managing multiple identities is extremely important for providing real value added to the user/identity. The IPTV system (and thus QoE) should be accompanied with various *means/mechanisms of adapting the service* (quality, content, etc.) to the context of the viewer as it can have significant influence on one's evaluation of the service (QoE). In addition to the notion of identities, a viewer can be classified into one of the three groups: (a) primary user (direct viewer), (b) secondary user (parent), and (c) traveler/guest (someone temporarily far from one's home environment). This led to the analogy with the technical issue of "sensing the spectrum" with cognitive radio networks— in particular the (IP)TV should now be *sensing the environment* in which it is operating in order to optimize and adapt to the (needs and experience of) feasible set of identities in front of it. Similarly, as (media) networks are becoming more (pro) active in a process of delivering optimized service (through cognition) also the IPTV services are on their way to "understand" and "fit to the purpose" of our daily living. This is an important challenge for emerging services like an IPTV as it is becoming a "multimodal" (multitude of service types—games, communication, fun, learning), "multidevice" (switching between rooms triggers a change of end device type but also a media network that serves the session) and "multistage" while accompanying us throughout our lives. For instance, it is realistic assumption (from technical point of view) that IPTV system senses our mood and adapts the service to the state we are in. On the contrary, it should also be able to "sense and adapt" to the stage of life we are in—for example, a child needs to watch tales but it should not do so all the day (impersonation of a parents' will), whereas an elderly person should be watching one's soap operas but also needs to have one's brain challenged (with quizzes, lifelong learning, and other brain-activities)—impersonation of child's expectations toward older father/mother. We are living in times where the only evident thing is the change and convergence—technologies, media networks, services evolve toward symbiosis with external environment (users or systems). The Internet as a concept is also in the process of addressing the need for an upgrade to Internet2, where the difference is in the fact that Internet is more a closed-loop system with the feedback mechanisms (cognition, sensing). Through this evolution toward "systems with a feedback" we are coming closer to the (ultimate goal) use of automation and (machine) learning of systems for a benefit of users. Here the analogy is the Google page rank algorithm that has changed (through the intelligent automation/improvement of search process) the way people use Internet today, improving the efficiency and reliability of information search—and thus quality of life.

This chapter introduces information about *how* services evolve (proposing new metrics needed to quantify that) and *what modifications are needed* in current measurement methodologies (architectures, algorithms) to keep pace with that change (user centric, feedback driven).

12.2 Framing the Environment

The user perception of audiovisual services is usually assessed by applying metrics focused on evaluation of quality of audio and video. In general, the notion of user's audiovisual perception measure is referred to in the literature as Quality of (user) Experience. Further extensions to this notion may include several additional parameters like channel zapping time, Electronic Program Guide (EPG) access time, and so on. Quality assessment of audiovisual content, as perceived by the end user, is well suited for traditional media (e.g., linear TV, Video on Demand (VoD)), where content is consumed by viewers in a rather passive way. The adoption of IPTV, with its intrinsic capability of two-way communication channel, is changing the landscape of networked media and therefore requiring a more comprehensive approach to evaluate user QoE. Emerging services introduce real convergence of services and allow users communicating with each other and interacting with the service at the same time, thus exploiting the new social aspect of entertainment in a completely redefined way. In the following sections authors will present an overview of new IPTV service trends, focusing on the aspect of service quality perception and evaluation. In particular, the following issues will be discussed:

- Factors, that impact the perceived quality of new convergent and interactive IPTV services (e.g., synchronized TV games, "virtual room" services, etc.), with new metrics that are important for quality assessment of such services.
- New media usage trends and new audiovisual technologies (e.g., scalable coding, recommender systems, etc.) which alter the user perception of audiovisual quality.

A multitude of well-recognized methods for QoE evaluation is available spreading from purely subjective tests (e.g., ITU-R BT.500-12), through black-box objective methods (models like Perceptual Evaluation of Speech Quality (PESQ)/ Perceptual Evaluation of Video Quality (PEVQ)) and eventually glass-box techniques, where QoE is derived from the network Quality of Service (QoS) parameters (e.g., International Telecommunication Union-Telecommunication Standardization Sector (ITU-T G.107)). However, all these methods have shortcomings that prevent it from providing the comprehensive user experience assessment for future emerging services. Relying on a set of techniques for evaluation of single media (video or audio) solely, not considering multimodality of multimedia and devices throughout the session, and especially not relying on the "user-content" dynamics classifies such an approach as incomplete or limited. Often various stakeholders in the value chain (e.g., mobile operators) complain about the unavailability of reliable tools that could offer complete and reliable testing facility of the user perceived quality considering multimedia or other emerging services. What is evidently on the span currently is the aspect of end-user perception of the content (whether mobile World Wide Web (WWW), IPTV video or gaming). However, it is often not satisfactory to rely on just measuring network level QoS to estimate customer satisfaction [1]. Mimicking end-user sensing capabilities using hardware (dedicated devices) coupled with special purpose software for data analysis (algorithmic approach)—for example, camera for visual perception with image processing algorithms or Electroencephalography (EEG) headset equipped with brainwave data mining algorithms—introduces greatest benefit for assessing user perspective in delivering accurate QoE measures and predictions. Thus, unobtrusive, user-centric, perception-driven, automated methods are sought to fill the gap between signal-driven perception (video, audio), network-derived parameters (QoS) and high fidelity but also expensive and not convenient subjective assessment methods.

The remainder of this chapter will focus on the overview of emerging methods that try to complement the current toolkit of QoE methods. Among all effective assessment of content perception, image processing-based assessment and automation techniques will be presented and evaluated with respect to their current state, challenges addressed and applications perspectives toward emerging services. The next section will shed some light on the perception of QoE and definitions of services in the changing environment of modern media networks from the perspective of standards and regulations.

12.3 Regulations and Standards: Current Duties of Operators and Service Providers

The European Telecommunications Standards Institute (ETSI) delivers a document TS 181 016 v.3.3.1 describing the Service Layer requirements for IPTV services within Next Generation Networking (NGN). The document is complementary to other documents about IPTV released by ETSI. Additional emphasis is put on the integration of IPTV services with the communication services defined in Telecoms & Internet converged Services & Protocols for Advanced Networks (TISPAN). The variety of services in IPTV shall not exclude services already available in traditional cable, satellite TV but also focus on new services that will attract new clients. Many service providers believe that to create a competitive TV service bundle the IPTV services should be integrated with other services like voice and data to provide unique, innovative applications [2]. Service providers integrate data, voice, and video services in their network in order to deliver a bundle of service to their customers which can be installed and accessed right away at the customer's premises. Currently, more and more service providers are deploying IPTV services in order to compete with traditional cable TV [3] and increase their revenue by providing enhanced TV experience. The necessary condition for the service provider is to ensure that the quality experienced by IPTV customers is similar to the quality of cable or satellite TV services. This section reviews some of the IPTV services described in ref. [4] and points out parameters that influence the overall user experience and thus service acceptability.

12.3.1 Changing QoE Perception in Transforming Audiovisual Environment

The adoption of IPTV technology is currently changing the landscape of networked media and entertainment. New convergent and social TV services will allow users for communicating with each other and interacting with the service at the same time. Although, these new services are still considered as supplementary to main content distribution services (TV and VoD), their quality impairments would impact on the overall acceptability of the offer and thus should not be ignored by IPTV service providers.

There is currently wide literature available (as pointed out in Section 12.3) on different aspects of audiovisual quality, but little knowledge is established on the QoE requirements, metrics and assessment methodologies of interactive and social TV services. ITU-T recommendation G.1010 [5] gives some guidelines on the classification of different network applications regarding their end-to-end quality metrics and major impairment effects (with special focus on delay characteristics and error rate). However, in the case of convergent TV services, one should additionally take into account cross-effects of different services used at the same time, network performance for simultaneous transport of different types of traffic (real time and nonreal time, data and signaling), as well as application and terminal performance for rendering the user interfaces of distinct service components.

In subsequent sections, example end-to-end quality requirements and impairment factors are discussed on the service-by-service basis, in the case of several prominent future IPTV service categories.

12.3.2 Anywhere, Anytime TV Services

Anywhere, anytime TV allows users for consuming content on any device in a convenient time, breaking the linear programming principle of traditional TV. The so-called *place-shifting TV* assumes consuming purchased content on different types of devices (TV set, PC, mobile phone). *Time-shifted TV* allows users for accessing previously recorded TV content at any convenient time. If the recorded content is stored on operator's servers (the so-called Network Personal Video Recorder (PVR)), the delay to access recorded item will be a relevant end-to-end quality indicator for the user, apart from audiovisual quality of the content itself. This delay will be determined by bandwidth constraints, but also by storage system access efficiency and network performance (packet transfer delay) for carrying IPTV control signaling traffic. With additional *bookmarking* capability, the user can set a bookmark while watching content and resume it in a later time or on another terminal. Thus, bookmark setting precision and retrieval accuracy can be considered as additional quality metrics from user point of view.

Session continuity services [6] allow for transferring active IPTV session between multiple terminals and access networks. With session mobility feature, user can start a media session on one terminal (e.g., Set-Top Box (STB)) and continue on another one (e.g., smartphone), attached to the same or another access network (e.g., the FP7 project OMEGA proves to deliver real seamless handover at home between Power Line Communication (PLC), Wireless Local Area Network (WLAN), Ethernet and Visual Light networks). With terminal mobility feature, the device can change access network attachment without considerable service disruption. In both cases, from user perspective, the session transfer time (impacted by session control signaling transport and processing delays, media buffering in the terminal, codec and application implementation efficiency), as well as the audiovisual quality degradation level (e.g., in the case of transfer from STB to mobile phone), will be important quality characteristics.

In the case of *remote control of IPTV services* (e.g., browsing EPG, zapping between TV channels, and commanding PVR recording from a smartphone or tablet), the responsiveness of mobile phone application, remote command execution delay (related with signaling transport delays and processing times), will affect the quality perception.

12.3.3 Interactive TV Services

Bidirectional communication between service platform and terminal device enables services, in which user interaction with the system can impact on the content received. In this case, the network level transport performance of traffic related with interactive component (which is usually of data type), as well as application-level efficiency and ergonomics will play an important role for an overall perception of the service.

In the case of *search, discovery, and content recommendation* services (extension of EPG, essential for usability of modern IPTV services, where multitude of available content makes the selection more and more troublesome for users), the page loading responsiveness and ease of Graphical User Interface (GUI) navigation will impact on the overall QoE assessment. Similar factors will be important for *interactive Web-like services on TV* (e.g., TV shopping, retrieving additional information about the program, news, gambling, access to Facebook, Twitter, etc.). In addition, one should take into account the smoothness of switching display between TV content and interactive service.

For *interactive synchronized services*, quality of synchronization between TV content and interactive features will be a crucial parameter. For example, consider a TV show, where a user in front of the TV can answer to the same quiz questions as participant in studio, using STB remote controller or dedicated mobile phone/tablet application. The quiz question has to be activated on user device exactly at the same moment as it is presented in the televised show, otherwise the contest would be unfair. This requires deploying a mechanism for timely triggering game events (e.g., using markers in streaming video content, or asynchronous signaling protocol), carefully taking into account effects of time synchronization, network delays, and content buffering in end devices.

Targeted advertising and personalized advertisement insertion requires synchronization of advertisement clip with the commercial slots in the original content. The smoothness of transition between main content and advertisement clips may impact on the overall perception of the service. Similar factor will be important in the case of *personalized channel service* [4], where users can create their own programming by merging content items from different sources (a kind of personalized playlist).

In *streaming games* on IPTV the game logic is implemented on a central server, and the game view is dynamically rendered and delivered as streaming video to each player's terminal device. This type of game implementation removes the computational burden from STB and thus is suitable for IPTV environment. User perception of the service will strongly depend on smoothness of gaming experience, determined by performance of game control signaling. As it represents a highly interactive type of traffic, the network transfer delays from clients to the game server will be essential for quality assurance. Since the game picture is distributed in real time as video streaming, its perception will be impacted by known factors of audiovisual quality like coding efficiency and network capabilities for carrying streaming traffic.

In the case of *user-generated content services* (e.g., YouTube on TV), the operator has little control over the quality of the content itself. However, transcoding operation may be required to adapt the content format for displaying on IPTV device and it should be performed with possibly minor effect on content perceived quality.

New *parental control services* [4] assume that parents may control (by sending Short Message Service (SMS) from remote device, or by other means) the content that is being watched by their children. Although quality parameters are difficult to identify in an explicit way, the level of assurance provided and friendliness of user interface will impact on the overall perception of the service.

Lastly, for *interactive services launched as widgets or applets on connected TVs*, additional quality factors should be studied like application loading time.

12.3.4 Social TV Services

TV watching experience can be enriched with social exchanges, thanks to blending content with interpersonal communication services. Perception of communication component and cross-effects of multiple services interactions will certainly impact on perceived quality of converged service as a whole.

IPTV presence [4] assumes enriching user's reachability information with indication of currently watched content (TV channel or VoD item). Presence information should be distributed to subscribed peers in a timely manner. Thus, presence update delay, affected by signaling processing efficiency and network transport performance, will be an important quality indicator from the user's point of view. Similar factors of signaling performance will be applicable for *CallerID on TV,* which presents information about an incoming phone call and is an instance of a more general category of notification services.

Voicemail on TV is another example of blending TV with telephone network and it allows users for listening to their voicemail messages while watching TV. The aspects of end-to-end voice quality (see, e.g., ref. [7]) will play an essential role for QoE perception, with TV set acting as end device for voice transmission.

In *videoconference on TV*, the TV set will act as terminal device for audiovisual communication and the user will be concerned with combined quality of several media components, including real-time bidirectional audio and video (synchronized with each other to avoid lip sync effects). In addition to pure communication service quality aspects (see ref. [8]), the application-level factors like quality and smoothness of mixing communication with TV content (e.g., display of videoconference as a picture-in-picture on TV, mixing voice with TV audio), will impact on the overall QoE.

Instant messaging on TV is a text-based interactive communication service and timing aspects of text entry and display on TV, as well as messaging delays in the network, will contribute to its quality assessment.

Shared Service Control [4] denotes a new type of convergent service, allowing multiple users for watching the same content at the same time, with the impression of watching together despite being in distinct physical locations. Several factors will impact on overall perception of this kind of service. These factors are as follows:

- Interdestination media synchronization—imagine watching a live football match together—it would not be acceptable if one participant sees the video shifted in time comparing to others. Fulfilling this requirement involves mechanisms in media source (synchronization of streams destined to multiple end devices) and in the receivers (adjusting the buffering time), which is not trivial in the presence of different network delays to considered end points.
- Media playback—could be controlled by any participant (e.g., if one participant presses pause, this is reflected on the screens of all other participants), so precise synchronization of media control commands (affected by application processing and IPTV control signaling delays in the network) is required to maintain smooth user experience.
- Multiparty communication sessions (messaging, voice, videoconference)—can be established between participants of the shared TV service. In this case, the quality impact of communication component and smooth mixing of different real-time media will play an important role for overall QoE perception.

Parameters that influence the perceived service quality are of key importance for service acceptability. For example, during a trip to Oslo one of the authors stayed in a hotel that offered IPTV system in the room. However, the interaction with the system was frustrating as the channel zapping time was extremely long (order of tens of seconds) resulting in bad perception of the service. As a result, the decision was obvious to switch the TV set off. This is only one example of how seriously a service (basic) QoS parameter can influence the QoE of a user and in turn affects the user's attitude toward a service (the TV set was switched off because it was *annoying to the user*). As more IPTV systems are being deployed more work should be conducted during the predeployment phase of the system [9]. IPTV services should be tested in comprehensive testbeds where conditions are similar to those in commercial IPTV deployments. In this way, QoE of end customers could be assessed before deploying the services in the real network (this is similar to recommendations and best practices provided by the Information Technology Infrastructure Library (ITIL) *de facto* standard). The next step toward a scalable user-centric quality measurement system would be to build QoE monitoring mechanisms into the IPTV architecture, which can adapt to changing network conditions (as well as user context) and evaluate user quality perception in real time. The next section describes some relevant work conducted toward service quality evaluation as perceived by the customer.

12.4 QoE of Audiovisual Services

There is a growing interest in QoE in ITPV and IPTV itself as being the mainstream technology to replace traditional TV broadcast systems and introduces really convergent services. In Figure 12.1 authors have collected the number of

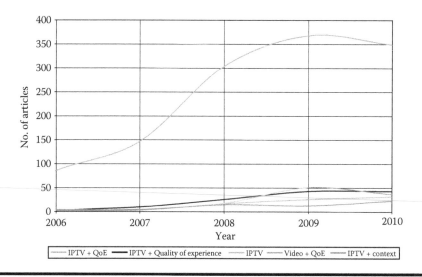

Figure 12.1 The number of articles over the last few years in IEEE database according to different search criteria [own elaboration].

research publications over the last 4 years in the Institute of Electrical and Electronics Engineers (IEEE) database according to different search criteria related to QoE. It can be seen, that IPTV has gained a lot of attention since the year 2006. In parallel, a considerable amount of work (nearly 15%) on IPTV research was carried out toward assessing QoE in IPTV. However, as depicted in Figure 12.1, this work is mostly targeted at assessing video/voice quality as perceived by the end user and little work was done toward assessment of other nonaudiovisual IPTV services (e.g., Parental Control). One should take into account that assessing only video or voice quality is not enough to deliver proper estimates of user's QoE as different factors (like context, user's attitude) can also influence the way a user perceives the overall IPTV service quality.

Current efforts of the International Telecommunication Union (ITU) are focusing on providing a definition of QoE which should take into account the future landscape of interactive services. To date QoE is defined by ITU as: *"the overall acceptability of an application or service, as perceived subjectively by the end-user"* [10:5]. However, ITU states that the term QoE *"... includes the complete end-to-end system effects (client, terminal, network, service infrastructure, etc.)"* and that the *"overall acceptability may be influenced by user experience and context."* The later statement provides an extension to the traditional QoE notion, stating that the end-user's QoE in not only directly influenced by QoS parameters of the network, but also by end-user's previous experience with a similar system or one's experience in general. Thus, user's QoE should be assessed at different layers of abstraction, and this is highlighted in the scientific literature from this domain.

12.4.1 State-of-the-Art QoE Measurement

In recent years, many researches have tried to modify existing QoE methodologies, which try to evaluate quality as perceived by end users, in order to provide a most comprehensive approach for QoE assessment in the entertainment domain. Pereira [11] proposes the QoE model that consists of three major components: sensorial, perceptual, and emotion evaluation. According to the article, these three components are key aspects in the process of video service evaluation and determination of general user satisfaction. The traditional approach to perceptual evaluation of audiovisual services uses subjective or objective assessment methods. The subjective tests provide the research community with valuable information on how users perceive video/voice quality but the tests have to be performed following ITU-T recommendations [12] to produce most reliable (and comparable) results. An interesting elaboration about common mistakes repeated by researchers during preparation of subjective video assessments can be found in ref. [13]. Authors point out that some tests are conducted under uncontrolled conditions and thus the results may be influenced by poor subjective test preparation. Some examples of common mistakes are: the number of subjective test participants is below the required minimum [12]; the video test sequences are obtained from low-quality Video Home System (VHS) tapes; or there is a too high percentage of removed outliers which indicates a mistake during test preparation. On the other hand, the subjective methodologies do not take the user context into account. For example—there is a substantial difference between a subjective test participant viewing a movie at home and in a controlled lab environment. An interesting approach to assess subjective QoE can be found in ref. [14]. Test participants were given a full-length movie on DVD to be viewed in their home environment. The movie had several scenes inserted with degraded quality and the test subjects were not aware of the existence and timing of such scenes. During traditional subjective test users were told to rate the video quality which implies that they should focus on the quality of particular scenes. This, however, stimulates the participants to concentrate on the video quality rather than the content. On the opposite, the use of full-length movies is more realistic approach (context-aware) that may cause subjects not to notice an impaired scene during movie consumption because, that is, they might be emotionally affected by the movie plot and thus not paying so much attention to the video artifacts. Thus, the impaired scenes might have no influence on a user's perceived QoE. Consequently, authors in ref. [14] imply that user expectation and context have influence on perceived video quality. Similar research was conducted and described in ref. [15], where authors tried to assess the influence of sensory effects (wind, heat, etc.) on perceived quality. The results show that user tend to rate the quality of same video samples higher when sensory effects are present.

For practical considerations it is, however, advised to minimize the user involvement directly in the quality-rating process (during tests and trials) as it is a time-consuming and usually costly task. That is why assessments methods, that rate quality of video in real time without direct feedback from the user, are needed. In particular, such mechanisms should be aware of the set of potential impairments that can influence video quality and, if possible, should be evaluated in terms of its correlation with human perception, network QoS and codec parameters. Several objective Video Quality Assessment (VQA) algorithms have been proposed—each of them is classified into one of the following categories:

■ Full-reference—original/reference source signal known at the input of the receiver (e.g., PEVQ).
■ Reduced-reference (or partial-reference)—only some information about source signal is given, like the codec used to compress the video, frame sequence number, and so on.
■ No-reference—no information available at the receiver about the source signal, quality metrics are solely calculated based on the received signal (e.g., E-model for Voice over Internet Protocol (VoIP)).

The different models for objective video assessment are depicted in Figure 12.2. In general, VQA algorithms are more complex and computationally demanding

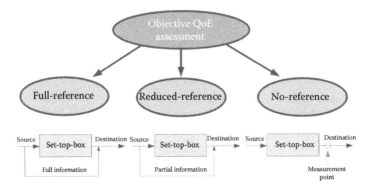

Figure 12.2   Objective QoE assessment models [own elaboration].

than Image Quality Assessment (IQA) algorithms. Although, the video signal could be decomposed to image frames and traditional IQA algorithms could be applied as it is much more difficult to assess video when we take the problem of, for example, perceived-motion into account. Improving the performance of VQA would require implementing some motion modeling.

However, it is rather difficult to model perceived-motion as the studies toward this direction are incomplete and the knowledge about the way the human brain processes moving images is still insufficient. Thus, we have to rely on the work considering full-reference and reduced-reference approaches. Both models, however, require the presence of the original source signal which in practical implementation cannot be delivered. As a result, full reference methods can be only used to evaluate the performance of single devices in a controlled test lab environment. No-reference algorithms can be applied to deliver QoE estimates at user's home IPTV set. However, the accuracy of such algorithms is not yet satisfying. An interesting approach to assess user QoE that merges subjective and objective methods was presented in ref. [1]. Authors try to fill the gap between both methodologies by proposing an automated tool for the assessment of mobile WWW service. With a use of a camera and image processing software authors are able to assess QoE in the exact form as subjectively perceived by service end users. The camera's role is that of pretending user sight to an extent possible, it "observes" the progress of Web pages being loaded and evaluates relevant key performance indicators (availability, accessibility, download time, etc.). Several tests were carried out and the results proved that it is possible to assess subjective user QoE in an automated fashion especially that results show that QoS measure fail to properly indicate user perspective (QoS results are more optimistic, for example, lower delay—than QoE indicators). The quality measures of several mobile telecom provider's WWW services was performed in several regions in Poland in order to evaluate mobile customer satisfaction.

12.4.2 QoE Measurements: Practical Perspective of a Provider

Apart from the scientific literature there is a rich market of devices capable of emulating an IPTV system with end users in order to assess a user's perceived QoE [3]. The market leading testbed equipment developer IXIA comes with their flag product IxLoad [16]. IxLoad is a full integrated test solution for evaluating the performance of multiplay (voice, video, data) networks and devices. The testbed has the capability to emulate subscribers and associated protocols enabling realistic assessment of QoS-based QoE. Another product is the IxN2X ITPV QoE solution which is IXIA's most comprehensive testbed to validate IPTV QoE under realistic conditions [17]. There are also other similar tools on the market like Agilent's N2X IPTV QoE test solution. It has the ability to dynamically simulate tens of thousands subscribers per system using multiple access protocols. Additionally, each subscriber may have a unique channel zapping profile assigned. Both producers, IXIA and Agilent, provide a comprehensive test equipment allowing service providers to build a fully integrated and scalable network testbed, that enables QoE assessment in the predeployment phase. Still as the size of the service provider's network grows and more users are being served some issues may still arise and the QoE of the customers could fall below an accepted threshold. A solution to this problem could be the Service Assurance Platform xVu from Mariner [18]. The tool has the ability to proactively monitor and diagnose network-related issues which have an impact on the customers QoE. In turn, a service provider can target and isolate the issue before it has a negative influence on customer experience. Such solution allows the service provider to minimize incoming help desk call volumes, reduce rework and maintain a satisfactory level of service quality. However, such techniques can only probe part of the real user experience spectrum, that is, audiovisual quality, service initiation time, channel zapping time, and so on. There is more into QoE than only visual quality than measured service quality.

12.4.3 Conclusions

The landscape of multimedia services is still changing therefore requiring new methods and models with the ability to realistically assess user QoE. This, however, requires the model to be aware of the user and one's context. As can be seen, there is rich scientific literature on QoE and a growing market with out-of-the-box testbed solutions that can probe a user's QoE. YouTube is a good example of how user's perception has changed during the last decade. People are more into YouTube because it allows them to share and watch video content in a manner most convenient to them. Usually, the service provides some level of QoS but it is far from excellent. Still hundred thousands of users use the service on a daily basis. We will not make any serious mistake if we say that this is due to the fact that the service satisfies their customer's (visitors) particular needs even though sometimes it can take seconds to load a video stream. This observation about how the perception of QoE has changed during the last decade will be discussed in the next sections and related to current IPTV services and their evolution.

12.5 Redefined Approach to QoE Assessment of IPTV Services

New interactive services available in emerging IPTV systems have redefined the way QoE should be understood, and thus a different approach to assess user satisfaction is needed to provide feedback to the IPTV service provider. Techniques to assess QoE described in the previous chapter are still valid for most of IPTV services. However, there are services available (e.g., Parental Control, Targeted Advertisement, CallerID on TV) that are not only relying on proper QoS handling but also may benefit from high service personalization and matching individual user preferences. According to CISCO the IPTV service evolution can be decomposed into three explicit stages [2]. **The early IPTV** services provide access to many Digital Television (DTV) channels, enable High-Definition Television (HDTV) content and allow the user to record content. Thus, the end user could benefit from an enhanced TV experience when compared with traditional broadcast TV but still the user was considered a passive consumer. **In the next stage** of IPTV service evolution interoperability and sharing between the IPTV system and other customer equipment was needed. The result was an interconnected home IP network coupled with IPTV system. This initiated a roll out of new services in IPTV (on-screen Interactive Program Guide (IPG), SurfWeb TV, etc.) but there was still room for improvement and the system could benefit even more if additional information related to the end user was present. That is why, in the **third stage of IPTV service evolution**, personalization of services have created the opportunity to provide interactive service that utilize user related information to provide better QoE for IPTV customers. The three stages of IPTV service evolution are depicted in Figure 12.3. The evolution of IPTV services has created many opportunities, but also gaps, in the way QoE should be evaluated. In ref. [19] author shows that good QoS does not always guarantee good QoE. Traditional, nonpersonalized services are mostly influenced by QoS parameters of the network, and the network parameters have influence on the proper delivery of QoE. This applies to audiovisual services (e.g., Linear TV) where the user is a passive consumer and has limited control on where and what he wants to watch. This is not longer valid when we take current IPTV systems into account, where the user has more control on the content itself (e.g., introduction of Trick Play functionalities). Thus, the landscape of customers has shifted from a pure passive to a mix of more active consumers that participate in IPTV service creation by directly interacting with the IPTV platform.

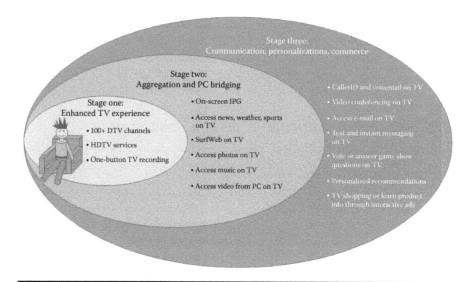

Figure 12.3   The evolution of IPTV services. (Adapted from White Paper; *The Evolving IPTV Service Architecture*; Cisco, 2007.)

Additionally, more services, like, for example, content recommendation, require user personalization mechanisms that utilize the information about the user, either extracted from the user's IPTV profile or derived from direct user feedback (e.g., user ratings, comments or knowledge extracted from the social networks). These services require more in-depth QoE mechanisms to measure user satisfaction. A customer plays a key role in current and future IPTV systems and more emphasis should be put on how to tailor a service to individual user needs. This is why a personalized IPTV service's QoE should not only be measured in terms of QoS, but also needs to take into account the user's opinion of a system, one's expectations, previous experience with a similar system, and user's attitude towards a system. In Figure 12.4 we have depicted under which conditions, in general, QoE is affected at different stages of IPTV service evolution. In first deployments of IPTV platform the service provider's scope was to deliver high QoS in order to ensure a right level of QoE. Thus, traditionally QoS had high influence on the perception of a service in the early stage of IPTV evolution. However, thanks to the increasing deployment of QoS mechanisms, increasing network capabilities and throughputs, service providers can maintain a good level of service quality for the customer. Yet, personalized services require more than just maintaining QoS constraints in order to provide high QoE. At some level of IPTV services evolution, personalization is starting to play an important role with respect to the end-user's satisfaction. It is evident that matching individual user preferences in emerging IPTV services is of considerable importance for the customer. Nevertheless, the service provider has to provide an appropriate QoS level

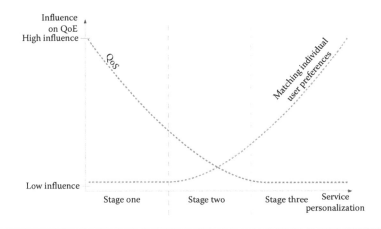

Figure 12.4 QoE versus service personalization [own elaboration].

but this, however, has a small impact on the service perception and acceptability. Let us consider the evolution from a perspective of Parental Control services for example. Originally this service was designed to allow parents to restrict access to specific channels for children. Nowadays, Parental Control service enhances their capabilities by SMS notification, or allowing parents to incorporate learning time into child's TV watching time [4]. This is a good example of how a service, well known from traditional broadcast TV, has evolved during last years. What is even more important to the user is the fact that he/she can choose how to use a service as different options of usage are available to him/her. Additionally, QoS constraints are less relevant with the original definition of Parental Control and do not influence the service acceptability. The key for high acceptability here is to assure that the service can be personalized and tailored to individual customer needs. As a conclusion, in today's quickly evolving IPTV environment focusing only on satisfying the QoS metrics has decreasing influence on the service acceptability and user satisfaction in general. More importantly, it is required to make the user feel he/she is in control of the system and the system itself is configured according to one's preferences. That is why personalization mechanisms should be applied to match end-users expectations and satisfy one's QoE.

Based on the above elaboration it is straightforward to come to the conclusion that current and future IPTV services could be divided into three groups of services according to their QoE requirements:

■ Services where QoE is strongly dependent on QoS (voice communication)
■ Services where matching individual preferences have high influence on QoE (personalized EPG)
■ Services where a balance between QoS constraints and personalization is of key importance (watching VoD based on system recommendation)

The challenge of measuring what is the level user satisfaction with a personalized service should be given additional attention as the research toward this issue is insufficient according to the authors' best knowledge. In particular, the question arises how to measure user's QoE when interacting with audiovisual and nonaudiovisual (Parental Control, CallerID on TV, etc.) IPTV services. One solution would be to gather user-generated feedback through subjective assessment (either by direct questions or ratings). However, TV consumption should be a rather unwinding and relaxing activity and should not involve the user in a tiresome rating process. That is why the next section reviews some interesting research toward unobtrusive user feedback gathering and shows the unexplored potential of such approach in QoE domain. In particular, we review the scientific literature to see which methods can be used to gather user feedback.

12.5.1 Role of the Human in the QoE Holistic View: Change of Paradigm

Preliminary work has been done toward unobtrusive user's QoE assessment during human–system interaction. In particular, the assessment of this interaction could be exploited to provide feedback information on system acceptability, usability, and accessibility. These factors are also important and influence the users' opinion of a system. A visible shift toward the user-centric approach in QoE dimension is reflected in ITU-T Recommendation G.1080 [20]. The QoE dimensions there are divided into two groups—network-related (QoS) and user-related (human component). The formal division is presented in Figure 12.5. It can be seen that emotions and user experience (which can be defined as user context) are expected to influence

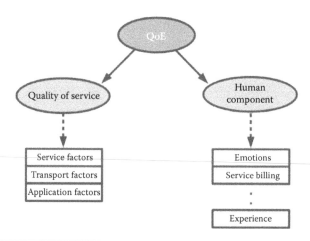

Figure 12.5 QoE dimensions. (Adapted from ITU-T; *G.1080: Quality of Experience Requirements for IPTV Services*; May 2008.)

QoE. The question is how to measure those parameters without involving the user in tiresome subjective tests.

There is still a gap between what is defined here as human component and QoS. That is why current research should focus on how to relate those two worlds in a big picture view. Using subjective methods is not always possible and impractical in some cases (due to its high cost, time, availability of test sample, etc.). However, some services will still require the use of subjective techniques (either in the form of direct questions or user ratings) but recent studies have shown that part of that process can be automated, thus not requiring continuous conscious feedback from the user [21]. The conclusion is that assessing user's QoE requires measuring the user response to the system he/she is interacting with (either it is a Virtual Environment, IPTV or VoIP, etc.). Three methods (see Figure 12.6) for assessing human response to the systems can be identified [22]:

- Subjective (by direct questions)
- Performance (observing how people behave)
- Physiological (measuring nonvoluntary responses of the body)

Users' subjective impressions are usually collected via surveys (questionnaires or interview).

At this point the user opinion on the system is collected in order to estimate if the user finds the system useful, if he can benefit from using it, or if it satisfies one's expectations. Gathering the user response seems to be important in terms of future system adaptation and enhancement to satisfy the user QoE. The surveys can include three kinds of questions—factual-type (responders age, sex, experience with computers, education level); opinion-type (ask about the responders thoughts about external stimuli); and attitude (ask about responders internal response, users attitude can be divided as depicted in Figure 12.7). Indeed, the current technique

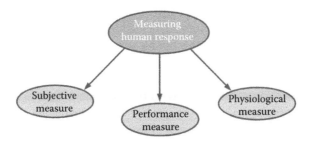

Figure 12.6 Measuring human response based on ref. [22]. (Adapted from Whalen, T.E.; Noel, S.; Stewart, J.; *International Symposium on Virtual Environments, Human–Computer Interface, and Measurements Systems;* **VECIMS, Lugano, Switzerland, pp. 8–12, July 27–29, 2003.)**

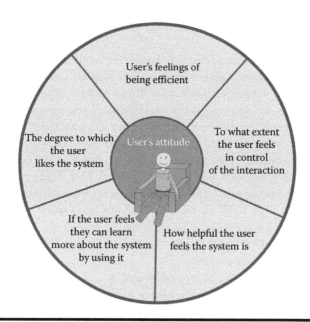

Figure 12.7 User's attitude based on ref. [22]. (Adapted from Whalen, T.E.; Noel, S.; Stewart, J.; *International Symposium on Virtual Environments, Human–Computer Interface, and Measurements Systems*; VECIMS, Lugano, Switzerland, pp. 8–12, July 27–29, 2003.)

for IPTV services personalization is still in its infancy. While watching a program the user can personalize the stream by selecting a different point of interest (if available, e.g., the visual quality level), a different live channel or program and using trick functions, for example, PAUSE, PLAY, re-PLAY, and STOP. Such approach demands the user intervention in a continuous manner and has a limited personalization means through failing to consider the exact user and environment context as well as the user special preferences and level of satisfaction with what they are watching. In this context Motorola proposes another approach to make individualized recommendations on a multiuser device, such as a STB, without asking users for authentication [23,24], that is through grouping anonymous family members' preferences into separate clusters based on their similarities, and providing user(s) with recommendations matching the tastes of all family members. We notice that this approach does not require the user(s) authentication to the system, but involves them in a short interactive filtering process for recommendation list, which could be seen a bit tedious by certain users for whom watching TV should remain a lazy and unengaged activity.

Another way to collect the information on a user's opinion of a system would be to observe the user or measure physiological responses of one's body. The latter can be performed by measuring galvanic skin response (GSR), blood pressure, heartbeat rate, face recognition techniques. The most relevant techniques used by the research community to collect information about user reaction to a given stimuli are presented in Figure 12.8. In particular, the question arises on how to assess the psychological aspects of a user in a possibly unobtrusive way. Interaction with a system should be as pleasant as possible for the user and should not interrupt one's experience with the system. Furthermore, this interaction can usually trigger positive or negative user emotion.

Many researches, projects, and experiments prove that analysis of human emotions may be helpful in the area of services' content evaluation, particularly in the entertainment services domain (i.e., for pictures, clips, movies, and music QoE assessment). Such analysis can be used for automated tagging of content in order to enhance capabilities of content recommendation systems based on individual and personalized users' preferences [21]. Efficient methods for data mining/correlating user emotions (against audio visual content) in a possible unobtrusive way are sought. In the literature several methods are enumerated to capture emotions online. These methods include GSR, face recognition, and

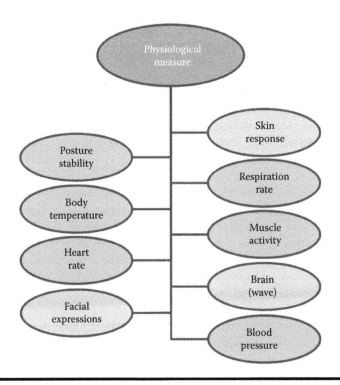

Figure 12.8 **User psychological measure. (Adapted from Ferre, M.;** *6th International Conference Proceedings; EuroHapitics 2008,* **LNCS 5042, pp. 129–138.)**

mapping user brainwaves into emotions. The authors in ref. [26] propose tracking of emotional states based on face recognition approach and indirectly based on assessment of attention, perception, interaction or skills level in order to achieve QoE adaptation in the e-learning systems. In ref. [21] author proposes utilizing a user's emotions to build content-based recommender for multimedia resources which are described by affective (emotional) metadata acquired in a noninvasive manner using face recognition through a video camera. This work has shown that awareness of user emotional response to a given stimuli, together with appropriate labeled content can increase the accuracy of the content recommendation system and thus influences the overall QoE of the service. In particular, the user emotional feedback can be used to solve the "cold start problem" of collaborative filtering-based recommender for images [27]. However, this approach still focuses on images only. Human mental states, which are detected by analyzing subjects face expression recordings are also considered in ref. [28]. Research documented in ref. [29] presents results of experiment, in which face expression detection and automated analysis of body language are used for emotion recognition. Detected emotions are induced by video clips characterized by different emotional features (e.g., causing anger, sadness, happiness, joy, or neutral). On the opposite to the above mentioned work, the authors in ref. [30] note the fact that popular methods for emotions detection based on face recognition techniques may be insufficient, because face expressions are relatively easy to control, regardless of perception and emotions. For more reliable and adequate results of emotion assessment, signals which are more spontaneous and less vulnerable to intentional control should be considered. Examples of such signals are those derived from the peripheral nervous system (e.g., heartbeat rate changes, EMG—electromyography, GSR—Galvanic Skin Response, etc.). In ref. [11], the author notes that emotion may be characterized both, by positive and negative features; however, such simplification may determine inappropriate results of content evaluation due to differentiation of target emotions related to various content (e.g., fear which is negative emotion should be desired brain response for horror movie). The system should take into account the type of video content and level of emotions which are adequate for this particular type of content and which are adequate to user satisfaction. Author considers parameters, such as heartbeat rate, muscles activity, skin conductance, and temperature, to measure user emotional response to content. However, specific metrics for QoE estimation are not defined, but general equations with specific weights for sensorial, perceptual, and emotional dimensions are included.

The above-mentioned techniques for emotion recognition provide only one-dimensional emotional information on a two-dimensional emotional space model proposed by Russel [31]. Mapping EEG signals into emotions, however, provides information on the type of emotion (valence axis) and the intensity of the emotion (arousal axis). The comparison between different methods is depicted in Figure 12.9. Authors in ref. [30] describe an experiment, in which methods for collecting

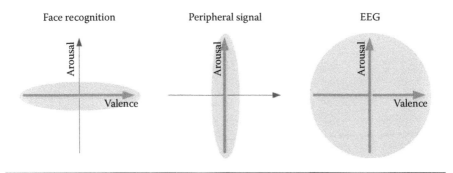

Figure 12.9   Three dimensions of emotions measurement in the two-dimensional space.

nonvoluntary responses of the human body, like EEG, GSR, skin temperature and blood pressure, are used to recognize emotions when viewing pictures from International Affective Picture System (IAPS) database [32]. After collecting data from EEG device, participants performed a self-assessment and evaluation of emotional valence and arousal level based on numerical (with five possible numerical judgments) scale in order to validate results from EEG.

The authors in ref. [33] also propose emotions detection based on EEG analysis. In the described experiment, EEG signal was a starting point for detection of eye blinking frequency and intensity during playing a racing video game. The purpose of the experiment was to prove that EEG provides capabilities for determination of logical interdependencies between EEG signals and level of concentration during video game.

To conclude this section let us recall the movie "The Game" [34]. In the movie "a game" refers to a notion of interactive event (kind of so called "exergame"), taking place in the real world, prepared for a participant and tailored to one's personality (to trick one's intelligence). Building the behavioral model of the main character, however, requires multitude of tests to be performed to understand the personality of the participant and one's preferences (but also behaviors, decision-making process, etc.). The profiling of the participant takes place through one's participation in several tests, one of which includes analyzing the response to video stimulus. In the movie Nicholas Van Orton (played by Michael Douglas, see Figure 12.10) sits in front of a TV screen and is presented with a set of short video sequences. One's response to the moving images is collected and analyzed by the company performing the test. The data were then used to prepare a real-life game setup for Van Orton. By then, 1997, this was a little bit futuristic, but today's technology has reached a very mature level and devices capable to read user brainwaves and translate them to emotions are accessible for everyone.

One such device is the Emotiv EPOC [35] presented in Figure 12.11. Other devices of this type are available on the market, like the NeuroSky Mindwave [36]

Figure 12.10 Scene from David Fincher's *The Game* (starring Michael Douglas).

or Neural Impulse Actuator [37], however, their performance is not satisfying, nor do they provide the necessary applications, like the Emotiv does (Software Development Kit).

According to the survey performed within the UP-TO-US project [38] all participants agreed that their emotional state influences their selection and assessment of video and audio content and also that movies should trigger emotions. The participants were generally in favor of motion-controlled interaction, which would mean not having to use yet another remote control. However, their sine qua

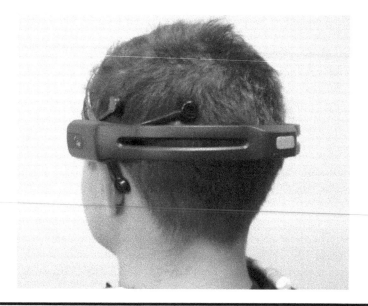

Figure 12.11 Light-weight EPOC emotive headset.

nonconditions were: the simplicity of the required motions, the performance level of the motion recognition technology (multiuser situation, parasitic movements) and effectiveness of the system as a whole (compared to a conventional remote control). The opinions are not sharply contrasted on the idea to carry a helmet that could read wave's brain in order to recognize moods. It is more difficult for the people to pronounce on this kind of process, that they know little the principle (measurement, performance, and reliability of this helmet) and that it is about a little captivating projection of use. Despite the consensus on the importance of mood in their criterion of selection, the process of mood recognition by a helmet, and especially the relevant association with the good TV program, makes the participants skeptical. Users could not imagine the impacts of emotion in IPTV service choosing or viewing. Above-stated facts show that there is room for (research on and deployment off) emotion-centric user profiling (algorithms) and enhancement of the Return on Investment (ROI) of IPTV systems with emotion reading devices in the loop.

12.6 A Redefined Approach to QoE Assessment of IPTV Services

There is a relationship between a service and individual user preferences and thus additional factors could be identified in order to provide a full perspective of QoE in the IPTV service domain. We gathered the factors that influence perceived service quality (QoS and individual preferences) in Table 12.1 to show that it is possible to assess QoE at different layers (based on QoS parameters and individual preferences). IPTV services were already described in Section 12.2 of the following chapter. Here we present some additional indicators that may influence service acceptability and customer QoE. Furthermore, it is interesting to note that emerging IPTV services would mostly benefit from high personalization.

However, media networks, where voice, video, and data traffic are present, are limited by the bandwidth capacity which impacts on QoE. One can simply increase the capacity of the network but this solution has proven to be highly ineffective and costly. On the other hand proper traffic management could solve some of the issues associated with IP networks. Peer-to-peer traffic is soaking most of the available bandwidth leaving a small portion of it for voice/video streaming. In the end this could affect the IPTV end-user's experience. In this light a key requirement for achieving IPTV service success is to enable QoE measurement that will monitor and benchmark the perception that an IPTV end-user has of the system. This is not a trivial task when we take into account that QoE not only relates to picture quality but also accounts for usability, responsiveness, and the user's opinion about the service. This approach is complementary to QoS, where measurements are based on collected network parameters (jitter, packet loss, and delays).

Table 12.1 IPTV Services and Factors that Influence Perceived Service Quality

Service	QoS-Based QoE	Individual Preferences
Time-shifted TV	Audiovisual quality, delay to access the recorded item (when stored on external server)	Define recording time and content to be recorded, consume content at a convenient time
Bookmarking	Setting precision, retrieval accuracy	User can set a bookmark while watching content and resume it in later time or on another device
Session continuity services	Session transfer time, quality degradation level	No service disruption, service can be adjusted to user context, lifestyle, and daily habits
Remote control of IPTV services	Responsiveness of mobile phone application, remote command execution delay	User can use a mobile handheld, or any other device capable to send remote commands, to control IPTV services
Search, discovery content recommendation	Page loading responsiveness, ease of GUI navigation	Content recommended according to user consumption trends, one's preferences, context, and lifestyle
Interactive Web-live services on TV	Page loading responsiveness, ease of GUI navigation, smoothness of switching display between TV content and interactive service	User may benefit from retrieving additional information about the program, news. User can access one's social Web accounts without the need to switch to a computer
Interactive synchronized services	Quality of synchronization between TV show and content depends on signaling protocol performance, timely triggering game events	User can actively participate in a quiz in a remote fashion

Table 12.1 (continued) IPTV Services and Factors that Influence Perceived Service Quality

Service	QoS-Based QoE	Individual Preferences
Targeted advertising and personalized advertisement insertion	Smoothness of transition between main content and advertisement clips	Takes into account individual user preferences, advertisement tailored to user expectations and needs
Personalized channel service	Performance of transport protocols and signaling, smoothness of transition between user selected content	Creation of individual personalized channel
Streaming games	Performance of transport protocols and signaling, coding efficiency	Games selected according to user gaming experience and preferences
User-generated content	Transcoding	User can create and share the content of one's own and access the other user's content
Parental control services	Ease of use	Level of assurance
IPTV presence	Presence update delay	
CallerID on TV	Signaling performance	
Voicemail on TV	Voice quality	Ease of use
Videoconference on TV	Audiovisual quality	Ease of use
Instant messaging on TV	Messaging delay	Ease of use
Shared service control	Synchronization accuracy, transport protocol performance	Impression of watching together despite being in distinct physical locations, personalization by utilizing user context information (friends, family, watching habits, etc.)

12.7 Final Conclusions

As mentioned in this chapter, emerging services, and especially IPTV require defining new metrics and do introduce new challenges for quality measurement methods to be well representing the user perspective. It is important to see that the landscape of services is changing and probably stakeholders should reengineer the way we perceive the new service capabilities. We need to focus especially on providing efficient measurement of services that includes the cross-layering of the measurement process with human component in the center. QoE is not easy challenge as what we experience is truly dependent on the (user) context, content, and (quality) of networks. Current research(ers) needs to rely on availability of novel sensors that help measuring and in turn study the (user's) feedback to the IPTV system (the portfolio of services and their key parameters). This will enable collection of new data that can bring into life new correlations on the relation between services, content, and eventually the user. It seems possible and beneficial with current devices to include them in the process of end-to-end quality management to identify new echelons of user perception. Yet, the research (on emerging services) and commerce (deploying it) cannot forget that the feedback between the user and service should bring the value that is not only visible in the third parties' ROI from TV commercials but which is value added to the primary and secondary users of the emerging services (like IPTV). People spending more time on watching IPTV should be benefiting from the IPTV system in the way they relax/have fun (emotional assessment) but also, for example, lifelong learning strategy (brain power is not negatively affected by being a passive user).

Acronyms

DTV	Digital Television
EEG	Electroencephalography
EMG	Electromyography
EPG	Electronic Program Guide
ETSI	European Telecommunications Standards Institute
GSR	Galvanic Skin Response
GUI	Graphical User Interface
HDTV	High-definition Television
IAPS	International Affective Picture System
IEEE	Institute of Electrical and Electronics Engineers
IPG	Interactive Program Guide
IPTV	Internet Protocol Television
IQA	Image Quality Assessment
ITIL	Information Technology Infrastructure Library
ITU-T	International Telecommunication Union—Telecommunication Standardization Sector

NGN	Next Generation Networking
PESQ	Perceptual Evaluation of Speech Quality
PEVQ	Perceptual Evaluation of Video Quality
PVR	Personal Video Recorder
QoE	Quality of Experience
QoS	Quality of Service
ROI	Return on Investment
SMS	Short Message Service
STB	Set-top Box
TISPAN	Telecoms & Internet converged Services & Protocols for Advanced Networks
VoD	Video on Demand
VoIP	Voice over Internet Protocol
VQA	Video Quality Assessment
WLAN	Wireless Local Area Network
WWW	World Wide Web

References

1. Flizikowski, A.; Puchalski, D.; Sachajdak, K.; Hołubowicz, W.; The gap between packet level QoS and objective QoE assessment of WWW on mobile devices; *Image Processing and Communications Challenges 2; Advances in Intelligent and Soft Computing*, 2010, 84/2010, 461–468.
2. White Paper; *The Evolving IPTV Service Architecture*; Cisco, 2007.
3. White Paper; *Ensure IPTV Quality of Experience*; Agilent Technologies, 2005.
4. ETSI TISPAN Technical Specification 182 027; *IPTV Architecture; IPTV Functions Supported by the IMS Subsystem*, V3.5.1, 03/2011.
5. ITU-T Recommendation G.1010; *End-user Multimedia QoS Categories*; 11/2001.
6. 3GPP, *IP Multimedia Subsystem (IMS) Service Continuity*; Stage 2+, TS 23.237, V9.6.0 Release 9, 3GPP, 10/2010.
7. ITU-T Recommendation G.1020; *Performance Parameter Definitions for Quality of Speech and Other Voiceband Applications Utilizing IP Networks*; 07/2006.
8. ITU-T Recommendation G.1070; *Opinion Model for Video-Telephony Applications*; 04/2007.
9. Online resource: http://www.youtube.com/watch?v=G6oMy63WGWA (last visited 10.05.2011).
10. Recommendation ITU-T P.10/G.100 (2006), Amd.1 (2007), New Appendix I—Definition of Quality of Experience (QoE).
11. Pereira, F.; A triple user characterization model for video adaptation and quality of experience Evaluation; *IEEE MMSP 2005, IEEE International Workshop on Multimedia Signal Processing*, October–November 2005, Shanghai, China.
12. ITU-R; *BT.500-12: Methodology for the Subjective Assessment of the Quality of Television Pictures*; September 2009.
13. Keimel, C.; Oelbaum, T.; Diepold, K.; Improving the verification process of video quality metrics; *First International Workshop on Quality of Multimedia Experience*; pp. 121–126, July 29–31, 2009, San Diego, California, USA.

14. Staelens, N.; Moens, S.; Van den Broeck, W.; Marien, I.; Vermeulen, B.; Lambert, P.; Van de Walle, R.; Demeester, P.; Assessing the perceptual influence of H.264/SVC signal-to-noise ratio and temporal scalability on full length movies; *First International Workshop on Quality of Multimedia Experience*; pp. 29–34, July 29–31, 2009, San Diego, California, USA.

15. Waltl, M.; Timmerer, C.; Hellwagner, H.; A test-bed for quality of multimedia experience evaluation of sensory effects; *First International Workshop on Quality of Multimedia Experience*; pp. 145–150, July 29–31, 2009, San Diego, California, USA.

16. Online resource: http://www.ixiacom.com/solutions/testing_video/index.php (last visited: 25.07.2011).

17. Online resource: http://www.ixiacom.com/products/display?skey=ixn2x_iptv_qoa_pa (last visited 10.05.2011).

18. Online resource: http://www.marinerpartners.com/en/home/newsevents/news/marine rsiptvqoesolutionimprovesservicequalityandre.aspx (last visited 10.05.2011).

19. Soldani, D.; Li, M.; Laiho, J.; QoS provisioning, in Soldani, D.; Li, M.; Cuny, R. (eds), *QoS and QoE Management in UMTS Cellular Systems*; John Wiley & Sons, Ltd, Chichester, UK, November 6, 2006.

20. ITU-T; *G.1080: Quality of Experience Requirements for IPTV Services*; May 2008.

21. Marko, T.; *Recognition and Usage of Emotive Parameters in Recommender Systems*; PhD Thesis, University of Ljubljana, 2010.

22. Whalen, T.E.; Noel, S.; Stewart, J.; Measuring the human side of virtual reality; *International Symposium on Virtual Environments, Human–Computer Interface, and Measurements Systems*; VECIMS, Lugano, Switzerland, pp. 8–12, July 27–29, 2003.

23. Bonnefoy, D.; Bouzid, M.; L'huillier, N.; Mercer, K.; 'More Like This' or 'Not for Me': Delivering personalized recommendations in multi-user environments; *User Modeling Conference*; Springer, 2007, 4511, pp. 87–96.

24. Lhuillier, N.; Bouzid, M.; Mercer, K.; Picault, J.; *System for Content Item Recommendation*; Great Britain patent GB2438645 (A), 2007.

25. Ferre, M.; Haptics: Perception, devices and scenarios, *6th International Conference Proceedings*; EuroHaptics 2008, LNCS 5042, pp. 129–138.

26. Moebs, S.; McManis, J.; A learner, is a learner, is a user, is a customer—So what exactly do you mean by quality of experience?; *2004 IEEE International Conference on Systems, Man and Cybernetics – SMC*, October 2004, Hague, Netherlands.

27. Online resource: http://www.readwriteweb.com/archives/5_problems_of_recommender_systems.php (last visited: 09.05.2011).

28. El Kaliouby, R.; Robinson, P.; Mind reading machines: Automated inference of cognitive mental states from video; *IEEE International Conference on Systems, Man and Cybernetics*; 2004.

29. Crane, E.; Gross, M.; Motion capture and emotion: Affect detection in whole body movement; *Affective Computing and Intelligent Interaction*; Springer 2007, 4738, 95–101.

30. Chanel, G.; Kronegg, J.; Grandjean, D.; Pun, T.; Emotion assessment: Arousal evaluation using EEG's and peripheral physiological signals; *Multimedia Content Representation, Classification and Security; Lecture Notes in Computer Science*; 2006, 4105/2006, 530–537.

31. Russel, J.A.; A circumplex model of affect; *Journal of Personality and Social Psychology*; 1980, 39(6), 1161–1178.

32. Lang, P.J.; Bradley; M.M.; Cuthbert, B.N.; International Affective Picture System (IAPS): Technical manual and affective ratings; *NIMH Center for the Study of Emotion and Attention*; 1997.
33. van Galen Last, N.; van Zandbrink, H.; *Emotion Detection Using EEG Analysis*; Delft University of Technology, 2009.
34. Online resource: http://www.imdb.com/title/tt0119174/ (last visited: 22.07.2011).
35. Online resource: http://emotiv.com/ (last visited: 22.07.2011).
36. Online resource: http://www.neurosky.com/Products/MindWave.aspx (last visited: 25.07.2011).
37. Online resource: http://gear.ocztechnology.com/products/description/OCZ_Neural_Impulse_Actuator/index.html (last visited: 25.07.2011).
38. Online resource: http://www.celtic-initiative.org/Projects/Celtic-projects/Call7/UP-TO-US/uptous-default.asp (last visited: 25.07.2011).

Chapter 13

QoE-Based Routing for Content Distribution Network Architecture

Hai Anh Tran, Abdelhamid Mellouk,
and Said Hoceini

Contents

13.1 Content Distribution Network: Context and Challenges 256
    13.1.1 Overview 256
    13.1.2 Content Delivery Processes...257
13.2 Quality of Experience Paradigm 258
    13.2.1 Introduction 258
    13.2.2 QoE/QoS Relationship...259
    13.2.3 QoE Measurement Methods...261
13.3 Proposed Approach: CDA-QQAR 261
    13.3.1 QQAR Algorithm...262
    13.3.2 Proposed Architecture and Server Selection Method 263
    13.3.3 Data Routing...265
13.4 Validation Results 265
    13.4.1 Testbeds for PSQA...266
    13.4.2 Simulation Results 267
13.5 Conclusion 270
Acknowledgments 271
References 271

13.1 Content Distribution Network: Context and Challenges

With the rapid growth and evolution of Internet today, the lack of a central coordination is critically important. However, without management makes it very difficult to maintain proper performance. At the same time, the Internet utilization and the accelerating growth of bandwidth intensive content continue to overpower the available network bandwidth and server capacity. The service quality perceived by end users is accordingly largely unpredictable and unsatisfactory. One proposes Content Distribution Network (CDN) as an effective approach to improve Internet service quality [1,2]. CDN replicates the content from origin servers and serves a request from a replica server close to requested end users. As CDN service includes two important operations: (1) Replica server selection used to select an appropriate replicated server holding the requested content and (2) Data routing used to choose the best path to serve data to the end user, we focus our attention in a first part to give a brief overview of CDN. We then present the content delivery processes including sever selection and data routing method in a second part.

13.1.1 Overview

Nowadays, the poor Internet service quality with the long content delivery delay is due to two problems: First, it is the lack of overall management for Internet. The absence of overall administration makes it very complicated to guarantee performance and deal systematically with performance problems. Second, the load on Internet and the diverse content continue to increase. These two problems make the access latency unpredictable and increase the delay in accessing content. With the diversity of Internet content, the Internet performance problem will become even worse. The basic approach to resolve this problem is to move the content from the places of origin servers to the places at the network edge.

In fact, serving content from a local replica server typically has better performance than from the origin server, and using multiple replica servers for servicing requests costs less than using only the data communication network does. How to select the most appropriate replica server having the desired content and how to route the content data in the core network from replica server to end user are the key challenges in designing an effective CDN, and are the major topics we will discuss in this chapter.

Different papers [3] describe the CDN general architecture. The architecture used here is depicted in Figure 13.1. The CDN is focused on serving both Web content and streaming media. We present now some main components of the architecture are: Origin Server, Surrogates, Clients, Access Network, Distribution Network, Content Manager, and Redirector.

Origin server contains the information to be distributed or accessed by clients. It distributes the delegated content to the surrogates of the CDN through the

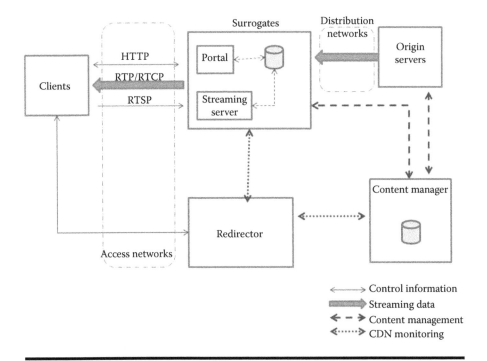

Figure 13.1 CDN architecture.

Distribution Network. We consider surrogate servers as replica servers of the origin servers. They act as proxy/cache servers with the ability to store and deliver content. We based on the structure, number of surrogate servers, location of these servers to classify the CDN and the algorithm executed to determine the server who serves each issued request. Clients are represented by PC or special set-top boxes which request and download a particular piece of content stored somewhere in the CDN. The CDN usually works with clusters of clients rather than individual clients. Clients access the service provided by the CDN through different Access Networks, depending on the ISP.

Distribution network interconnects the Origin Servers with the Multimedia Streaming Servers (MSSs) to deliver media objects within the streaming media CDN. The Content Manager controls the media objects stored in each Surrogate, providing this information to the Redirector in order to get each client served by the most suitable Surrogate. The Redirector module provides intelligence to the system, because it estimates the most adequate surrogate server for each different request and client.

13.1.2 Content Delivery Processes

CDN has many distributed components collaborating to distribute content across different servers and networks. There are commercial and research CDN, but all of them use common approaches to the problem of delivering content and services to the users. The common processes CDN researchers and operators have to solve are:

- *Content placement*: The problem of content placement involves how many servers to use, where to position them, and how to distribute content across them.
- *Surrogate placement*: Many researchers have focused on surrogate server placement and proposed several placement algorithms. There are different approaches to the problem. In ref. [4], the authors try to solve the problem by minimizing the number of traversed autonomous systems by when clients issue a request for a specific content. Other authors propose a technique using an unrealistic tree topology network [5], a greedy solution [6] or try to profit P2P content placement solutions to the CDN environment [7].
- *Content discovery*: is the process of identifying surrogates that store the desired content or a relevant part of it. P2P networks use mechanisms based on flooding and forward routing algorithms [8].
- *Content management*: There are different techniques to redirect client requests to the objects stored in a CDN [9]. Although most CDNs use DNS-based redirection schemes [10,11], others use URL rewriting, in which the origin server redirects clients to different surrogate servers by rewriting the dynamically generated pages.
- *Server selection*: This mechanism tries to estimate the most appropriate surrogate server to serve each individual client request. The best server does not always mean the one nearest to the client, the fastest or the most reliable but a combination of all these features.
- *Data routing*: The routing mechanism tries to find the best path to route data packet in the core network from chosen surrogate to end users.

In this chapter, we focus on the last two mechanisms: server selection and data routing.

13.2 Quality of Experience Paradigm

13.2.1 Introduction

Nowadays, users use the Internet not only to read news or send emails but also to use real-time applications (audio/video streaming services, VoD, video conference, etc.). Basic connectivity is a given and availability is no longer an effective differentiator for a broadband provider. To differentiate against competition, Internet service providers (ISPs) must differentiate themselves using QoE that means to provide fast and efficient downloads as well as immediate and uninterrupted streaming of video/audio.

The theory of quality of experience (QoE) has become commonly used to represent user perception. To evaluate a network service, one has to measure, monitor, quantify, and analyze the QoE in order to characterize, evaluate, and manage services offered over this network. In fact, the network provider aim is to provide a good user experience at minimal network resource usage. It is important for the network operator to be aware of the degree of influence of each networks factor on the user perception.

In fact, the end-to-end (e2e) quality is the most important aspect to be achieved for operators, service providers, and also for users. QoE has recently been introduced, which is a new concept to make e2e QoS more clearly taking into account the user's experience. QoE also takes into account the satisfaction and the perception of the users when using network services. Recently, many researches and proposals have been made in order to measure, evaluate, and improve QoE in networks. We need a definition of new effective parameters and values, new measurement methodologies, and new quality prediction systems.

U2U-QoE (User-to-User QoE) represents the overall satisfaction and acceptability of an application or service as perceived by the end user. U2U-QoE expresses user satisfaction both objectively and subjectively. It has become a major issue for telecommunication companies because their businesses are highly dependent on customer satisfaction and the average revenue per user can only be increased by value-added services taking into account the complete e2e system effects. U2U-QoE takes into account the entire e2e path chain including the Quality of Design of terminal equipment and Quality of Service supported by network components. In the context of communication services, it is influenced by content, network, device, application, user expectations and goals, and context of use. Other external factors that can also have an impact on U2U-QoE include user's terminal hardware, mobility and the importance of the application.

Technical metrics cannot prevent the customer churn problem if a subscriber is having a problem that is getting him or her dissatisfied to change service providers. So one may use QoE assessment to prevent this churn problem. Network providers that provide good QoE has a significant competitive advantage, while providers that ignore the importance of QoE may suffer unnecessary costs, lost revenue, and diminished market perception. In [12], the authors confirm this point with statistical results: around 82% of customer defections (churning) are due to frustration over the product or service and the inability of the provider/operator to deal with this effectively. More dangerously, this leads to a chain reaction, an unsatisfied customer will tell other people about their bad experience. An operator cannot wait for user complaints to react or to improve its service quality. 90% of users will not complain before defecting. So, the only way to overcome this situation is to devise a strategy to manage and improve QoE proactively.

13.2.2 QoE/QoS Relationship

The notion of Quality of Service has been discussed for more than a decade, but it concerns essentially the technical view on service quality. The QoS architectures (e.g., Integrated Services or Differentiated Services) are used to pave the way for high-quality real-time services like VoIP or video streaming. Recently, in order to redirect the focus toward the end user and quantify the subjective experience gained from using a service, the notion of QoE has emerged, describing quality as perceived by the human user.

QoS is defined early as "collective effect of service performance which determines the degree of satisfaction of a user of the service" [13:3], or "a set of qualities related to the collective behavior of one or more objects" [14:5], or "the networks capability to meet the requirements of users and applications" [15:1].

The notion QoE means the user perception about the quality of a particular service or network. It is expressed by human feelings like good, excellent, bad, and so on. In addition, QoS is fundamentally a technical concept, which usually has little meaning to users.

Many people think that QoE is a part of QoS but it is a wrong perception. In reality, QoE covers the QoS concept. In order to demonstrate this theoretical point, we observe a quality chain of user-to-user services (Figure 13.2). The quality perceived by end user when using end devices is called QoD (Quality of Design). The quality of network service including core network and access network is determined by QoS access and QoS backbone. QoE is satisfied only in the case that QoD and QoS are satisfied. So we can see that the user-to-user QoE is a conjunction of QoD and QoS. Indeed, the QoE is the most important factor for the operator to properly design and manage its network.

Even if a better network QoS in many cases will result in better QoE, maintaining all the QoS parameters will not ensure a satisfied user. For example, in a wireless network, if there is no coverage a short distance away, this causes a bad QoE although there is an excellent throughput in one part of a network.

Furthermore, in e2e QoS, the main challenge lies in how to take the dynamic changes in the resources of communication networks into account to provide e2e QoS for individual flows. Developing an efficient and accurate QoS mechanism is the focus of many challenges, mostly in complexity: e2e QoS with more than two noncorrelated criteria is NP-complete.

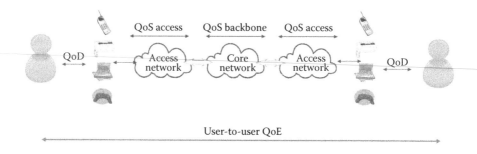

Figure 13.2 Quality chain in an e2e service.

Therefore, integrating the QoE concept in a QoS system is a solution to reduce this complexity. Indeed, this combination can always maintain satisfaction degree of end users, purpose of all network systems, without optimizing all QoS parameters.

13.2.3 QoE Measurement Methods

It is not easy to evaluate the perceived quality of a multimedia stream. In fact, assessing perceived quality requires real people to evaluate it subjectively. There are standard methods for organizing subjective quality assessments, such as the ITU-P.800 [16] recommendation for telephony, or the ITU-R BT.500-10 [17] for video. However, subjective evaluations are very expensive and cannot be a part of an automatic process. As subjective assessment is useless for real-time evaluations, a significant research effort has been done to obtain similar evaluations by objective methods. We cite here some commonly used objective measurement methods for audio: Signal-to-Noise Ratio (SNR), Segmental SNR (SNRseg), Perceptual Speech Quality Measure (PSQM) [18], Measuring Normalizing Blocks (MNB) [19], ITU E−model [20], and so on. Regarding video quality, there are some measurement methods such as: the ITS' Video Quality Metric (VQM) [21], Color Moving Picture Quality Metric (CMPQM) [22], and Normalization Video Fidelity Metric (NVFM) [23]. However, these methods cannot replace subjective methods because their provided assessments do not correlate with human perception, so their use as quality measurement method is limited.

In our approach, we decide to use PSQA tool [24], a hybrid between subjective and objective evaluation. PSQA tool measures QoE in an automatic and efficient way, such that it can be done in real time. It consists of training a Random Neural Network (RNN) to behave as a human observer and to deliver a numerical evaluation of quality, which must be close to the average value that a set of real human observers would give to the received streams.

13.3 Proposed Approach: CDA-QQAR

In this section, we present our protocol QQAR (QoE Q-Learning-based Adaptive Routing) and the idea to apply this protocol to a CDN, namely CDA-QQAR (Content Distribution Architecture using QQAR protocol). We consider QoE measured at end-user station in the system as metric to choose the "best" replicated server as well as route data content from these servers to end users. This architecture (Figure 13.3) requires replicated server placed at different locations.

These servers having different IP addresses are divided into different groups. Each group has a private anycast address. CDA-QQAR also needs network components placed at client side. This component works as a set-top box which we named here as iBox (Intelligent Box) (Figure 13.4).

This iBox component has the functionality to select the server based on a server selection table. Thus, all anycast requests are directed to the local iBox, which uses

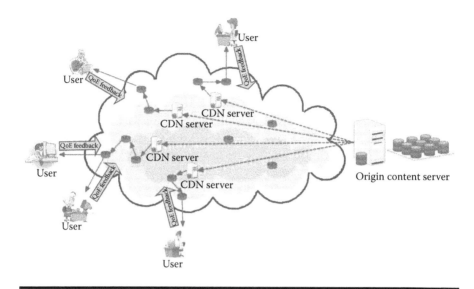

Figure 13.3   QoE-aware CDN architecture.

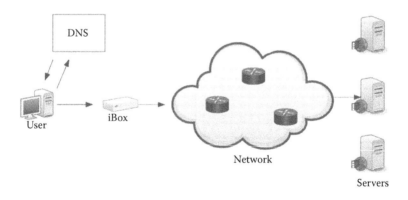

Figure 13.4   CDA-QQAR architecture.

request routing table to choose the appropriate replicated server. In the opposite direction, the data routing uses modified QQAR protocol (presented in Section 13.3.1) to route requested data to end users.

In the following subsections, we will present in detail our CDA-QQAR approach.

13.3.1 QQAR Algorithm

In this section, we briefly present our earlier QQAR [25], a QoE-based routing protocol. In order to evaluate QoE at the end user, we have applied the PSQA tool. The learning method of QQAR is based on Q-learning algorithm [26]. The latter representing one of the most important breakthroughs in Reinforcement Learning (RL) is an off-policy temporal difference (TD) control algorithm. The optimal action-value function is approximated directly by Q-learning, independent of the policy being followed. After executing action *a*, the agent receives an immediate reward *r* from the environment. By using this reward and the long-term reward, it then updates the *Q*-values influencing future action selection. We have modified our QQAR in applying Least-Squares Policy Iteration (LSPI) [27] which is a recently introduced RL method. With LSPI, the next-hop choice of routing protocol is based on the fact that this technique learns the weights of the linear functions, thus can update the *Q*-values based on the most updated information regarding the QoS criteria. Based on LSPI, the value function is approximated as a linear weighted combination $\phi_{xy}^T\omega$, where ϕ_{xy} is the QoS criteria vector of link *xy* and ω is weight vector. Following some LSPI mathematical transformations, the weight ω is extracted:

$$\omega = (\Phi^T(\Phi - \gamma P^\pi \Phi))^{-1} \times \Phi^T R \qquad (13.1)$$

where γ is discount factor. Φ is matrix representing QoS criteria for all links in the routing path. *P* is probability of entering neighbor node. In our approach we assume that $P = 1$. *R* is a matrix of reward values calculated by Equation 13.2. The proposed routing mechanism can be formulated as follows:

- *First step—Data packet flow*: the provider sends data packet to end user. When a packet is forwarded from a node to his neighbor, the QoS criteria vector ϕ is appended to the packet. The neighbor selection is based on softmax method with Boltzmann distribution [28].
- *Second step—At end-user side:* After data reach end-user station, the latter has to realize three tasks. First, it evaluates QoE by using PSQA tool to give an MOS. Second, with certain amount of collected samples ϕ, we determine reward values based on the equation

$$r_{xy} = \alpha MOS + \beta \phi_{xy}^T \omega \qquad (13.2)$$

Finally, we invoke the LSPI procedure to estimate new weights ω_{new}.
- *Third step—ACK message flow*: Each time a node receives an ACK message, it updates the *Q*-value of the link that this ACK message just passes through. The update process is introduced in Section 13.3.3.

13.3.2 Proposed Architecture and Server Selection Method

Figure 13.4 illustrates CDA-QQAR architecture, which consists of replicated servers, end-user station, and iBox. The iBoxes placing at client side stocks the server

Table 13.1 Server Selection Table

Anycast Address	Path	QoE Feedback	Server IP Address
Add 1	Path 1	q1	IPadd1
	Path 2	q2	IPadd2
	...	...	...
	Path n	qn	IPaddn
Add 2			
...			
Add m			

selection table (Table 13.1). The PSQA tool (cf. Section 13.2.3) is integrated into the iBoxes. Therefore, based on the information gathered of routing path, the iBox can give a QoE feedback as a MOS score. Then it stocks this MOS in the server selection table and sends it back to the routing path.

We present now a step-by-step description of the server selection operations of CDA-QQAR:

1. The user sends a domain name resolution query to the anycast-enabled DNS. The latter returns the anycast address to this user.
2. The user sends this obtained anycast address to his iBox, which realizes a table lookup to find the associated IP addressed of this anycast address. If the requested anycast address does not exist, iBox sends a failure message to the user and makes an entry for this anycast in the server selection table. In other cases, iBox makes a choice of replicated server basing on Softmax Action Selection method, which assigns a choice probability to each server:

$$p_i = \frac{e^{q_i/\tau}}{\sum_{j=1}^{n} e^{q_j/\tau}} \tag{13.3}$$

where p_i is the probability of choosing server i. q_i is the QoE feedback (MOS score) assigned to server i. τ represents a temperature parameter of Boltzmann distribution.

3. After choosing the replicated server, the request is sent to this chosen server using the unicast routing technique.

13.3.3 Data Routing

After receiving the request, the server serves requested data to the user using modified QQAR protocol. Concerning data routing problem, we analyze it in three steps:

1. *Data flow:* Each node (router) in the routing system has a routing table that indicates the Q-values of links emerging from this router. Based on softmax action selection rules, after receiving a packet, node x chooses its neighbor y_k among its n neighbors $y_i (i = 1::n)$ with probability presented in

$$P_{xy_k} = \frac{e^{Q_{xy_k}/\tau}}{\sum_{i=1}^{n} e^{Q_{xy_i}/\tau}} \tag{13.4}$$

   where Q_{xy_i} represents the Q-value of link xy_i and τ represents a temperature parameter of Boltzmann distribution. We apply Q-learning algorithm and LSPI technique to the problem of learning an optimal routing strategy. We consider the following QoS criteria for each link between node x and y in the system: (1) The delay:time to transmit a packet from x to y; (2) The distance (hop count) of y to the destination; (3) The bandwidth of link xy. We normalize all these values to the range of $[-1,1]$. When a packet is forwarded from node x to y, the vector $\omega\{d,c,b\}$ is appended to this packet, where d, c, b are delay, hop count, and bandwidth, respectively.
2. *At end-user side:* when a packet arrives to the iBox, the latter evaluates QoE using the PSQA tool to give an MOS score. It then determines rewards r_{xy} for each link xy of the routing path based on Equation 13.2. Finally, iBox calculates the new value of ω_{new} based on Equation 13.1 and assigns it with the MOS score into an ACK message. This ACK message is sent back to the routing path for updating Q-values of routers in this path.
3. *ACK message flow:* Once node x receives an ACK message from y, it calculates the reward r_{xy} in using Equation 13.2. Node x updates then the Q-value of link xy with equation follows:

$$Q_{xy} \leftarrow Q_{xy} + \varphi(r_{xy} + \gamma \max Q_y - Q_{xy}) \tag{13.5}$$

where Q_{xy} is the Q-value of link xy. Q_y is the maximum Q-value of node y. φ is the learning rate, which models the rate updating Q-value. γ is the discount factor.

13.4 Validation Results

In order to validate our proposed approach, this section presents our real testbed for collecting dataset to the PSQA tool. We then describe our simulation results using the OPNET simulator.

13.4.1 Testbeds for PSQA

Training RNN for PSQA tool needs a real dataset of the impact of the network on the perceived video quality. To construct this dataset, we conducted an experiment which consist to selecting a set of students and ask them to score the perceived quality of video using the MOS score. The testbed (Figure 13.5) is composed by client–server architecture and a network emulator. The client is VLC video client [29] and the server is VLC video streaming server [29]. The traffic between client and server is forwarded by the network emulator NetEm [30]. NetEm provides the way to reproduce a real network in a lab environment.

The current version of NetEm can emulate variable delay, loss, duplication, and reordering. The experimental setup consists of forwarding video traffic between server and client. Then, we introduce artificial fixed delay, variable delay, and loss on the link to disturb the video signal.

According to ITU-R [31], the length of the video should be at least 5 s. We choose the *SINTEL* video trailer [32]. This video is of 52 s duration, 1280×720 dimensions and 24 frames/s cadence and uses H.264 codec. This video was chosen because it alternates high and slow movements. Experiments were conducted with fixed delay values of [25, 50, 75, and 100 ms], variable delays of [0, 2, 4, 6, 8, 16, 32 ms], loss rate values of [0%, 2%, 4%, 6%, 10%, 15%, 20%, 25%, 30%], and

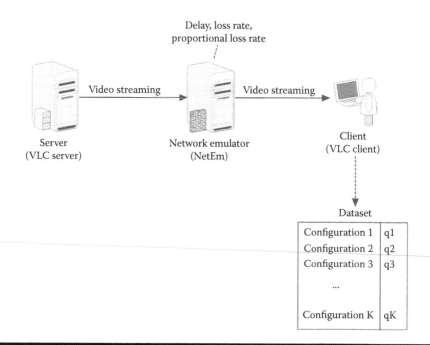

Figure 13.5   Testbed for the PSQA tool.

successive loss probability of [0%, 30%, 60%, 90%]. These values were chosen to cover the maximum of QoE range.

To collect data, we choose viewers with a strong cinematic experience. Nowadays, as a major part of monitors are LCD, we used the same ones. The particular screen used is a 19″ size screen "LG flatron L194wt–SF" which has 1440×900 resolution. The obtained dataset of this testbed is used for the learning process of PSQA tool. We then obtain the function f in Equation 13.1 and apply the latter to our system as QoE measurement tool (PSQA).

13.4.2 Simulation Results

OPNET simulator version 14.0 has been used to implement our approach. Regarding network topology (Figure 13.6), we have simulated three end users (u0, u40, u39(red color)), three iboxes (i41, i42, i43(yellow color)), and five replicated servers (sv10, sv44, sv45, sv46, sv47(green color)). Between users and the server's side, we have implemented an irregular network with 3 separated areas including 38 routers, each area is connected to each other by one link, all links are similar. The network system is dynamically changed in an average period of 200 s.

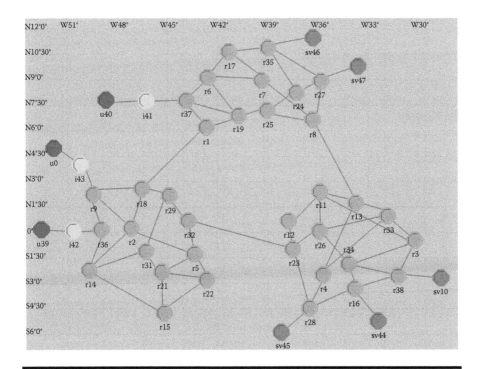

Figure 13.6   Simulation network topology.

In our simulation model, the frequency of sending request of end users is varied randomly from 100 to 300 s. Once the ibox receives a request, it realizes server selection process. To facilitate the simulation method, we assign all of five servers to the same anycast group. The training process of PSQA is realized in offline to give a function f_{PSQA} as output. We have integrated this output function f_{PSQA} in three iboxes.

We compare our approach with three kinds of algorithm in order to validate our results. All these algorithms are based only on QoS criteria.

1. *Those based on Distance-Vector (DV) algorithm.* In this algorithm, server selection and data routing are based on hop-count number.
2. *Those based on Link-State algorithm*: SPF (Short Path First). This algorithm is based on link cost.
3. *Those based on Standard Optimal QoS Multi-Path Routing (SOMR) algorithm* where sever selection and data routing are based on finding K Best Optimal Paths and used a composite function to optimize delay and link cost criteria simultaneously.

Figures 13.7 and 13.8 illustrate the result of average MOS score obtained by our four approaches in different levels of traffic network. To create these scenarios, we have generated different levels of traffic that stress the network. Each level represents the rate between the number of stressed links and the number of total links:

$$\text{level} = \frac{n_s}{N} \tag{13.6}$$

where n_s is the number of stressed links and N is the number of total links in the network.

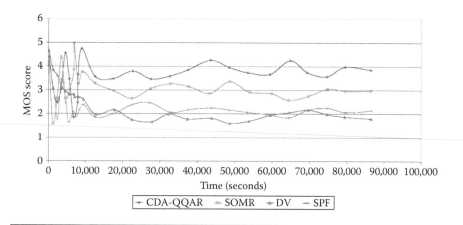

Figure 13.7   User perception in low-load traffic network.

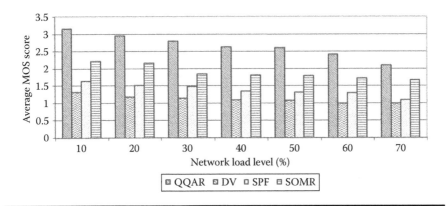

Figure 13.8   User perception in different load traffic levels of network.

Figure 13.7 shows that results of all tested algorithms fluctuate very much in the first 10,000 s. That is explained by the execution of initialization process. In other words, these four algorithms do not have any information about the system in this first period, they try to explore it. In these first 10,000 s, the MOS score of CDA-QQAR varies between 1.8 and 4.8, SOMR between 1.5 and 5, SPF between 1.7 and 3.5, DV between 2.5 and 4.5. After the first 10,000 s, algorithms gradually become stable. QQAR converges to the value of 3.8, SOMR to 3.0, DV and SPF to about 2.0.

Figure 13.8 gives us the average of these four algorithms for different traffic levels formulated in Equation 13.6: from 10% to 70%. We can see that more the system is charged, higher the average score decreases. However, at any level, the average MOS score of CDA-QQAR is better than all other algorithms. Regarding the other traditional protocols, the maximum value obtained by SOMR is just 2.3 with 10% for traffic level.

These results confirm that our approach CDA-QQAR takes into account the user perception in both cases than three other approaches and gives better results. So with our approach, despite network environment changes, we can maintain a better QoE from the user satisfaction point of view. Thus, CDA-QQAR is able to adapt its decisions rapidly in response to changes in the network dynamics.

Our experiment works consist also to analyze the overheads generated by these protocols. The type of overhead we observe in these experiments is the control overhead that is determined by the proportion of control packet number and the number of all packets emitted. To monitor this overhead value, we have varied node number in adding routers to network system. So the observed node numbers are [38, 50, 60, 70, 80]. The obtained result is showed in Figure 13.9.

We can see that the control overhead of our approach is constant (50%). That is explained by the equality between data packet and control packet in CDA-QQAR. Each generated data packet leads to an acknowledgement packet generated by the destination node. The control packet rates of DV, SPF, and SOMR are, respectively,

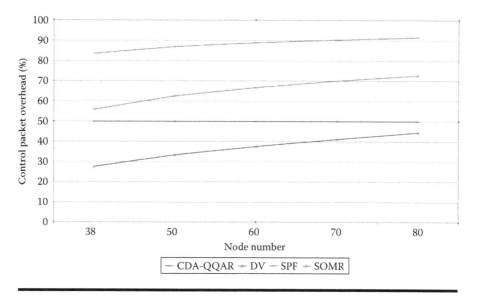

Figure 13.9 **Control overhead.**

0.03, 0.4, and 0.1 (packet/s). This order explains the control overhead order in Figure 13.7. Whereas the SPF algorithm has the highest value because of the highest control packet rate (0.4 packet/s) with multiple type of packet such as Hello packet, Link State (LS) Acknowledgement packet, LS Update packet, LS State request packet, etc. In the same time, DV algorithm has the smallest overhead value with a control packet rate value of 0.03. We can see also that the higher the number of node implies, the higher the overhead is. So with a stable overhead, our approach is more scalable than these three others.

13.5 Conclusion

There are currently several CDN systems created by different service providers such as Akamai, Speedera, Digital Island, and so on. However, most approaches existing in the literature realized these two operations based on only QoS criteria such as hop counts, round-trip time, bandwidth, and so on. The presented idea in this chapter is to work on the terminal user perception and not only on QoS criteria in the process of replica server selection. In fact, as e2e QoS optimization is an NP-complete problem regarding two or more nonadditive criteria, we try here to maintain QoE criteria instead of optimizing multiple QoS criteria and to optimize user's satisfaction. This idea starts from the point of view that improving the quality of the services as perceived by the users has a great effect as well as a significant challenge to the service providers with a goal to minimize the customer churn yet maintaining their competitive edge. Based on this kind of quality competition, QoE has been introduced, combining user perception, experience, and expectations without technical parameters. QoE takes into account the needs and the desires of the subscribers when using network services, while the concept of QoS just attempts to objectively measure the service delivered. As an important measure of the e2e performance at the service level from the user's perspective, the QoE is an important metric for the design of systems and engineering processes.

In this chapter, we propose a new routing process for content distribution architecture based on end-users perception. Our proposal, named Content Distribution Architecture using modified QQAR (CDA-QQAR), uses QoE measured by PSQA tool as a metric for choosing replica server and routes data to end users. To choose the "best" replica server, we propose to use network layer redirection based on QoE evaluation. This deduces the idea of anycast to determine a highest QoE server. Based on anycast communication method, one assigns the same anycast address to more than one node and a packet sent to this address is routed to the "best" node. CDA-QQAR requires replicated servers placed at different locations. These servers have different IP addresses but the same anycast address. The performance results we obtained with OPNET simulations demonstrate that our proposal yields significant performance improvements over other traditional approaches.

Acknowledgments

A large part of the work presented here was done in the framework of the European CELTIC IPNQSIS project.

References

1. Akamai. Cambridge, MA, USA: http://www.akamai.com
2. Digital island: http://www.digitalisland.com
3. G. Peng, *CDN: Content Distribution Network*, Technical Report TR-125, Experimental Computer Systems Lab, Department of Computer Science, State University of New York, Stony Brook, NY, 2003.
4. J. Kangasharju, J. Roberts, K. Ross, Object replication strategies in content distribution networks, in: *Proceedings of the Sixth International Web Content Caching and Distribution Workshop*, Boston (USA), June 2001.
5. B. Li, X. Deng, M.J. Golin, K. Sohrabi, On the optimal placement of web proxies on the internet: The linear topology, in: *Proceedings of the IFIP TC-6 Eight International Conference on High Performance Networking*, 1998, pp. 485–495.
6. L. Qiu, V. Padmanabham, G. Voelker, On the placement of web server replicas, in: *Proceedings of the IEEE INFOCOM 2001*, Anchorage (Alaska), April 2001.
7. I. Clarke, O. Sandberg, B. Wiley, T. Hong, Freenet: A distributed anonymous information storage and retrieval system, designing privacy enhancing technologies, in: *International Workshop on Design Issues in Anonymity and Unobservability*, Berkeley, CA (USA), Lujy, 2000.

8. D. Liben-Nowell, H. Balakrishnan, D. Karger, Analysis of the evolution of peer-to-peer systems, in: *ACM Conference on Principles of Distributed Computing*, Monterrey (USA), July 2002.

9. M. Kabir, E. Manning, G. Shoja, Request-routing trends and techniques in content distribution networks, in: *Proceedings of the ICCIT*, Dhaka, December 2002.

10. K. Park, V.S. Pai, L. Peterson, Z. Wang, CoDNS: Improving DNS performance and reliability via cooperative lookups, in: *Proceedings of the Sixth Symposium on Operating Systems Design and Implementation*, San Francisco CA (USA), December 2004.

11. Z. Mao, C. Cranor, F. Douglis, M. Rabinovich, A precise and efficient evaluation of the proximity of web clients and their local DNS servers, *USENIX'02*, Monterrey CA (USA), June 2002.

12. D. Soldani, M. Li, R. Cuny, *QoS and QoE Management in UMTS Cellular Systems*. Wiley Online Library, 2006.

13. International Telecommunication Union E. *800: Terms and Definitions Related to Quality of Service and Network Performance Including Dependability*, ITU-T Rec E.800, 1994.

14. ISO/IEC 13236: Quality of service: Framework, ISO/IEC 13236, 1998.

15. K. Kilkki, Quality of experience in communications ecosystem, *Journal of Universal Computer Science*, 14(5), 615–624, 2008.

16. ITU-T P.800. Methods for subjective determination of transmission quality—Series P: Telephone transmission quality methods for objective and subjective assessment of quality, August 1996.

17. Methodology for the Subjective Assessment of the Quality of Television Pictures, Recommendation ITU-R BT.500-10, ITU Telecom. Standardization Sector of ITU, August 2000.

18. J. Beerends and J. Stemerdink, A perceptual audio quality measure based on a psycho-acoustic sound representation, *Journal-Audio Engineering Society*, 40, 963–963, 1992.

19. S. Voran, Estimation of perceived speech quality using measuring normalizing blocks, in *1997 IEEE Workshop on Speech Coding For Telecommunications Proceeding*, 1997, IEEE, 2002, pp. 83–84.

20. ITU-T Recommendation G.107: The E-model, a computation model for use in transmission planning, *International Telecommunication Union*, August 2008.

21. S. Voran, The development of objective video quality measures that emulate human perception, in *Proceedings of IEEE Global Telecommunications Conference: GLOBECOM '91*, Phoenix, AZ, December 1991. vol.3. pp 1776–1781.

22. C. van den Branden Lambrecht, Color moving pictures quality metric, in *International Conference on Image Processing*, Vol. 1. IEEE, 2002, pp. 885–888.

23. C. van den Lambrecht, *Perceptual Models and Architectures for Video Coding Applications*, PhD dissertation, EPFL, Switzerland, 1996.

24. G. Rubino, Quantifying the quality of audio and video transmissions over the internet: The PSQA approach, *Design and Operations of Communication Networks: A Review of Wired and Wireless Modeling and Management Challenges*. Imperial College Press, London, 2005.

25. H. A. Tran and A. Mellouk, Real-time state-dependent routing based on user perception, in *The International Conference on Communications and Information Technology—ICCIT 2011*, IEEE, 2011.

26. Watkins and Daylan, Technical note: Q-learning, *Machine Learning*, 8(3), 279–292, 1992.

27. M. G. Lagoudakis and R. Parr, Least-squares policy iteration, *Journal of Machine Learning Research*, 4, 1149, 2003.
28. R. S. Sutton and A. G. Barto, *Reinforcement Learning: An Introduction,* MIT Press, Cambridge, MA, 1998.
29. Videolan. [Online]. Available: http://www.videolan.org/
30. S. Hemminger, Network emulation with netem, in *Linux Conf Au*, April 2005.
31. Recommendation 500-10: Methodology for the subjective assessment of the quality of television pictures, ITU-R Rec. BT.500, 2000.
32. Sintel video trailer. [Online]. Available: http://www.sintel.org/

Chapter 14

QoE of 3D Media Delivery Systems

Varuna De Silva, Gokce Nur, Erhan Ekmekcioglu, and Ahmet M. Kondoz

Contents

14.1 Introduction to QoE in 3D Video 276
    14.1.1 QoE of 3D Video 277
14.2 Elements of QoE of 3D Video 277
    14.2.1 Psyco-Physiological Analysis of Depth Perception
        in 3D Video 277
        14.2.1.1 Oculomotor Cues 277
        14.2.1.2 Visual Cues 278
    14.2.2 Geometry of Binocular Stereopsis 278
    14.2.3 Sensitivity of the HVS to Depth Perceived by Different Cues 279
        14.2.3.1 Sensitivity of the HVS for Binocular Disparity 281
        14.2.3.2 Sensitivity of the HVS for Retinal Blur 282
        14.2.3.3 Sensitivity of the HVS for Relative Size Cue 283
    14.2.4 Factors Affecting Visual Fatigue in 3D Viewing 284
    14.2.5 Effects of Ambient Lighting on Perceived 3D Video Quality 285
    14.2.6 Other Factors Affecting QoE of 3D-TV 287
14.3 Metrics for Evaluation of QoE Elements of 3D Video 288
    14.3.1 Stereoscopic Image/Video Quality Metrics 288
    14.3.2 Metrics of Depth Sensation 289
    14.3.3 Metrics for Prediction of Visual Fatigue 289

14.4 Conclusion 290
    14.4.1 Challenges in QoE of 3D Video 290
    14.4.2 Outline of a Future QoE Metric for 3D Video 290
References 291

14.1 Introduction to QoE in 3D Video

QoE indicates the overall user satisfaction gained by using a particular system that performs a defined function. While there is no universally accepted definition of QoE, it generally is a measure of the level to which a particular system meets the expectations of its users [1]. As QoE indicates user satisfaction, it needs to be evaluated subjectively by the users.

For the consistency of technical vocabulary, telecommunication standardization sector of the International Telecommunication Union—(ITU-T) has defined the term QoE as [2:1], *"The overall acceptability of an application or service, as perceived subjectively by the end-user."* The definition in Ref. [2] also indicated that the QoE includes the complete end-to-end system effects such as the terminal, network, and services infrastructure and noted that the overall acceptability may be influenced by user expectations and the context of usage.

In ITU-T Rec. G.1080, the elements of QoE in an IPTV scenario is depicted as in Figure 14.1.

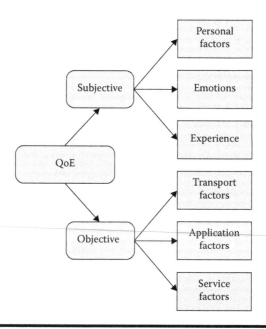

Figure 14.1   The elements of QoE in an IPTV scenario.

Quality of Service (QoS) provides an objective virtual measurement of performance criteria of a system, whereas QoE is a subjective measurement of performance.

14.1.1 QoE of 3D Video

3D video is an emerging form of media that gives its viewers an added dimension of experience (i.e., depth). 3D provides a sensation of depth by exploiting the properties of HVS. Thus, 3D video delivery systems have become an interesting application of QoE. This chapter introduces QoE in the context of 3D video delivery systems. 3D video provides a sensation of depth by providing two disparate views of the same scene to the two eyes. The HVS, which consists of the eyes, the neuronal pathway to the brain, and the brain, analyzes the binocular sensation and perceives a 3D view. Thus, 3D video provides a more natural experience than traditional 2D video. However, there has been research indicating that users sometimes experience visual discomfort while watching 3D video. This visual discomfort is mainly attributed to excessive disparities and a physiological phenomenon known as vergence–accommodation mismatch. Furthermore, recent research has shown contextual elements such as ambient lighting condition of the viewing environment have effect on the experience of 3D viewing.

14.2 Elements of QoE of 3D Video

14.2.1 Psyco-Physiological Analysis of Depth Perception in 3D Video

Several different cues are made use of by humans to perceive the depth of different objects in a scene. The depth cues can be classified mainly into two categories, namely, oculomotor cues and visual cues. While an extensive coverage of these cues is beyond the scope of this book, for the purpose of completeness, some of these cues are briefly described as follows.

14.2.1.1 Oculomotor Cues

Accommodation: When the eye fixates on some object, the optical power (ability to converge or diverge light) of its lens is adjusted so that the image of the fixated object is focused on the center of the fovea. This process is known as accommodation.

Convergence: When the eyes do not gaze at a particular object, the axes of the two eyes stay parallel to each other. To gaze on an object or to locate an object of interest on the fovea, the two eyes will move inwards from its parallel positioning. This process is known as convergence [3]. The brain interprets the muscular tension required for inward movement of eyes as a depth cue.

14.2.1.2 Visual Cues

Binocular Stereopsis: This is a cue by which the brain interprets a relative depth making use of the two slightly different perspectives seen by the two eyes. As this is the most important cue in 3D display, this factor will be described in detail later.

Relative Size: The same object when placed nearer to a viewer will appear bigger than when it is placed at a far away point. This cue is made use of by the brain to infer relative depth details.

Occlusion: When one object is occluded by another, the occluded image is perceived to be behind the other object.

Retinal Blur: When one object focused by the eye is kept sharp and another object is blurred, a sense of depth relative to the sharp object is perceived by the viewer.

Motion Parallax: This cue enables the users to see different views of the same scene while moving the head sideways.

Apart from these cues, there are many more cues that are used by the humans to perceive depth. The readers are referred to Ref. [3] for an extensive coverage of these cues.

3D video provides an additional experience of depth while watching 3D video as compared to 2D video. 3D display systems provide additional cues to its viewers that enhance the viewers' perception of depth in a video scene. The most important one of these additional cues is the binocular disparity, which is obtained by providing two views of the same scene, from slightly different perspectives, to each eye of a viewer. Head motion parallax is another additional cue provided by modern 3D display systems, which enhances depth perception.

14.2.2 Geometry of Binocular Stereopsis

The geometry of binocular stereopsis is illustrated in Figure 14.2, the binocular geometry projects slightly different images on the retinas of the two eyes. When eyes fixate on point P, then that point falls on the center of the fovea of each eye. Fovea is the most sensitive part of the retina. When the eyes fixate on point P, point Q casts the image α degrees away in one eye's fovea and β degrees away from the other. The binocular disparity η in this case is given as $(\beta - \alpha)$, measured in degrees of visual angle [3,4]. Making use of this stimulation the brain interprets depth of objects relative to the fixated point. The binocular disparity has both magnitude and direction.

Let the angular disparity due to point Q be η_Q,

$$\eta_Q = (\beta - \alpha) \tag{14.1}$$

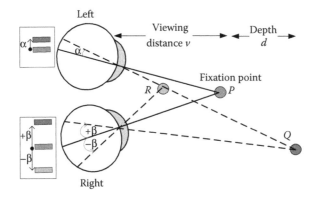

Figure 14.2 Geometry of binocular stereopsis. (From V. De Silva et al. Sensitivity analysis of the human visual system for depth cues in stereoscopic 3-D displays, *IEEE Transactions on Multimedia,* **13(3), 498–506, Jun 2011. © (2011) IEEE.)**

Geometrically, the binocular disparity is directly proportional to the depth (d) of the point Q relative to P, and inversely proportional to the square of the viewing distance v [3].

$$\eta \propto \frac{d}{v^2} \tag{14.2}$$

Furthermore, the angular disparity due to point R (η_R) could be given as in

$$\eta_R = (-\beta) - \alpha = -(\beta + \alpha) \tag{14.3}$$

According to Equations 14.1 and 14.3, $\eta_Q > 0$ and $\eta_R < 0$. This sign difference is interpreted by the brain as the relative positioning of points Q and R relative to point P. When the binocular disparity is greater than zero, the brain interprets that the point is behind the fixation point and vice versa. In this manner, the binocular disparity provides stimulation to the brain to perceive the depth of objects, relative to a fixation point [4]. This cue is known as the binocular stereopsis.

Figure 14.3 illustrates how a sensation of depth is provided by stereoscopic displays by way of pixel parallax (p).

14.2.3 Sensitivity of the HVS to Depth Perceived by Different Cues

This section gives an overview of recent results published regarding the sensitivity of the HVS for different depth cues in 3D video. The theoretical derivation and additional experimental details are found in Ref. [4]. The aim of the experiments in Ref. [4]

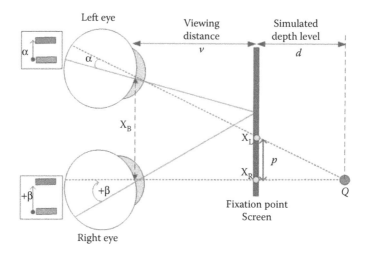

Figure 14.3 Eye positioning while 3D viewing. (V. De Silva et al. Sensitivity analysis of the human visual system for depth cues in stereoscopic 3-D displays, *IEEE Transactions on Multimedia*, **13(3), 498–506, Jun 2011. © (2011) IEEE.)**

is to model how much sensitive the humans are for depth cues such as binocular disparity, retinal blur, and relative size.

The results are based on a simple subjective experiment performed on a passive stereoscopic display where each viewer watches a synthetic image sequence with two objects as shown in Figure 14.4, both representing a synthetic image of a car. Initially, both the cars are placed at the same disparity level. A particular depth cue of the right side object is gradually changed (increased or decreased), while the depth cue of the left side object is kept unchanged. The subjects need to signal the coordinator, just when they sense a change in the depth-level difference between the two objects.

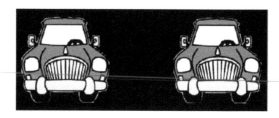

Figure 14.4 An example of a test stimuli used for the JNDD experiment. (V. De Silva et al. Sensitivity analysis of the human visual system for depth cues in stereoscopic 3-D displays, *IEEE Transactions on Multimedia*, **13(3), 498–506, Jun 2011. © (2011) IEEE.)**

14.2.3.1 Sensitivity of the HVS for Binocular Disparity

The sensitivity with binocular disparity is measured at five selected testing disparity levels (−16, −8, 0, 8, 16). Further, sensitivity is also measured by increasing and decreasing the disparity at each level. The results obtained from the experiment are presented in Figure 14.5, where positive disparity in Figure 14.5 corresponds to objects seen in front of the screen and vice versa. These results illustrate a maximum sensitivity at zero disparity and decreasing sensitivity with increasing disparity.

Weber's law provides a relationship between an initial stimulus and the difference of that stimulus to perceive a just noticeable change. According to Weber's law [3], at a larger initial stimulus, a larger stimulus difference is required for a subject to perceive a change in the initial stimulus. When binocular disparity is the initial stimulation, the change in binocular disparity is the stimulus difference. When objects are on the screen, the binocular disparity is zero, and hence the lowest just noticeable difference or highest sensitivity. When objects are in front or behind the screen, binocular disparity is proportional to the distance the objects are simulated to be in front or behind. Therefore, as the binocular disparity increases, the sensitivity decreases or the just noticeable difference increases. The

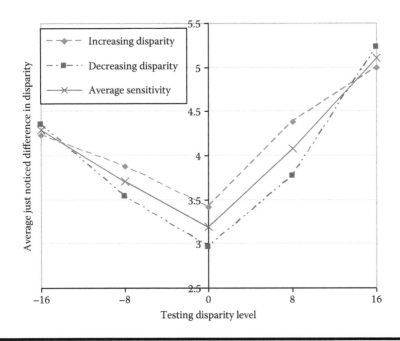

Figure 14.5 **Average of just noticed difference in depth at various testing disparity levels (viewing distance = 2 m). (V. De Silva et al. Sensitivity analysis of the human visual system for depth cues in stereoscopic 3-D displays,** *IEEE Transactions on Multimedia,* **13(3), 498–506, Jun 2011. © (2011) IEEE.)**

same experiment is also performed on an auto-stereoscopic display, for which the sensitivity results showed a similar trend as in Figure 14.5 [5].

14.2.3.2 Sensitivity of the HVS for Retinal Blur

The sensitivity with retinal blur as simulated by artificial image blur is measured at three testing disparity levels (−16, 0, 16) to reflect different positions relative to the screen. Furthermore, to provide useful insights, the sensitivity with blur is measured at three different rates of blur change. Note that the disparity of both the objects is kept at its original disparity level, and the only change within the experiment is that the right object is blurred at a particular rate.

To measure the sensitivity with retinal blur, in the first experiment, each subject is shown nine different sequences (three each at three different rates of change of blur) that use artificial image blur to simulate depth changes. The results indicate that 26 out of 28 subjects (93%) identified artificial image blur as a depth cue in at least one of the nine sequences. Here, 18 of the subjects (70%) identified blur as a depth cue in at least five out of nine sequences. Out of the 26 subjects who identified blur as a depth cue, 24 of them (92%) interpreted that the blurred object is moving backwards of the sharp object. In these experiments, we use Gaussian blur to simulate depth changes and the extent of Gaussian blur (σ) is varied at three different rates as shown in Figures 14.6 and 14.7.

Furthermore, subjects tend to identify blur as depth cue especially at low change of rates of blur (70% at 0.047 σ/s to 50% at 0.186 σ/s). The reason for this

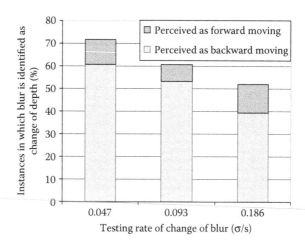

Figure 14.6 Subjective results for identifying blur as a depth cue. (V. De Silva et al. Sensitivity analysis of the human visual system for depth cues in stereoscopic 3-D displays, *IEEE Transactions on Multimedia,* **13(3), 498–506, Jun 2011. © (2011) IEEE.)**

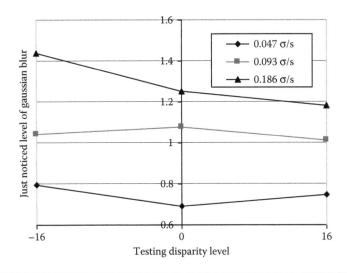

Figure 14.7   Just noticed level of Gaussian image blur for three rates of blur variation. (V. De Silva et al. Sensitivity analysis of the human visual system for depth cues in stereoscopic 3-D displays, *IEEE Transactions on Multimedia*, 13(3), 498–506, Jun 2011. © (2011) IEEE.)

identification rate difference is that since blur is based on the accommodation response of the eye, full accommodation response requires a minimum fixation time of 1 s or longer.

The sensitivity of the HVS in understanding image blur as a depth cue is summarized in Figure 14.7. Accordingly, the subjects were highly sensitive to blur as a depth cue when it is changed at a lower rate. Further, it should be noted that the sensitivity is almost constant irrespective of the testing disparity level.

14.2.3.3 Sensitivity of the HVS for Relative Size Cue

To measure the sensitivity for relative size cue, the size of the right-side object is changed to simulate a depth change, while the left-side object is kept at the original size. For this experiment, six (i.e., 3 disparity levels × 2 increasing and decreasing sizes) test sequences are used.

All the subjects who participated in the test perceived an object that is increasing its size to be moving forward and vice versa. Generally, most participants were slightly more sensitive to increasing size than to decreasing size, especially at screen level or in front of the screen. As shown in Figure 14.8, the just noticeable difference in object size that is required to perceive a change of depth is almost equal at all the testing disparity. The object size should be changed by about 4% of the original size for the subject to perceive a depth difference.

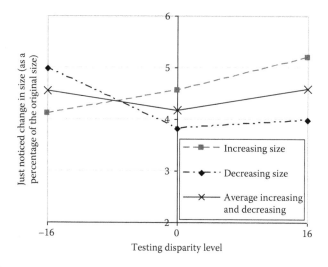

Figure 14.8 **Sensitivity of relative size as a depth cue. (V. De Silva et al. Sensitivity analysis of the human visual system for depth cues in stereoscopic 3-D displays,** *IEEE Transactions on Multimedia*, **13(3), 498–506, Jun 2011. © (2011) IEEE.)**

14.2.4 Factors Affecting Visual Fatigue in 3D Viewing

It was introduced earlier that accommodation is the process where the optical power of the eye is adjusted to locate a point of interest on the fovea. However, the optical power of the lens fluctuates at a certain frequency, and the eye tolerates a certain amount of retinal blur without readjusting accommodation [6]. This optical power difference is known as the ocular depth of focus and is presented in dioptres (D) (or m^{-1}). In other words, objects within a certain span of the object of interest will be seen sharp without readjusting accommodation. This span is known as the Depth of Field (dof).

For natural viewing, accommodation and convergence are reflexively linked [7]. In stereoscopic 3D viewing, this link is disturbed since the eyes accommodate on the screen level no matter where the object is projected. This is often known as the "accommodation–convergence mismatch," and is considered to be a major reason for visual discomfort in stereoscopic displays, if the objects are projected to be beyond the *dof.* Therefore, stereoscopic display systems are designed to minimize the visual discomfort due to accommodation–convergence [8], by limiting the amount of binocular disparity [7] and thus, simulating the objects to be within the *dof.*

In Ref. [9], Yano et al. presents results for subjective and objective visual fatigue for stereoscopic viewing. In the experiments presented in Ref. [9], the subjects are required to read a set of Japanese characters placed at some point in a stereoscopic display. After each session, the visual fatigue was measured both by measuring the accommodation response and a subjective questionnaire. The results presented in the paper can be illustrated as in Figure 14.9.

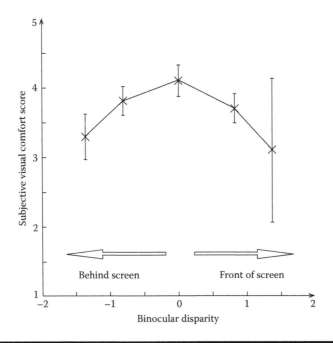

Figure 14.9 Variation of subjective visual fatigue with increasing binocular disparity. (Adapted from S. Yano, M. Emoto, and T. Mitsuhashi, *Displays*, 25(4), 141–150, Nov 2004.)

As the binocular disparity is increased, the visual fatigue seems to be increasing. This is in correspondence with the accommodation–convergence mismatch. When the simulated depth is increased, the difference between the convergence distance and accommodation distance increases, as the eyes are always accommodated at screen level. This will make the brain to resolve two conflicting sets of information, which results in a discomfort.

Yano et al. [9] also report the motion factor that causes visual discomfort. It is reported in Ref. [9] that subjects are more comfortable with objects with lateral motion in stereoscopic video than with objects that have motion in the depth direction. This fact could be due to the constant change of convergence of the eyes.

14.2.5 Effects of Ambient Lighting on Perceived 3D Video Quality

Experiments are carried out in Ref. [10] to find the relationship between the background illumination and its effect upon perceptual 3D video quality and subjective depth perception in 3D video. Extensive subjective experiments are carried out for assessing the 3D video quality and depth perception with six video sequences at four different illumination levels at four channel bandwidths. The perceptual 3D

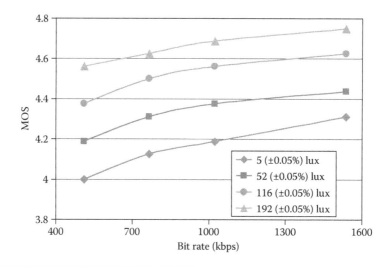

Figure 14.10   Subjective Mean Opinion Score (MOS) for perceived image quality vs. the channel bandwidth, at four distinct illumination levels.

video quality results of the performed experiments can be summarized as given in Figure 14.10. In Figure 14.11, the subjective MOS score for the perceived depth in the 3D video sequences shown are illustrated. In both the figures, 5 lux is the darkest environment whereas 192 lux is the brightest environment in which the experiments were performed.

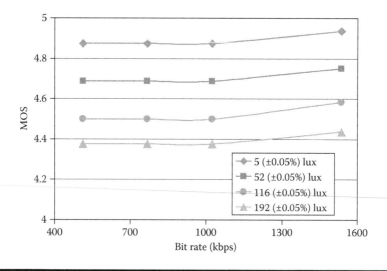

Figure 14.11   Subjective MOS for perceived depth in 3D video vs. the channel bandwidth, at four distinct illumination levels.

To summarize the results, when the ambient illumination in the usage environment for 3D video consumption increases, the video quality perception ratings of the viewers also increase. This is due to the fact that in a well-lit environment, the visual artifacts (due to compression, transmission, rendering, etc.) in the 3D content cannot be easily detected by the human eye as much as they can be noticed while viewing the same content in a dark environment.

In contrast to the perceived image quality, as the ambient illumination in the environment increases, the depth perception ratings of the viewers decrease. These facts result as brighter light conditions hinder the clear visibility of the essential cues (e.g., sharpness, shadowing, reflection, contrast, etc.) that enhance the overall depth sensation in 3D content.

14.2.6 Other Factors Affecting QoE of 3D-TV

Apart from the latest results outlined in earlier subsections, there has also been research covering certain other aspects of QoE of 3D video. In this section, these factors are briefly introduced.

Naturalness of 3D video: In an initial effort to characterize naturalness in 3D video, Seuntiens et al. [11] displayed few scenes in both 2D and 3D modes to be subjectively evaluated by the viewers. Each of the scenes was corrupted with six levels of white Gaussian noise. Results illustrated that the naturalness was rated higher in 3D than in 2D at the same Gaussian noise level. The added dimension of depth clearly demonstrates an improvement in the naturalness of the content that is being viewed. The added value of 3D over 2D in terms of naturalness is around 4 dB (expressed in noise level). This indicates that the naturalness plays a major role in the psychological impact of 3D viewing.

Feeling of presence in 3D video: In a recent research studying the effects of 3D viewing on partially sighted people, the results indicated that the users, irrespective of them being partially or fully sighted, tend to feel more engaged with 3D viewing rather than with 2D viewing.

Effects of visual acuity of the viewers: Recently, the Royal National Institute of Blind People (RNIB), which is the leading organization representing the interests of 2 million people living with sight loss in the United Kingdom carried out a research on what 3D-TV means for those partially sighted people who rely on their remaining sight to watch TV [12]. Results indicated that across both partially and fully sighted participants, 3D was preferred to 2D to view the films. However, this preference was not significant for the partially sighted people to the extent it was for the fully sighted sample. As for the viewing experience, most participants did not experience adverse effects of 3D viewing. Sighted participants were more likely to report negative effects of 3D viewing than the partially sighted participants, even for the relatively short duration of presentation of the 3D film clip.

14.3 Metrics for Evaluation of QoE Elements of 3D Video

One of the major challenges associated with subjective evaluation of QoE is that it is very time consuming. Moreover, the subjective evaluation of QoE is expensive and sometimes requires multidisciplinary knowledge. Therefore, objective quality metrics that measure various aspects of QoE are currently of utmost importance. This section will outline recent efforts within research communities that consider the development of QoE metrics.

14.3.1 Stereoscopic Image/Video Quality Metrics

3D visual metrics is a key enabling element for 3D distribution system deployment. However, due to its complexity, an overall 3D QoE metric that includes multiple experience factors could only be a long-term goal. It is envisaged in the literature that a 3D visual experience metric will have at least two key performance indicators, as shown in Figure 14.12.

It is reported in Ref. [13] that the perceived image quality of 3D video does not show a considerable change as compared to 2D video. The human visual system is able to locate errors in either one of the stereoscopic views successfully. Therefore, as for the image quality, existing state-of-the art video quality metric could be effectively used. However, the measure of depth perception is not as straightforward as

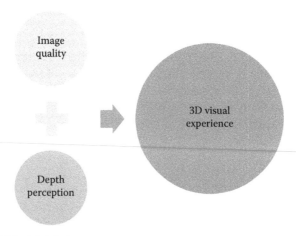

Figure 14.12   A representation technique of 3D visual experience.

the image quality measurement. An initial effort to measure the depth perception of 3D video is described briefly in the next section.

14.3.2 Metrics of Depth Sensation

At the beginning of this chapter, it was explained that the human brain uses several cues for depth perception. It was shown in Ref. [14] that the human brain independently perceives the degradation in monocular and binocular cues. Thus, the depth perception could be modeled as follows:

$$\text{Depth\_perception} = D_M^\alpha D_B^\beta \tag{14.4}$$

where D_M is the contribution of monocular cues, D_B is the contribution of binocular cues and α and β are positive constants illustrating the relative weights. In the method proposed in Ref. [14], contribution from the monocular cues was measured with the quality of the color texture video measured with VQM. The contribution from the binocular cues was measured with a novel metric described as follows.

In the proposed Depth Distortion Model (DDM), the depth map was segmented based on its histogram. Thereafter, three factors as listed here was measured based on the segmented depth map.

- Distortion of the relative distance in depth axis among depth planes: Measured as the average change of significant depth luminance.
- Distortion of the consistency in perceived depth within the depth planes: Measured as the variance of the error in each depth plane.
- Structural comparison (SC) of the depth maps: Borrowed from SSIM, the covariance of the two depth maps.

Finally, a single metric was derived to measure the depth perception as follows:

$$\text{Depth\_perception} = (1 - \text{VQM})^\alpha \cdot \text{MDDM}^\beta \tag{14.5}$$

MDDM in Equation 14.5 represents the DDM of each of the depth planes, and $\alpha = 1.5$ and $\beta = 1$, which were experimentally derived. Results of the proposed method are given in Table 14.1, and as illustrated, the proposed method performs the best in terms of correlation coefficient between the objective scores and MOSs.

14.3.3 Metrics for Prediction of Visual Fatigue

Even though 3D video brings great entertainment, it can cause visual fatigue or headache. Choi et al. [15] propose a metric for visual fatigue measurement. Spatial

Table 14.1 Results of the Proposed DPM

Objective Quality Model	Depth Perception		
	CC	RMSE	SSE
Average PSNR of the rendered left and right videos	0.7788	0.0737	0.0579
Average SSIM of the rendered left and right videos	0.8065	0.0674	0.0547
Average VQM of the rendered left and right videos	0.7753	0.0739	0.0603
Proposed DPM	0.8716	0.0325	0.0379

complexity (i.e. the complexity of scene composition), temporal complexity (i.e., the amount of motion in a scene), depth position (i.e., average distance from the display plane), and scene movement (i.e., the overall speed of the 3D video due to object and camera movements) are the elements of this visual fatigue metric. Using the proposed metric, the depth sequence can be adjusted to lower the visual fatigue.

14.4 Conclusion

14.4.1 Challenges in QoE of 3D Video

Future research could determine the extent to which the underlying aspects of viewing experience and naturalness are accountable for the difference in the assessment of 2D and 3D images. Moreover, the viewing experience and naturalness in combination with different 2D and 3D artifacts should be monitored for a possible QoE metric. Effect of age in 3D viewing is also an important factor that should be investigated for developing a 3D QoE metric.

14.4.2 Outline of a Future QoE Metric for 3D Video

The diagram of a future QoE metric for 3D video can be seen in Figure 14.13. As shown in the figure, the dimensions of a future QoE metric is a function of QoE of audio (i.e., QoE_{Audio}), QoE of video (i.e., QoE_{Video}), comfort, and context. Audiovisual content consists of both audio and video contents. Thus, QoE metrics should be developed both for audio and video, and then they should be combined. Comfort of viewing users is also a key element of a future QoE metric. Moreover, contextual factors effecting 3D video and audio quality should be included in a future QoE metric.

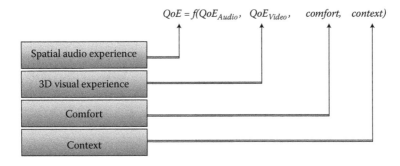

Figure 14.13   The diagram of a future QoE metric for 3D video.

References

1. H. Batteram, G. Damm, A. Mukhopadhyay, L. Philippart, R. Odysseos, and C. Urrutia-Valdés, Delivering quality of experience in multimedia networks, *Bell Labs Technical Journal*, 15(1), 175–193, May 2010.
2. ITU-T P.10/G.100 Ammendment 1, *Vocabulary for Performance and Quality of Service*, Jan 2007.
3. V. Bruce, P. R. Green, and M. A. Georgeson, *Visual Perception: Physiology, Psychology, & Ecology*, 4th ed. Psychology Press, Hove, NY, 2003.
4. V. De Silva, A. Fernando, S. Worrall, H. K. Arachchi, and A. Kondoz, Sensitivity analysis of the human visual system for depth cues in stereoscopic 3-D displays, *IEEE Transactions on Multimedia*, 13(3), 498–506, Jun 2011.
5. D. V. S. X. De Silva, E. Ekmekcioglu, W. A. C. Fernando, and S. T. Worrall, Display dependent preprocessing of depth maps based on just noticeable depth difference modeling, *IEEE Journal of Selected Topics in Signal Processing*, 5(2), 335–351, Apr 2011.
6. S. Reichelt, R. Häussler, G. Fütterer, and N. Leister, Depth cues in human visual perception and their realization in 3D displays, in *Society of Photo-Optical Instrumentation Engineers (SPIE) Conference Series*, 2010, Vol. 7690, p. 10.
7. M. T. M. Lambooij, W. A. IJsselsteijn, and I. Heynderickx, Visual discomfort in stereoscopic displays: A review, in *Stereoscopic Displays and Virtual Reality Systems XIV*, San Jose, CA, USA, 2007, Vol. 6490, p. 64900I-13.
8. T. Inoue and H. Ohzu, Accommodative responses to stereoscopic three-dimensional display, *Applied Optics*, 36(19), 4509–4515, 1997.
9. S. Yano, M. Emoto, and T. Mitsuhashi, Two factors in visual fatigue caused by stereoscopic HDTV images, *Displays*, 25(4), 141–150, Nov 2004.
10. G. Nur, S. Dogan, H. Kodikara Arachchi, and A. Kondoz, *Assessing the Effects of Ambient Illumination Change in Usage Environment on 3D Video Perception for User Centric Media Access and Consumption*, Spain, 2010.
11. P. Seuntiens et al., Viewing experience and naturalness of 3D images, in *Proceedings of the SPIE*, 2005, Vol. 6016, pp. 43–49.
12. J. Freeman and J. Lessiter, *Evaluation of Partially Sighted People's Viewing Experiences of 3D Relative to 2D TV: Scoping Study Report*. Royal National Institute of Blind People (RNIB), Dec 2010.

13. B. Julesz, Cyclopean perception and neurophysiology, *Investigative Ophthalmology & Visual Science*, 11(6), 540, 1972.
14. S. L. P. Yasakethu, D. V. S. X. De Silva, W. A. C. Fernando, and A. Kondoz, Predicting sensation of depth in 3D video, *Electronics Letters*, 46(12), 837–839, Jun 2010.
15. J. Choi, D. Kim, B. Ham, S. Choi, and K. Sohn, Visual fatigue evaluation and enhancement for 2D-plus-depth video, in *Image Processing (ICIP), 2010 17th IEEE International Conference on*, pp. 2981–2984.

USER-CENTRICITY AND IMMERSIVE TECHNOLOGIES

Chapter 15

Perceived QoE for User-Centric Multimedia Services

Nikolaos Zotos, Jose Oscar Fajardo, Harilaos Koumaras, Lemonia Boula, Fidel Liberal, Ianire Taboada, and Monica Gorricho

Contents

15.1 Introduction 296
15.2 Overview on Adaptive IPTV Components 297
    15.2.1 QoE-Enhanced IP Multimedia Subsystem 298
    15.2.2 Quality of Experience/Perceived Quality-of-Service
        Mechanism 298
    15.2.3 Perceptual Cross-Layer Management/Monitoring 299
        15.2.3.1 Advances in Cross-Layer Adaptation 300
15.3 User-Centric IPTV Architecture 301
    15.3.1 Enhancing IMS 303
        15.3.1.1 Specifications of the Proposed MCMS 305
        15.3.1.2 IMS Service Control Interface 305
        15.3.1.3 Policy and Charging Control Function 306
    15.3.2 Multimedia Content Management System 307
    15.3.3 Monitoring and Adaptation Engines 311
        15.3.3.1 Monitoring 311
        15.3.3.2 Adaptation 313

15.3.4 QoS Support and Mapping..316
      15.3.4.1 TN, AN, and TAM MCMS-Driven Adaptation
            Mapping...316
      15.3.4.2 TN/AN NQoS Mapping.................................317
15.4 QoE/PQoS Model 317
    15.4.1 Video Service Adaptation...322
      15.4.1.1 Multiuser Cross-Layer Adaptations...............324
15.5 Conclusions...343
Acronyms...343
Acknowledgment 345
References 345

15.1 Introduction

The convergence of multimedia services with mobile/fixed networks and broadcast-interactive applications is creating new demands for high-quality and user-responsive service provision management. To face the challenges of defining and developing the next generation of ubiquitous and converged network, together with the respective service infrastructures for communication, computing, and media, the industry has launched various initiatives in order to design the reference network architecture for user-centric multimedia services provision and standardize the various modules/interfaces that are necessary for delivering the expected multimedia services in Next Generation Network (NGN) networks, such as Internet Protocol TV (IPTV) (e.g., Live multicast IPTV or Video-on-demand) [1].

However, this strong commercial interest may be hindered by the lack of efficient user-centric network management mechanisms, which will dynamically adapt/optimize network traffic policy to maximize perceived user satisfaction. One of the visions of future multimedia communication networks is that services will be sold in a consumer mass market based on the provision of content that meets various perceptual quality requirements (e.g., video and/or voice quality). There are numerous approaches to this marketing model, but the most important of them is the Perceived Quality of Service (PQoS) or Quality of Experience (QoE) concept because it provides a direct link to user satisfaction. The requirements for PQoS for multimedia services and applications with differing bandwidth demands should provide the user with a wide range of potential quality choices, including, for example, cases of low, medium or high perceptual levels, and an indication of service availability and cost. However, existing multimedia services and network operators' infrastructures do not provide any PQoS-aware management mechanism within its service provision control system [1].

Given the need for perceived Quality of Service (QoS) provisioning, it is expected that the success of new business opportunities and innovative multimedia services within the new convergent environment will be significantly based on novel user-centric network management solutions that employ cross-layer-adaptive techniques in order to (a) compensate for network impairments (Network QoS— NQoS) according to the time-varying conditions of the network delivery chain, (b) perform a content-dependent optimization of the encoding and/or streaming parameters, and to (c) improve the end-user experience/satisfaction by maximizing the delivered PQoS level.

This chapter discusses the QoE consideration for user-centric services and presents an architecture for serving adaptive IPTV to mobile and fixed network consumers based at the real-time quality that the end-user experiences. The core component of this architecture is an innovative Multimedia Content Management System (MCMS) focused on performing a dynamic cross-layer adaptation for the optimization of the user experience in terms of the delivered PQoS level for IPTV. This multimodal management system makes use of the current IP Multimedia Subsystem (IMS) management functions by providing necessary perceptual awareness capability. A fully functional small-scale prototype is presented in order to demonstrate the added value of MCMS in terms of perceptual enhancement. The main scope of the chapter is to present the overall QoE-driven IPTV architecture and the respective perceptual benchmarking of it via a subjective evaluation procedure.

The chapter is organized as follows: Section 15.2 presents an overview of all the core components of the adaptive IPTV system (IMS, QoE, Perceptual Cross layer adaptation, network management, etc.). In Section 15.3, the user-centric IPTV architecture is presented, containing NQoS to PQoS mapping operations, the Multimedia Content Management System, the monitoring and adaptation modules presentation, the enhancements of the current IMS functions, and UMTS/WiMAX technologies QoE/QoS mapping. Section 15.4 analyzes in depth the QoE model used for IPTV cross-layer adaptation. Section 15.5 concludes the benchmarking of the whole architecture, and finally Section 15.6 draws the conclusions.

15.2 Overview on Adaptive IPTV Components

This section deals with the description of the basic high-level functions and components, which comprises an adaptive IPTV service provisioning platform in NGN networks. The main components of the system are: (a) the IP Multimedia Subsystem (IMS) which were designed for providing integrated multimedia services combining mobility and IP network connectivity, (b) the Quality of Experience (QoE) giving to the end user the capability to set by himself the quality thresholds for the multimedia service one experiences, and (c) the Perceptual Cross-layer management/monitoring.

15.2.1 QoE-Enhanced IP Multimedia Subsystem

The IMS infrastructure is creating novel business opportunities for new and emerging multimedia services, such as Voice over IP (VoIP) (e.g., voice/video call) and Internet Protocol TV (IPTV) (e.g., Live multicast IPTV or Video-on-demand) services. However, this strong commercial interest in the IMS environment is hindered by the lack of efficient user-centric network management mechanisms, which will dynamically adapt/optimize network traffic policy to maximize perceived user satisfaction. One of the visions of future mobile communication networks is that services will be sold in a consumer mass market based on the provision of content that meets various perceptual quality requirements (e.g., video and/or voice quality). There are numerous approaches to this marketing model, but the most important is the Perceived Quality of Service (PQoS) concept because it provides a direct link to user satisfaction. The requirements for PQoS for multimedia services and applications with differing bandwidth demands should provide the user with a wide range of potential quality choices, including, for example, cases of low, medium, or high perceptual levels, and an indication of service availability and cost. However, existing IMS infrastructure does not provide any PQoS-aware management mechanism within its service provision control system [2].

15.2.2 Quality of Experience/Perceived Quality-of-Service Mechanism

As an important measure of the end-to-end performance at the services level from the user's perspective the QoE is an important metric for the design of systems and engineering processes. This is particularly relevant for video services because bad network performance may highly affect the user's experience, mainly because these services are compressed and have low entropy. So, when designing systems the expected output, that is, the expected QoE, is often taken into account also as a system output metric.

This QoE metric is often measured at the end devices and can conceptually be seen as the remaining quality after the distortion introduced during the preparation of the content and the delivery through the network until it reaches the decoder at the end device. There are several elements in the video preparation and delivery chain and some of them may introduce distortion. This causes the degradation of the content and several elements in this chain can be considered as "QoE relevant" for video services. These are the encoding system, transport network, access network, home network, and end device.

The concept of QoE in engineering is also known as Perceived Quality of Service (PQoS), in the sense of the QoS as it is finally perceived by the end user. The evaluation of the PQoS for audiovisual content will provide a user with a range of potential choices, covering the possibilities of low-, medium-, or high-quality levels. Moreover, the PQoS evaluation gives the service provider and network operator the capability to minimize the storage and network resources by allocating only the resources that are sufficient to maintain a specific level of user satisfaction.

Another approach for measuring QoE in Video content is using a referenceless analysis. In this case, the QoE is not measure comparing an original video to delivered one, but by trying to detect artifacts such as blockiness, blur, or jerkiness directly in the video. This approach is based on the idea that customers do not know the original content.

The evaluation of the PQoS is a matter of objective and subjective evaluation procedures, each time taking place after the encoding process (postencoding evaluation). Subjective quality evaluation processes (PQoS evaluation) require a large amount of human resources, establishing it as a time-consuming process. Objective evaluation methods, on the other hand, can provide PQoS evaluation results faster, but require large amount of machine resources and sophisticated apparatus configurations. Toward this, objective evaluation methods are based and make use of multiple metrics.

15.2.3 Perceptual Cross-Layer Management/Monitoring

Regarding the research in the area of the service management, the proposed approach goes beyond the state of the art, leaving back typical network-related management systems and proposes a service management extension to the upcoming IMS mobile communication platform, which uses PQoS monitoring and evaluations in order to adapt the service provision with twofold objectives:

- Maximization of the end-users' satisfaction by taking into account developed PQoS schemes.
- Minimization of the impact of possible service degradations by obtaining the greater number of satisfied users.

The exploitation of such dynamic cross-layer adaptation procedure combines different mechanisms that are expected to support a better PQoS management in the provision of multimedia services. The use and integration of these adapting mechanisms into a common PQoS-aware management system is a move beyond the current state of the art, which its novelty is further supported by the integration of the proposed management system within the IMS platform. Some of these service adapting mechanisms that will be integrated in the proposed management system are

- Content adaptation in the server with regard to a set of media parameters such as the codec and bit rate or spatio-temporal characteristics.
- More "classical" network resource management mechanisms (e.g., DiffServ/ MPLS) that guarantee specific NQoS levels associated to the traffic belonging to critical sessions, through traffic marking and classification of traffic flows into the deployed traffic engineering mechanisms.

◼ Error resilience adaptation, in order to improve the service robustness at the reception through the use of more efficient Forward Error Correction (FEC) schemes at the service generation.

The majority of the current research projects related to service management have been mainly restricted to technical issues rather than in analyzing their applicability to the management of users' satisfaction. Only limited research has been performed toward the use of PQoS evaluations as input for service management mechanisms, but even these cases are in general related to concrete mechanisms or they simply use a limited set of PQoS characteristics only for monitoring purposes. MCMS project with a pioneering way comes to integrate these two discrete research areas by defining novel distributed monitoring systems and feedback mechanisms in conjunction with the application of existing traffic models for the forecast of possible perceptual degradations using network measurements, which in turn will launch the most optimal adaptation mechanisms with the scope the maximization of the delivered PQoS level [1].

15.2.3.1 Advances in Cross-Layer Adaptation

Cross-layer adaptation schemes especially in the areas of streaming multimedia and wireless/mobile networks have attracted significant attention over the last couple of years. A lot of worldwide research and a significant number of EC-funded projects (ASTRALS, SUIT, ENTRONE II, porTiVity) aim at offering some kind of CLC. As a result, significant results have been reported at the physical layer, link adaptation and channel-aware scheduling. In parallel, wireless multimedia transmission robustness has been improved by means of application layer packetization, rate-distortion optimized scheduling, joint source-channel coding, error resilience, and concealment [3].

All the currently proposed schemes are network oriented by specializing to specific network technologies and specific transmission conditions. The proposed approach goes beyond all the current approaches, considering the following characteristics:

◼ It is not bounded to a specific network technology. On the contrary, it is applied across the network delivery chain of the requested multimedia service, covering heterogeneous network technologies (i.e., DiffServ/MPLS Core and UMTS, Access network).
◼ It is content-aware, considering in its adaptation actions the spatial and/or temporal dynamics of the media service under adaptation.
◼ It exceeds up to the application layer, integrating in its actions the dynamic adaptation of the encoding/streaming parameters. Thus, it moves beyond the state of the art, combining the advances in adaptation with the latest multimedia robust transmission research outcomes.

■ It is aiming at maximizing the user satisfaction by optimizing through adaptation actions the respective QoS-sensitive network parameters, showing the path toward the future intelligent user-centric adaptation techniques.

Cross-layer adaptation includes service layer adaptation (e.g., source/terminal coding parameters and FEC), network layer adaptation (e.g., traffic policies) and link layer adaptation (e.g., service classification), and is controlled by an intelligent action engine within the MCMS system. The two most promising multimedia services, that is, VoIP (voice/video call) and IPTV (live IPTV and Video on Demand (VoD)) will be used as a vehicle to demonstrate PQoS-aware MCMS system in the project.

The PQoS-aware dynamic cross layer adaptation will be performed according to

■ The monitored PQoS degradation at the end-user terminal device, which is also the triggering event for the initiation of the adaptation procedure.
■ The time-varying conditions of the Access and Transport network.
■ The type of the delivered service (i.e., Voice and/or Video for VoIP and IPTV applications).
■ The content dynamics (i.e., Action movie/Talk show or High/Low dynamic conversation).

15.3 User-Centric IPTV Architecture

Fixed/mobile convergence is a massive trend that requires adequate network and service infrastructures. One of the visions of this trend is that services will be sold in a consumer mass market based on the provision of content at a requested quality, exploiting the Perceived Quality of Service (PQoS) concept. The evaluation of the PQoS for multimedia content that have variable bandwidth demands, will provide a user with a range of potential choices covering, for example, the possibilities of low-, medium-, or high-quality levels, indication of service availability and costs. A user-centric IPTV system should consider a Perceived QoS-aware management mechanism within its service provision management system, eliminating its traffic policies to the network, which is a typical traffic differentiation mechanism that classifies the service bearers to different classes with specific QoS constraints [4,5].

A novel approach for such system is the adoption of the PQoS awareness into the current IMS management system, toward which the whole traffic engineering is not performed abruptly, but with scope the PQoS optimization at the user terminal. More specifically the system comprises of an IMS compatible PQoS-aware dynamic cross-layer adaptation management system, which in case of possible distorted playback of the multimedia service at the users' terminal, will be able to dynamically and in real time adapt the various NQoS parameters across the layers

(i.e., service, network, and link) of the network delivery chain (i.e., service generation, delivery, and consumption) with scope the optimization/maximization of the
delivered PQoS level and as such not interfering with the end-user experience of the
content being consumed. Moreover, the proposed management system does not
only improve abruptly the delivered PQoS level but also performs a sophisticated
reallocation of the occupied resources in order to keep the total traffic of the bearer
constant, indicating that the proposed management system uses more efficiently
the already utilized resources, without requesting/spending more resources in order
to perform the PQoS optimization.

Figure 15.1 depicts the overall architecture, which comprises:

■ The MCMS Module [1], which is the main entity of the user-centric IPTV
   system and focuses on monitoring the network statistics (i.e., core, access,
   terminal), the service generation (i.e., Media Server Resource Function
   (MSRF), VoIP terminals), and the service perceptual level at the end-user
   terminal in order to define and apply an optimal cross-layer adaptation action
   across the delivery network chain and media lifecycle (i.e., service generation
   node, core network, access network, and end-user terminal) for maximizing
   the user satisfaction.
■ The Multimedia Server and Resource Function (MSRF), which is an IMS-
   based MRF module with an additional fourfold responsibility:
   – The IPTV service generation, session management, and service streaming.
   – The VoIP service session management and signaling.

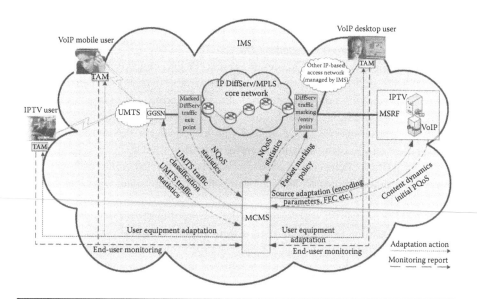

Figure 15.1   The overall proposed QoE-aware IMS architecture.

- The monitoring of the spatial and/or temporal content dynamics along with the selected encoding/streaming parameters.
- The adaptation, according to the MCMS commands, of the IPTV encoding parameters.

■ An IP Core—DiffServ/MPLS Compatible Transport Network.

In the proposed architecture it is considered a DiffServ/MPLS-enabled core network for the delivery of the requested multimedia service. IMS- and MCMS-compatible modules and interfaces are considered for the packet marking and traffic monitoring at the edges of the DiffServ/MPLS traffic network.

■ The Access Network

At the access side a wireless access network is considered (e.g., UMTS, WiMAX), which provides service/bearer classification mechanisms for providing QoS constraints on the delivered service type (e.g., video, voice, data, etc.).

■ The Terminal Adaptation Module (TAM)

At the QoE-enabled user terminals (e.g., 3G mobile handset, SIP voice/video phone (hardphone or softphone), PDA) a TAM will be integrated, which enables the terminal's interaction with the appropriate interfaces/modules of the MCMS.

■ The proposed MCMS and NGN IMS interaction

The IMS Application Servers (AS) for IPTV application is hosted within the Media Resource Function (MRF), which executes IMS applications and services by manipulating Session Initiation Protocol (SIP) and Session Description Protocol (SDP) signaling for interfacing with other systems. The proposed architecture considers that MRF and AS modules are combined into a single entity, which will be called as Media Server Resource Function (MSRF). The MSRF, besides being a media server, provides mechanisms for bearer-related services such as conferencing or bearer transcoding, through a controller Media Resource action Controller (MRFC) and a processor Media Resource Function Processor (MRFP) in compliance with the MCMS decisions [1,3].

15.3.1 Enhancing IMS

The predominant candidate for the convergence of an user-centric and adaptive IPTV system with NGN technologies is the IP Multimedia Subsystem (IMS), which is a standard (originally defined by the 3rd Generation Partnership Project 3GPP) for next-generation mobile networking applications based on Session Initiation Protocol (SIP) as the basic common signaling protocol. It is a specification framework that introduces the functional and network elements, as well as service platforms and the respective architecture, which enable real multimedia convergence on an IP-based infrastructure. Currently, IMS has been adopted as the standard control system in TISPAN NGN architectures [2].

The HSS (Home Subscriber Server) is the master user database that supports the IMS network entities that are actually exploited by the CSCF modules for handling the calls/sessions. The HSS contains the subscription-related information (user profiles), performs authentication and authorization of the user, and can provide information about the physical location of the user [1].

The control layer of the IMS infrastructure consists of nodes for managing call establishment, management, and release, which are called Call Session Control Functions (CSCF). The CSCF inspects each SIP/SDP message and determines if the signaling should visit one or more application servers en route to its final destination. In this multimodal management environment the MCMS modules comes to enhance the current IMS management capabilities by adding real-time dynamic cross-layer adaptation procedures for providing end-to-end perceptual optimization and therefore maximization of the user experience [2,6,7] (Figure 15.2).

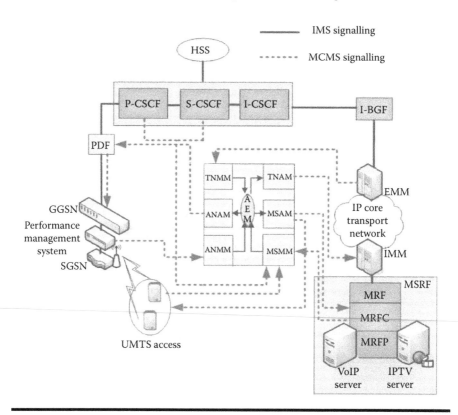

Figure 15.2   The proposed MCMS and NGN IMS interaction.

The MCMS architecture is based on a central decision module, the Action Engine Module (AEM), responsible for taking optimal adaptation decisions based on the monitoring of network and perceptual statistics, gathered by IMS-based monitoring and adaptation modules. Afterward, the AEM will process all the selected statistics and define a perceptually optimal cross-layer adaptation action.

15.3.1.1 Specifications of the Proposed MCMS

The MCMS is the central entity of the proposed QoE-aware and user-centric IMS architecture that seamlessly communicates with the already-existing IMS modules and interfaces. The MCMS interacts with each element in the service provision chain via real-time monitoring and adaptation/control mechanisms to achieve end-to-end Perceived QoE maximization. Thus, MCMS moves away from traditional NQoS-centric adaptation/management scheme and is able to achieve an end-to-end QoE optimization based on a user-/customer-centric approach. This provides an efficient solution/approach for future networked multimedia making it possible to maintain the quality of the media at every step of the media lifecycle from creation to consumption [8].

There are two functions associated between the MCMS and the TAM of the QoE-enabled user terminals:

Adaptation function: According to the control command/parameters received from the MCMS, the QoE-enabled user terminal adapts on the fly its VoIP/Videocall codec, encoding bit rate or mode, aiming at end-to-end perceived quality improvement.

Monitoring function: Reporting delivered QoE (e.g., Mean Opinion Score (MOS) score for voice and video) and relevant terminal parameters (e.g., codec type, bit rate, encoding mode, packetization parameters) to the MCMS module. These parameters are obtained by the TAM interface, which is implemented in the terminal device.

The MCMS waits in an idle mode which changes to active as soon as the TAM triggers an Alarm event to the MSMM. A two alarm method is chosen for more flexibility in experimenting and minimizing the amount of data exchanged. The alarms considered are two: *Warning Alarm* and *Red Alarm* (Figure 15.3).

15.3.1.2 IMS Service Control Interface

The interface between an SIP AS and the S-CSCF is called *ISC* (IMS Service Control) interface. Depending on the role of an AS, the procedures of *ISC* can be divided into two main categories:

1. For incoming session initiating SIP messages, the S-CSCF analyzes them based on initial filter criteria from the user profile as part of the HSS subscriber data

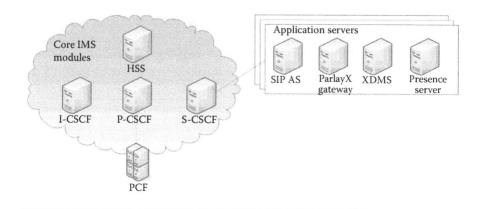

Figure 15.3   IMS architecture and elements.

and then routes them to the appropriate SIP Application Server for further processing. The AS then can act as User Agent Server (UAS), SIP Proxy or a Redirect Server.

2. The Sip Application Server may initiate own SIP requests and acts as User Agent Client (UAC) or Back to Back User Agent (B2BUA).

Sh *Interface*: 3GPP has defined the *Sh* interface as part of the IMS for Application Servers to access subscriber profile data in HSS. The protocol used for message exchange is Diameter. Subscriber related data from the HSS can be accessed from SIPSEE with help of a Diameter protocol adapter. Applications are also able to subscribe to special events in the HSS (e.g., change of a special item in user profile, user registration status, etc.) and receives notification events sent from HSS.

15.3.1.3 Policy and Charging Control Function

The Policy and Charging Control (PCC) Function is the component responsible for making policy decisions. PCRF decisions are based on session and media-related information taken from the P-CSCF. Some of the supported functionalities are

1. To store session and media-related information
2. To generate PCC rules to install at the PCEF, with session and authorized QoS information
3. To provide an authorization decision on receiving a bearer authorization request
4. To update the authorization decision at session modifications related to session and media-related information
5. The capability to recall the authorization decision at any time

6. The capability to enable/prevent the usage of an authorized bearer
7. To inform the P-CSCF when the bearer is lost or modified
8. To pass an IMS-charging identifier to the Gateway GPRS Support Node and to pass a GPRS-charging identifier to the P-CSCF

The user-centric IPTV system policy management functions be based on a Policy Control Framework and will include a Policy and Charging Rule Function (PCRF), a Policy and Charging Enforcement Point (PCEF), an XML Driven Data Repository and a Web Management Interface. This implementation follows 3GPP Release 7 Specifications for the IMS Policy Control Architecture [4].

The Rx interface between the P-CSCF and PCRF will be Diameter-based, following 3GPP TS 29.214 and passes session information to the PCRF. The Gx interface between the PCRF and PCEF will use Diameter and follow 3GPP TS 29.212 and passes enforcement policy rules to the PCEF. Based on Policy Control Framework, policies can be stored locally or remotely on an XDM Server—XCAP is used for remotely accessing the policy repository.

PCRF: Regarding policy control, the Policy Control and Charging Rules Function (PCRF) performs the logic for QoS authorization and for binding application-level sessions to network resources. PCRF communicates with P-CSCF via Rx interface for authorization responses or reauthorization requests triggered by network events. For QoS authorization, the PCRF uses the service information received from the P-CSCF and/or the subscription information received from the SPR to calculate the proper QoS authorization (QoS class identifier, bit rates). Policy decisions are based on the creation and management of PCC Rules. A PPC Rule is made up of a service data flow template, jointly with the related charging and QoS information [4].

PCEF: The Policy and Charging Enforcement Function (PCEF) is in charge of receiving policy decisions from the PCRF and translate them to access network-specific procedures to ensure the fulfillment of those requirements. Depending on the terminal and access network characteristics and capabilities, the IP-CAN bearers can be controlled by the User Equipment (UE) (UE-only mode), the IP-CAN (NW-only mode) or it can be possible that it can be managed by both of them (UE/NW mode). This characteristic determines where the bearer binding is performed. The bearer binding is the process by which the multimedia flow description is associated to the IP-CAN bearer that will support it [4].

15.3.2 Multimedia Content Management System

The user-centric IPTV system introduces an innovative IMS-compatible Multimedia Content Management System (MCMS) focused on performing a dynamic cross-layer adaptation for the optimization of the end-user experience in terms of the delivered PQoS (Perceived Quality of Service) level for IPTV (Internet Protocol TeleVision) and VoIP (Voice over Internet Protocol) services—two of the most important multimedia services with the potential to create significant wealth for Europe.

This multimodal management system is being applied in an integrated and coherent way across all the network layers and delivery-chain nodes based on a user-/customer-centric approach rather than the traditional network QoS-oriented one. Toward this, the management system makes use of advanced IMS compatible PQoS and NQoS monitoring and adaptation mechanisms across the network delivery chain, thereby significantly enhancing the current IMS management functions by providing necessary perceptual awareness capability. Novel perceptual mapping frameworks between the NQoS-related monitored parameters and the delivered PQoS level will be researched for both IPTV (live IPTV and VoD) and VoIP (voice/video call) services.

Based on the monitoring data, a decision engine initiates a dynamic cross-layer adaptation procedure, which extends from the service generation entity to the user terminals.

An overview of the PQoS-aware MCMS concept is depicted in Figure 15.4. On the left, a multimedia service provision full cycle from content creation, content transport (via core and access networks) to content consuming is shown (from top to bottom). On the right, Multimedia Content Management System (MCMS) interacts with each element in the service provision chain via real-time monitoring and adaptation/control mechanisms to achieve end-to-end Perceived Quality of Service (PQoS) maximization.

As shown in Figure 15.4, real-time monitoring provides important information, from source (e.g., content dynamics), core/access network (e.g., network/link QoS), and terminal (e.g., delivered PQoS). Cross-layer adaptation includes service layer adaptation (e.g., source/terminal coding parameters and FEC), network layer adaptation (e.g., traffic policies), and link layer adaptation (e.g., service classification), and is controlled by an intelligent action engine within the MCMS system. The two most promising multimedia services, that is, VoIP (voice/video call) and IPTV (live IPTV and VoD) will be used as a vehicle to demonstrate PQoS-aware MCMS system in the project.

The new MCMS management system moves away from traditional NQoS-centric adaptation/management scheme and will be able to achieve an end-to-end PQoS optimization based on a user-/customer-centric approach. This will provide an efficient solution/approach for future networked multimedia making it possible to maintain the quality of the media at every step of the media lifecycle from creation to consumption.

The AEM is the central decision module (Figure 15.5), responsible for taking optimal adaptation decisions based on the network monitoring and perceptual statistics. The AEM exploiting theoretical mapping frameworks between NQoS and PQoS will process all the selected statistics and define a perceptually optimal cross-layer adaptation action.

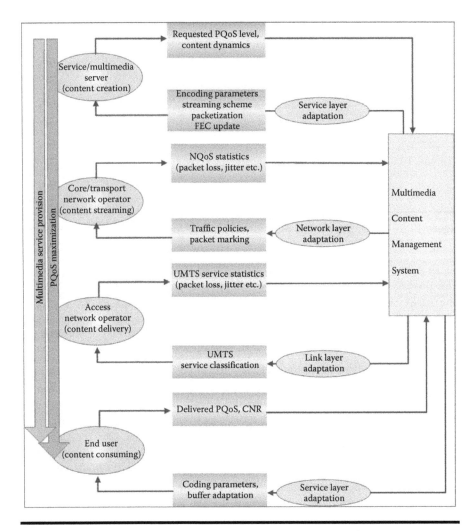

Figure 15.4 Conceptual diagram of the MCMS for optimizing the user experience.

The AEM will execute a sophisticated processing procedure and a decision algorithm, and it will decide the appropriate actions and adaptations across the network delivery chain, in order to optimize the delivered PQoS. The AEM will have interfaces with the rest of MCMS modules (described above) to get and provide the monitoring and adaptation. In the following subchapters these interfaces are specified. The interfaces hide the internal communication used between the different monitoring and adaptation modules and the AEM. These communication implementation details depend upon architecture decisions, like single/multiple hosts, single/multiple processes.

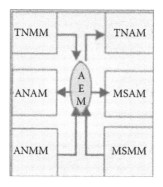

Figure 15.5 AEM, the central decision module of MCMS.

In a first step the interfaces that are defined will be simple ones with extension possibilities. During the project this extension will be defined if needed. The AEM shall be able to take QoS control decisions based on an advanced mechanism/algorithm (e.g., such as neuronal, genetic programming, or expert system), taking into account multiple and interrelated PQoS theoretical models.

The AEM needs to have some intelligence to process the information coming from the different Monitoring Modules, and make decisions about the best way to improve PQoS, sending instructions to the Adaptation Modules. In next subsection, a quick overview of the possible algorithms is presented, followed by the algorithm selection at the time being. Anyway, it must be taken into account that, during the implementation phase, this decision could be changed due, for example, performance reasons. The expert system resides within a network functional entity interacting with other network entities. There will be some interfaces to obtain the case-specific input data and to communicate the output decision. The modules providing the case-specific information to the AEM will be the MCMS monitoring modules. And the modules receiving the expert system output will be the MCMS adaptation modules.

Additionally, the base Expert System architecture has been extended to take care of other user-centric IPTV system specific tasks, such as the handling of IMS sessions. In general, the AEM is decomposed into the following main components:

- The *AEM logic*, which is the main component containing the system intelligence.
- The *database* where all data required by the logic are stored.
- The adapter between the logic and the database.
- The interface between the logic and the rest of MCMS modules.

15.3.3 *Monitoring and Adaptation Engines*

15.3.3.1 *Monitoring*

15.3.3.1.1 Multimedia Service Monitoring Module

The Multimedia Service Monitoring Module (MSMM) performs monitoring of the service session through the P-CSCF and S-CSCF modules of the IMS. This is the only module of the MCMS that is activated when a service is requested; while all the rest remain in idle mode until the active adaptation procedure starts. Then, the MSMM, except from simply informing the MCMS for the liability of the service, monitors the QoE and Carrier-to-Noise ratio (CNR) values at the end-user mobile terminal device, while at the service generation site, the content dynamics and the encoding parameters are also monitored through the TAM of the terminal device.

Concerning IPTV service, the MCMS requests the monitoring data related to the IPTV components, in order to use them as input for session adaptation decisions taking. To achieve this, it uses the MSMM, which requests monitoring data from the MSRF, the IPTV Soft terminal and the IPTV Test terminal at the end-user site through the TAM. The MSRF returns the current video streaming properties such as the video bit rate and the video frame rate. The test terminal is responsible for monitoring the PQoS value for the video the user is watching. Figure 15.6 depicts the flow of monitored data from MSMM to the AEM.

Before any session monitoring process, the MCMS should be aware of session instantiation on its successful initiation between the terminal and the MSRF. In a similar way, the MCMS should also be aware of the session teardown (ending). Finally, during the session streaming process, the monitoring operations are triggered at the MCMS on the reception of the "warning/alarm" message coming from

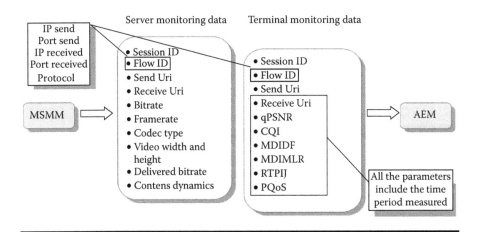

Figure 15.6 MS monitoring data and flow diagram.

the TAM. The related communication between the MCMS, TAM, and the MSRF are ensured through the MSMM module.

15.3.3.1.2 Transport Network Monitoring Module

The Transport Network Monitoring Module (TNMM) module is used during the dynamic cross-layer adaptation procedure for monitoring network statistics like packet loss, jitter, delay, and so on at the DiffServ/MPLS core transport network. Toward this, the appropriate External Marking Modules (EMM) interface will be developed and integrated at the egress edge router of the core transport network for enabling interaction and communication with the TNMM of the MCMS.

In order to perform *monitoring*, the following actions should take place:

1. Each router collects data about the traffic. (TC Monitoring Module).
2. Each router sends the collected data to the MonitoringSocketServer of the EMM, where the total data calculation takes place by the Total Data Calculator.
3. The AEM communicates with the EMM's TNMMSocketServer through the MCMS TNMM so as to get the needed information. This is done in three steps:
   a. The AEM calls the TNMM's function for data collection.
   b. The TNMM triggers the socket client which gets the data from the socket server on the EMM.
   c. The TNMM returns the data to the MCMS AEM through the function's result.

Steps 1 and 2 are continuously executed every 1 s. Figure 15.7 depicts the TN data flow diagram from the TNMM to the AEM.

15.3.3.1.3 Access Network Monitoring Module

The Access Network Monitoring Module (ANMM) module monitors the UMTS access network statistics, based on the exploitation of historical data, which are

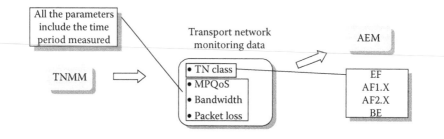

Figure 15.7   TN monitoring data and flow diagram.

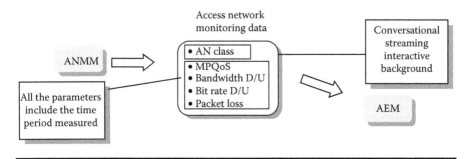

Figure 15.8   AN monitoring data and flow diagram.

updated in near real time by the UMTS performance management system at frequent time periods.

In order to perform *monitoring*, the following actions should take place:

1. The user-centric IPTV system-based PCEF monitors the access network traffic every 1 s. PCEF stores the monitored data in a file.
2. MCMS AEM receives a WARNING ALARM from the TAM.
3. The MCMS AEM communicates with the PCF so as to get the current status of the network traffic through the MCMS ANMM
   a. MCMS AEM communicates with the MCMS ANMM and calls its on ReceiveANMMXX function.
   b. MCMS ANMM communicates with the IMS-based PCEF so as to get the monitored data.
   c. The MCMS ANMM gets the monitored data from the IMS-based PCEF using the scp linux command and creates an ANMonitoredData object which contains the data in an AEM-understandable form.
   d. MCMS ANMM returns the AEM-understandable monitored data (as result of the onReceiveANMMonitoredData function) to the AEM.

Figure 15.8 depicts the AN data flow diagram from the ANMM to the AEM.

15.3.3.2 *Adaptation*

For adaptation purposes, MCMS considers the following modules, through which the optimal adaptation actions for QoE optimization are applied.

15.3.3.2.1 Multimedia Service Adaptation Module

The Multimedia Service Adaptation Module (MSAM) performs adaptation actions at the end-user terminal device and at the service generation entity (i.e., the MSRF for IPTV services or the end-user terminal for VoIP applications) relative to the decoding

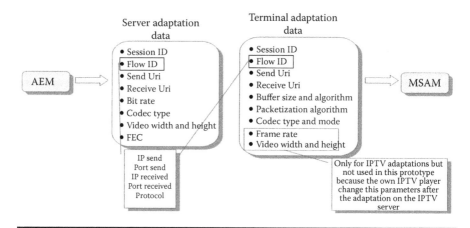

Figure 15.9 MS adaptation data and flow diagram.

(i.e., buffer scheme) and encoding (bit rate, packetization, etc.) parameters, as well as FEC value adaptation toward enhancing the error resilience of the delivered service.

The role of the MSAM in session adaptation operation is to forward the adaptation enforcement commands, when the AEM sends them to the IPC Agent. Then, these requests are encapsulated in SDP/SIP messages and sent to the MSRF in the case of IPTV service and TAM in the case of VoIP service via the SIP Agent. Figure 15.9 depicts the MS data flow diagram from the AEM to the MSAM.

15.3.3.2.2 Transport Network Adaptation Module

The Transport Network Adaptation Module (TNAM) applies the adaptation actions to the DiffServ/MPLS-enabled core transport network through the Internal Marking Module (IMM), which will be developed and integrated at the ingress router of the core network. The IMM receives the adaptation actions from the TNAM and translate them to DiffServ/MPLS compatible commands, which are finally applied by marking appropriately the incoming traffic.

In order to perform *adaptation*, the following actions should take place:

1. MCMS AEM (after decision according to the last version of the monitored data) communicates with the TNAMSocketServer on IMM by triggering the TNAMSocketClient.
   a. MCMS AEM calls the MCMS TNAM function to perform adaptation. AEM writes the new class of the session, the source IP and the destination IP in a file named TNADAPT.dat.
   b. MCMS TNAM triggers the TNAMSocketClient and sends the TN Adaptation command to the TNAMSocketServer on the IMM.

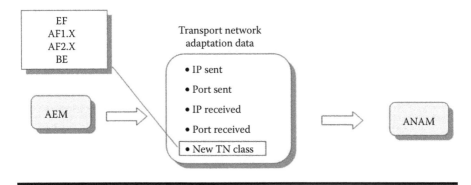

Figure 15.10   TN adaptation data and flow diagram.

In this step the TNAM executes the shell script which starts the TNAMSocket Client. The TNAMSocket Client reads the TNADAPT. dat file and sends its content to the TNAMSocketServer on the IMM.

2. The Transport Network Adaptation Module performs adaptation as soon as the TNAMSocketServer receives the Adaptation Data from the TNAMSocketClient. Figure 15.10 depicts the TN data flow diagram from the AEM to the TNAM.

15.3.3.2.3 Access Network Adaptation Module

The Access Network Adaptation Module (ANAM) applies the adaptation actions, decided by the AEM of the MCMS, to the UMTS access network through the IMS PDF module. The Policy Decision Function (PDF), in turn, applies them at the Gateway GPRS Support Node (GGSN) by performing service bearer classification in order to improve its QoS characteristics and therefore enhance the delivered QoE level. Figure 15.11 depicts the AN data flow diagram from the AEM to the TNMM.

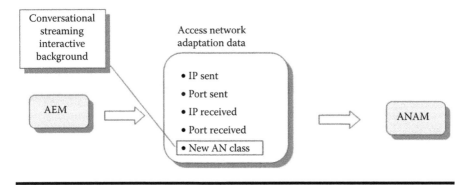

Figure 15.11   AN adaptation data and flow diagram.

In order to perform *adaptation*, the following actions should take place:

1. The MCMS AEM receives a RED ALARM from the TAM.
2. The MCMS AEM (after decision according to the last version of the monitored data) communicates with the PCF Adaptation Module through the MCMS ANAM.
   a. The AEM communicates with the ANAM and calls its adaptation-related function.
   b. ANAM sends the adaptation command to the PCF Adaptation Module (using the ANADAPT.dat file which contains the flow class and the source and destination IPs) using the scp linux command.
   c. The PCF Adaptation module performs adaptation. Through this action, the module creates new flows between the two clients which have the desired class.

15.3.4 QoS Support and Mapping

15.3.4.1 TN, AN, and TAM MCMS-Driven Adaptation Mapping

In IPTV sessions only quality video adaptations are performed.

- *QoS.* IPTV applications are very sensitive to loss/error packets but not to delays of the total stream.
  - *IPTD (IP packet Transfer Delay).* IPTV is not very sensitive to the time the data actually take to reach the user.
  - *IPDV (IP packet Delay Variation) or jitter.* IPTV is very sensitive to jitter during the decoding process.
  - *IPLR (IP packet Loss Ratio).* Direct impact on image and sound quality.
  - *IPER (IP packet Error Ratio).* Strong impact on image and sound quality.
- Depending on the type of service, the following considerations should be taken into account.
  - *Live IPTV.* A single audio/video stream is sent to many users and this saves bandwidth. Adaptation of this service produced by a degradation of the *PQoS* is performed only if there are a percentage of users that perceive it.
  - *VoD.* Higher bandwidth required as the audio/video stream is sent to a single user, and the PQoS adaptation is performed if this user detects degradation.
- Type of changes
  - *VideoFrameRate* changes (15–30).
    - FrameRate changes affect to the amount of data transmitted, and therefore to the bandwidth.
    - It affects the terminal and the server.

- *VideoBitRate* changes
  - Maintains video size and frame rate.
  - High DCT coefficients are removed and not transmitted.
  - Results: different amount of data transmitted (impact on bandwidth)
    - Low bit rate
      - Less bandwidth required
      - High DCT coefficients removed
      - Less image definition
      - Low video quality
    - High bit rate
      - More bandwidth required
      - More image definition
      - High video quality
- *Video Resolution* changes
  - Depends on the terminal. Typical: QCIF (176 × 144) Mobile and 320 × 240 PDA or Smartphone.
  - Affects to the bandwidth.
  - A higher resolution implies an increase of video quality and data transmitted (impact on bandwidth).
- *Video codec* changes
  - Terminal and server renegotiation required.
  - Implies a little delay and some messages interchange before the user can see the video with the new codec.
  - User receives a notified of the change.
  - Typical codecs: MPEG-4 (Table 15.1).

15.3.4.2 TN/AN NQoS Mapping

The NQoS mapping between heterogeneous networks is very important especially when the provisioned service is IPTV. The NQoS support should be at the same level for the whole End-to-End connection. Table 15.2 shows the NQoS mapping between heterogeneous networks used for a user-centric IPTV system.

15.4 QoE/PQoS Model

Two features relevant to the proposed approach should be presented:

- An enhanced network monitoring system, which provides detailed information of the QoS metrics for each network segment.
- The possibility to afford advanced service-level adaptations based on a combination of parameters, as well as to exploit network-level adaptation procedures.

Table 15.1   IPTV Service Adaptation Steps and NQoS Mapping

Type of Media	Parameter	Premises	Value	Actions	
IPTV	Any oher	TN class *BE*	Alarm	Flow reclassification	*New TN class:* AF23
		TN class *AF23*	Alarm	Flow reclassification	*New TN class:* AF22
		TN class *AF22*	Alarm	Flow reclassification	*New TN class:* AF21
		TN class *AF21*	Alarm	Flow reclassification	*New TN class:* AF13
		TN class *AF13*	Alarm	Flow reclassification	*New TN class:* AF12
		TN class *AF12*	Alarm	Flow reclassification	*New TN class:* AF11
		TN class *AF11*	Alarm	N/A	*No change available*
		AN class *background*	Alarm	Flow Reclassification	*New TN class:* Streaming
		AN class *streaming*	Alarm	N/A	*No change available*
		Video *CIF* (352 × 288)	Alarm	Terminal IPTV video change	*Codec:* H.264 (default no change)
					Codec mode: default
					Frame rate: default
					Packetization algorithm: default
					Buffer size: default
					Buffer algorithm: default
					Video settings: QCIF (176 × 144)

Table 15.1   (continued) IPTV Service Adaptation Steps and NQoS Mapping

Type of Media	Parameter	Premises	Value	Actions	
	Video QCIF (176 × 144)		Alarm	Server IPTV video change	Codec: H.264 (default no change)
					Bit rate: default
					Video settings: QCIF (176 × 144)
					FEC: default
				Terminal IPTV no change	Codec: H.264 (default no change)
					Codec mode: default
					Frame rate: default
					Packetization algorithm: default
					Buffer size: default
					Buffer algorithm: default
					Video settings: no change
				Server IPTV no change	Codec: H.264 (default no change)
					Bit rate: default
					Video settings: no change
					FEC: default

Under service degradations, the optimal multimedia adaptation for each case will depend not only on the bare end-to-end (e2e) NQoS values, but also on the source of these impairments. For example, depending on where the degraded network segment is, the modification of different service-level parameters will have a different impact on the final result.

Table 15.2   TN and AN NQoS Traffic Class Mapping

Traffic Class	Conversational (Real Time)	Streaming (Real Time)	Interactive (Best Effort)	Background (Best Effort)
Charac-teristics	Preserve time relation (varia-tion) between information entities of the stream	Preserve time relation (variation) between information entities of the stream	Request/ response pattern	Destination is not expecting the data with a stringent time
	Conversational pattern, there-fore, very low delay and jitter	Delay and jitter require-ments are not as strict as with the conversa-tional class	Retransmis-sion of payload content in-route	Retransmis-sion of payload content in-route might occur
Example Applica-tions	Voice over IP	Streaming audio and video (IPTV)	Web brows-ing	Downloading email
WiMAX class mapping/ DSCP based	UGS	rtPS	nrtPS	Best effort (BE)
Diffserv class/map to DSCP	Expedited forwarding class	Assured forwarding 2 class	Assured forwarding 3 class	Best effort

Figure 15.12 illustrates the basics of the general adaptation procedure here pro-posed. Each horizontal plane represents a combination of access network (AN) and core network (CN) performance states. For each coordinate within the plane, the vertical axis represents all the possible combinations at service level, each combina-tion with different values for the n-tuple of variable service-level parameters. Each combination in the 3D space can take an associated QoE value, which will lead the decision-making process for the adaptation system [9,10].

This proposal therefore tries to cope with the following objectives:

■ When a new combined AN/CN state is detected, the automated system infers the optimal service configuration for the new network state to ensure that the system is placed at the best scoring position in the vertical axis.

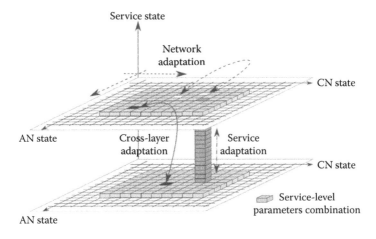

Figure 15.12   Cross-layer adaptation procedure.

For a new service configuration selected, the system estimates if the expected QoE fulfils the required quality.

- If the QoE is accurate enough for the user, and if the service configuration differs from the current one, then the system launches the procedure to take the required adaptation action, which could affect any or both of the endpoints. For example, in a VoIP service the sender endpoint can be affected if the AMR mode or the packetization scheme is modified, as well as the receiver endpoint if the de-jittering buffer size is changed.
- Otherwise, the only possible solution is to move to another AN/CN state where the service provision is able to comply with the quality requirements. This action is represented as network adaptation in Figure 15.12, and it likely entails a possible change in the network or in the QoS classification of the traffic involved. The system thus selects a new x–y coordinate able to cope with the QoE requirements.

If the new position in the horizontal plane requires modification of the multimedia session to achieve the corresponding best configuration, this will represent a cross-layer adaptation.

The former is basically a network-driven service adaptation approach. The latter becomes a more complex cross-layer service and network adaptation, and its real-world implementation involves the capability to request modifications in the network utilization as considered in the MCMS approach.

This section is focused on the adaptation of mobile video services. First of all, the combined impact of the UMTS transmission characteristics and the mobile video encoding is evaluated. From this study, the most suitable service-level configuration is selected for the different AN states. Afterwards, the more general problem of serving multiple mobile video users is studied. In this case, both the AN and the CN monitoring are considered as inputs to the cross-layer adaptation process.

15.4.1 Video Service Adaptation

Following the proposed approach, when a PQoS Alarm is received at the MCMS from a mobile video user, the AEM has to infer the set of adaptations that allows recovering an accurate QoE level. In this case, the influence of the adaptation on other users has to be considered. For example, promoting a user at the CN has a direct impact on the performance of the entire CoS. Thus, an intelligent decision making system is proposed that tried to solve this kind of problems with the additional constraint of the system responsiveness [11].

The specific settings are presented in Table 15.3. In this case, the considered service is a mobile video streaming displayed at a mobile handset, with a dejittering buffer of one second.

The considered video clips for this study are a set of reference video sequences at QCIF resolution, encoded at different frame-rate and bit-rate values (Table 15.4).

For each AN_state, which is determined by the BLER value, the experienced IP Packet Loss Ratio (IPLR) is a function of three variables: the Content Type (CT), the Frame Rate (FR), and the target Source Bitrate (SBR).

Table 15.3   UMTS Settings

BLER values	{0, 5, 10, 20, 30, 40, 50} %
MBL	1.75

Table 15.4   Mobile Video Encoding Settings

Video Sequences	FR (fps)	SBR (kbps)
Akiyo, Foreman	$i = 1 - 3$ for {7.5, 10, 15} fps	$j = 1 - 3$ for {48, 88, 128} kbps
Suzie, Carphone	$i = 1 - 2$ for {10, 15} fps	$j = 1 - 2$ for {90, 130} kbps
Stefan	$i = 1 - 3$ for {7.5, 10, 15} fps	$j = 1 - 3$ for {88, 128, 256} kbps
Football	$i = 1 - 2$ for {10, 15} fps	$j = 1 - 2$ for {130, 200} kbps

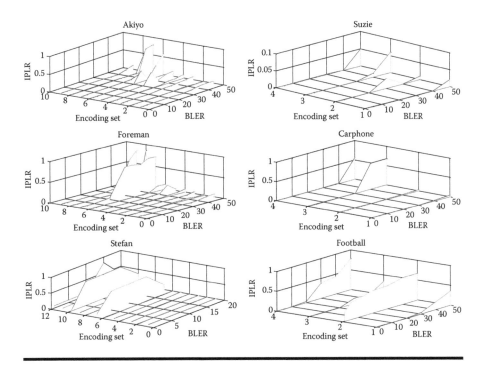

Figure 15.13   IPLR evolution of videos for different BLERs and encoding sets.

Figure 15.13 overviews the relations between BLER and IPLR found in the simulation experiments. As can be observed, for each Content Type, the expected IPLR at a BLER value is different for each encoding setting.

As well, the expected MOS for different SBR at no losses is obtained, and for different IPLR values, for the different video types.

Taking both mappings as inputs, the service-level adaptation logic in this case is defined as follows:

- The video service is being provided at the best encoding settings for the specific content type.
- A change in the AN state is detected, this is, the experienced BLER increases.
- If the resulting IPLR makes the predicted MOS decrease, a PQoS Alarm is generated by the endpoint and sent out to the MCMS.
- The MCMS uses the AN monitoring interface in order to infer the current AN state.
- The adaptation logic makes the decision of switching to the best performing encoding setting for the specific AN state.

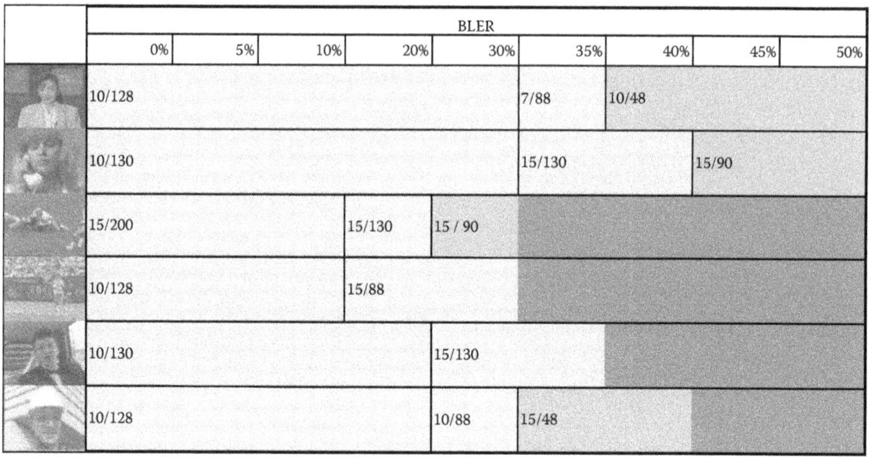

		BLER								
	0%	5%	10%	20%	30%	35%	40%	45%	50%	
	10/128					7/88	10/48			
	10/130					15/130		15/90		
	15/200			15/130	15 / 90					
	10/128			15/88						
	10/130				15/130					
	10/128				10/88	15/48				

Figure 15.14   Best-performing encoding sets (FR and SBR) for different AN states.

Figure 15.14 shows the best scoring encoding set (combined FR and SBR) parameters) for the different AN states considered and for each Content Type. The colors indicate the suitability of the service from a PQoS perspective.

15.4.1.1 Multiuser Cross-Layer Adaptations

When a PQoS Alarm is received at the MCMS, it invokes the AEM algorithm to solve the problem of the adaptation decision making. In this section, we analyze the complex problem of the possible adaptation of multiple IPTV flows at the same time, with the additional constraint of the system responsiveness [12].

The MCMS keeps a track of registered multimedia sessions currently being provisioned by the proposed system. Upon reception of a PQoS Alarm, the MCMS invokes the related Monitoring Modules in order to obtain an updated picture of the AN, and/or the CN.

After this process, the AEM is run in order to optimize the system performance, and thus maximize the PQoS levels of the multimedia sessions [13]. The set of service characteristics that are currently taken into consideration in the optimization problem are:

- For each IPTV session:
    - CT: Content Type
    - SBR: Source BitRate
    - SR: Spatial Resolution
    - FR: Frame Rate
    - DLBR: Maximum BitRate allowed to the Radio Bearer
    - BLER: Block Error Rate in the AN

   - CoS: Class of Service in the CN
   - IPLR: IP Packet Loss Ratio for the associated CoS
   - Alarm: Indicates if a PQoS Alarm has been generated by this session
▪ For the CN:
   - Initial assigned load to each CoS
   - Additional background traffic load associated to the BE

For the case study, two main variable parameters are considered: the SBR and the CoS for each multimedia flow. Based on this information, optimal values for the rest of variable parameters can be inferred. The optimization problem is decomposed into a global minimization of the objective function, which is based on expected MOS scores.

Due to the stringent requirements of the MCMS/AEM for solving adaptation problem in real time, the selected optimization approach is based on genetic algorithms. This approach allows us to obtain optimal or suboptimal solutions in lower execution times than other exact optimization approaches.

Since not all the alternatives are evaluated, the correct implementation of the problem constraints in the creation and reproduction functions is critical for the accuracy of the system. As well, the correct setting of the algorithm options has a great impact on the trade-off between optimality and responsiveness.

Additionally, in order to reduce the complexity of the system we limited the values that several parameters can take. This makes the adaptation system more implementable in an actual management system (Table 15.5).

Table 15.5 Allowed Values for Input Parameters

Parameter	Allowed Values
CT	1 (Low Motion), 2 (Medium Motion), 3 (High Motion)
SBR	{80, 130, 200, 256} kbps
SR	1 (320 × 240), 2 (176 × 128)
FR	{10, 12, 15} fps
DLBR	{384, 128} kbps
BLER	No Restrictions
CoS	1 (BE), 2 (AF22), 3 (AF11), 4 (EF)
IPLR	No Restrictions
Alarm	0 (Nondegraded Session), 1 (Degraded Session)

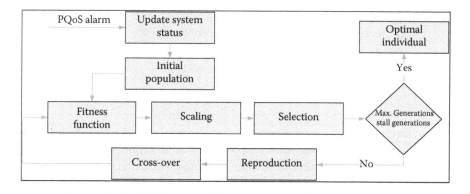

Figure 15.15   AEM GA-based logic.

15.4.1.1.1 Implementation Details

The AEM logic is illustrated in Figure 15.15.

In the following, the specific GA settings and the implemented functions are briefly described:

15.4.1.1.1.1 Initial Population — The initial population is generated from initial system status. The number of individuals to be generated must be established *a priori*. A higher number of individuals provides higher probabilities of finding the optimal solution. On the other hand, a high number of individuals increase the execution time for each generation. For the case studies considered in this document, a size of 20 individuals has been proven to be a suitable trade-off for the needs of the proposed user-centric IPTV system. Each individual represents a possible system state that can be reached by performing a set of adaptation actions over the registered multimedia sessions. In other words, this is the variable that can be modified to tests the accuracy of each alternative solution. Therefore, the individual genome is defined as a matrix that relates the SBR value and the associated CoS in the CN.

The implemented creation function follows the logic illustrated in Figure 15.16.

The Initial Individual is received as an input argument, and represents the system status at the moment of the invocation of the algorithm. This individual is included in the population. At a first stage, the creation function discards all the SBR values that are not suitable for the currently experienced BLER conditions. This is achieved by selecting the best scoring SBR value for the service conditions, following the evolution presented in Figure 15.17.

After the AN constraints are attended, the creation function generates different permutations varying the SBR and the CoS values for each of the IPTV sessions that generated the PQoS Alarm.

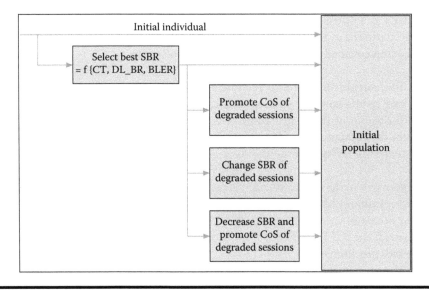

Figure 15.16   Creation function.

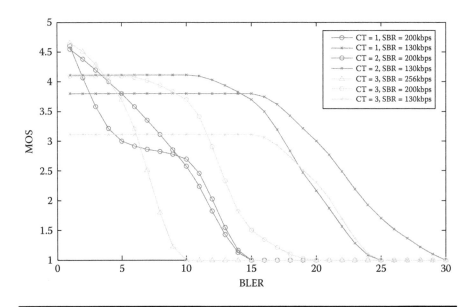

Figure 15.17   Evolution of MOS for different CT/SBRs.

15.4.1.1.1.2 Fitness Function — The fitness function provides an assessment of how accurate the evaluated solution is.

The implemented fitness function receives two inputs:

- The assignment matrix to be evaluated, which includes the selected SBR for each media flow and its selected CoS in the CN.
- The vector of IPTV structures, which contains all the information required for the computation of the expected MOS of each flow registered in the user-centric IPTV system.

The output of the fitness function is determined by a combination of the average and the minimum MOS values obtained for the whole set of the IPTV sessions.

The fitness function follows the algorithm illustrated in Figure 15.18.

First, all the IPTV session structures are updated with the new values proposed in the assignment matrix.

With this new traffic situation, the new conditions of the CN are estimated. Based on the assignment of SBR values to each CoS, and taking into account the additional background traffic present in the BE CoS, the new estimated IPLR values are computed. At this point, the fitness function keeps updated values of the AN and CN conditions, as well as of the service characteristic.

Based on previous works, the expected PQoS level for each media flow can be computed. Three impairment factors are considered:

- $MOS_{enc} = f\{CT, SBR, SR\}$
- $MOS_{CN} = f\{CT, SBR, IPLR_{CoS}\}$
- $MOS_{AN} = f\{CT, SBR, DL\_BR, BLER\}$

Thus, the expected PQoS level for each media flow can be obtained from the combination of these impairment factors as $MOS = f\{MOS_{enc}, MOS_{CN}, MOS_{AN}\}$.

Finally, all MOS values associated to each registered IPTV session are consolidated and the fitness score is computed based on the average and minimum values.

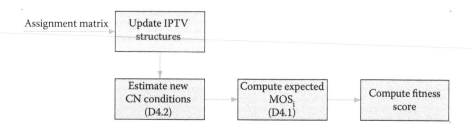

Figure 15.18   Fitness function.

15.4.1.1.1.3 Scaling — Once obtained, the fitness score for each of the individuals within the population, the scaling function provides a way of prioritizing individuals. The used function provides to each individual an expectation value proportional to its fitness score. This strategy can be used since the scale in the implemented fitness functions is representative of the accuracy of each solution.

15.4.1.1.1.4 Selection — In order to introduce a degree of randomness, the selection of parents for the next generation of individuals is based on a stochastic uniform function. In this way, the best scoring individuals have a higher priority of becoming a parent, but those individuals with lower scores also have their chance to be included in the reproduction.

From the selection stage, there are two alternative outcomes:

- If the algorithm has reached its final iteration, the best scoring individual from the current generation is returned. This may happen when the algorithm reaches the Maximum number of generations established, or when the best score has not changed over the last Stall generations. These values are set up to 5 and 2 respectively for the case study.
- Otherwise, the algorithm goes on to the reproduction phase.

15.4.1.1.1.5 Reproduction — The reproduction function determines how the next generations of individuals are generated. In the case study here presented, at each generation we assure the survival of the three best scoring individuals. The rest of individuals are generated by cross-over between their selected parents.

15.4.1.1.1.6 Cross-Over — The cross-over function receives as input the current generation of individuals, as well as a series of individual identifiers in order to select two parents for each new individual that has to be created.

Therefore, for each pair of parents, the implemented cross-over function iterates for each media flow and selects one of the parents to include the settings. The cross-over is made in a per-flow basis, in order to avoid duplications in the assignments of a flow to a CoS and to avoid as well the possible splitting of the traffic associated to a flow over different CoS.

15.4.1.1.2 Adaptation Actions Driven by AN Degradations

The impact of the UMTS link performance in the QoE of video sessions is analyzed hereafter. For this study, 20 different service conditions were considered, as presented in Table 15.6. In order to analyze the effects of different service parameters, for each considered Content Type different SBR values were included and the expected PQoS was evaluated at three different BLER conditions. The FR and SR values are set up at their default values: the FR is set up to 12 fps for Low Motion

Table 15.6 IPTV Service Conditions

Session ID	CT	SBR	FR	SR	BLER
1	1	200	12	320 × 240	0
2	1	200	12	320 × 240	10
3	1	200	12	320 × 240	25
4	1	130	12	320 × 240	0
5	1	130	12	320 × 240	10
6	1	130	12	320 × 240	25
7	2	200	12	320 × 240	0
8	2	200	12	320 × 240	10
9	2	200	12	320 × 240	25
10	2	130	12	320 × 240	0
11	2	130	12	320 × 240	10
12	2	130	12	320 × 240	25
13	3	256	15	320 × 240	0
14	3	256	15	320 × 240	10
15	3	200	15	320 × 240	0
16	3	200	15	320 × 240	10
17	3	200	15	320 × 240	25
18	3	130	15	320 × 240	0
19	3	130	15	320 × 240	10
20	3	130	15	320 × 240	25

and Medium Motion videos and 15 fps for High Motion videos; meanwhile, the SR is set up to 320 × 240 pixels (Table 15.7).

Apart from this series of IPTV sessions described, no additional background traffic traverses the CN. Thus, in this situation the CN does not experience any degradation.

The sources of impairment for the QoE are the encoding process and the conditions of the AN. From the results illustrated in Figure 15.17, the relationship between the expected MOS values and the experienced BLER for the different values of the content type and the encoding bit rate can be inferred.

Table 15.7 VoD Considered CoS

	CoS_1	CoS_2	CoS_3	CoS_4
Load (kbps)	3482	0	0	0
IPLR (%)	0.0562	0	0	0

As a result, the MOS values associated to these IPTV sessions are presented in Table 15.8.

As can be seen, 9 out of the 20 video sessions experienced bad quality levels. These sessions will generate a PQoS Alarm, which would be served by the MCMS/AEM.

At this stage, four different approaches are introduced:

■ No adaptation

After detecting a PQoS degradation, the IPTV session is not adapted at all.

■ SBR adaptation: MOS = f {CT, SBR/FR, BLER}

After detecting a PQoS degradation, the IPTV sessions adapts the SBR.

For a considered CT, the experienced BLER conditions imply an IPLR value and its associated MOS value. The behavior illustrated in Figure 15.17, indicates that for each combination of CT, SBR, and BLER, an expected MOS value can be inferred. Thus, the IPTV session is set up to the new SBR value that maximizes the expected MOS.

From a general perspective, this is a network-driven service adaptation.

In addition, this adaptation is considered as a standalone decision-making process. The BLER value is not modified by modifying the power control functions of the link layer, so the impact of the adaptation of a session on the performance experience by other users in the same cell is limited.

■ Cross-layer adaptation: MOS = f {CT, SBR/FR, DL_BR}

As a step forward, this alternative introduces the capability of modifying the Radio Bearer in function of the experienced service conditions.

If the service can be provided at one of the highest allowed SBR values, the UMTS DL Bearer is kept to 384 kbps in order to support the service. Otherwise, when the service is supported over the lowest SBR, it may be preferable to assure very low BLER values to guarantee no further impairments than the encoding itself. For this purpose, the maximum supported bit rate for the UMTS DL Bearer is decreased to 128 kbps, which exhibits a better resilience to noise and interference at the same transmission power levels.

Table 15.8   VoD MOS Values and Alarm State for Considered Configurations

Session ID	CT	SBR	FR	SR	BLER	MOS	Alarm
1	1	200	12	320 × 240	0	4.5491	0
2	1	200	12	320 × 240	10	2.6992	1
3	1	200	12	320 × 240	25	1	1
4	1	130	12	320 × 240	0	4.0634	0
5	1	130	12	320 × 240	10	4.0634	0
6	1	130	12	320 × 240	25	1	1
7	2	200	12	320 × 240	0	4.508	0
8	2	200	12	320 × 240	10	2.5796	1
9	2	200	12	320 × 240	25	1	1
10	2	130	12	320 × 240	0	3.7656	0
11	2	130	12	320 × 240	10	3.7656	0
12	2	130	12	320 × 240	25	1.6997	1
13	3	256	15	320 × 240	0	4.5946	0
14	3	256	15	320 × 240	10	1	1
15	3	200	15	320 × 240	0	4.0406	0
16	3	200	15	320 × 240	10	3.6963	0
17	3	200	15	320 × 240	25	1	1
18	3	130	15	320 × 240	0	3.0739	0
19	3	130	15	320 × 240	10	3.0739	0
20	3	130	15	320 × 240	25	1	1

Thus, this approach can be considered a combined service/network service adaptation. As in the previous case, the impact on other users is limited and each user can be adapted by itself.

■ MCMS adaptation: MOS = f {CT, SBR/FR, BLER, DL_BR, SR}

Finally, this approach introduces a new service-level parameter in the adaptation process, namely the Spatial Resolution. For some contents it is preferable to switch to a lower SR when the SBR is under a threshold.

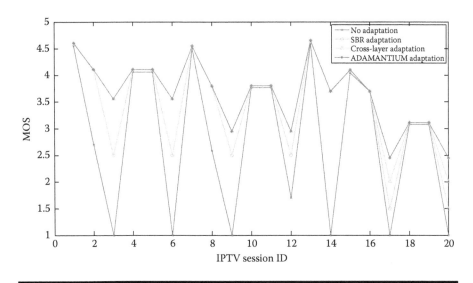

Figure 15.19 Comparative MOS achieved for different IPTV sessions and adaptation approaches.

As a result, the adaptation algorithm considered by the MCMS/AEM for improving the video service performance under BLER degradations considers different service-level parameters (SBR/FR and SR) as well as the AN adaptation (DL_BR).

Figure 15.19 shows the results obtained for each of the 20 IPTV service conditions analyzed in this section.

For a better understanding of the results, Table 15.9 shows the final service conditions after the MCMS/AEM determines the set of adaptation actions to perform in the service provision chain.

15.4.1.1.3 Adaptation Actions Driven by CN Degradations

This section focuses on the impact of the CN performance in the QoE of video sessions.

For this study, 20 different service conditions are considered, as presented in Table 15.10. In order to analyze the effects of different service parameters, for each considered Content Type different SBR values are included and the expected PQoS is evaluated.

Apart from this series of IPTV sessions described, no additional background traffic traverses the CN. Thus, in this situation the CN does not experience any degradation (Table 15.11).

Under no addition AN degradations, the evaluation result is PQoS levels for the different considered Content Types in function of the encoding SBR and

Table 15.9   Final Service Conditions and Achieved MOS

Session ID	CT	SBR	FR	SR	DL_BR	BLER	MOS
1	1	200	12	320 × 240	384	0	4.6034
2	1	130	12	320 × 240	384	10	4.1086
3	1	80	12	320 × 240	128	0	3.555
4	1	130	12	320 × 240	384	0	4.1119
5	1	130	12	320 × 240	384	10	4.1118
6	1	80	12	320 × 240	128	0	3.555
7	2	200	12	320 × 240	384	0	4.5488
8	2	130	12	320 × 240	384	10	3.7973
9	2	80	12	320 × 240	128	0	2.9535
10	2	130	12	320 × 240	384	0	3.7996
11	2	130	12	320 × 240	384	10	3.7996
12	2	80	12	320 × 240	128	0	2.9535
13	3	256	15	320 × 240	384	0	4.6557
14	3	200	15	320 × 240	384	10	3.7
15	3	200	15	320 × 240	384	0	4.0944
16	3	200	15	320 × 240	384	10	3.7
17	3	80	15	176 × 128	128	0	2.4454
18	3	130	15	320 × 240	384	0	3.1148
19	3	130	15	320 × 240	384	10	3.1147
20	3	80	15	176 × 128	128	0	2.4454

the experienced IPLR. As all the sessions are transmitted over the BE CoS, the expected IPLR value is the same for all the sessions. Thus, the High Motion video sessions experience further PQoS degradations than the other Content Types. As can be observed, at this specific system status the last four IPTV sessions detect the degradation of the PQoS below the threshold of 3 and generate a PQoS Alarm.

Table 15.10   VoD Considered Configurations

Session ID	CT	CoS_1	CoS_2	CoS_3	CoS_4	MOS	Alarm
1	1	200	0	0	0	3.0323	0
2	1	200	0	0	0	3.0323	0
3	1	200	0	0	0	3.0323	0
4	1	200	0	0	0	3.0323	0
5	1	200	0	0	0	3.0323	0
6	1	200	0	0	0	3.0323	0
7	1	200	0	0	0	3.0323	0
8	1	200	0	0	0	3.0323	0
9	1	200	0	0	0	3.0323	0
10	1	200	0	0	0	3.0323	0
11	2	200	0	0	0	3.0483	0
12	2	200	0	0	0	3.0483	0
13	2	200	0	0	0	3.0483	0
14	2	200	0	0	0	3.0483	0
15	2	200	0	0	0	3.0483	0
16	2	200	0	0	0	3.0483	0
17	3	200	0	0	0	2.3914	1
18	3	256	0	0	0	2.7192	1
19	3	256	0	0	0	2.7192	1
20	3	256	0	0	0	2.7192	1

Table 15.11   VoD Considered CoS

	CoS_1	CoS_2	CoS_3	CoS_4
Load (kbps)	4168	0	0	0
IPLR (%)	2.1966	0	0	0

From these conditions, the adaptation approaches proposed for study are:

■ Partial set multi-objective: Promotion of CoS

In this approach, the only adaptation action considered is the promotion of CoS within the CN. Upon the reception of the PQoS Alarms related to a set of video sessions, the MCMS/AEM makes the decision whether to promote the traffic associated to each flow in the CN to one of the upper CoS. The best-performing configuration is selected as the one that provides the highest average MOS values for the set of degraded sessions.

■ Partial set multi-objective: Change of SBR

The same selection criterion is implemented in this approach, selecting the best alternative configuration that with highest average MOS among all the degraded session. However, in this approach the only adaptation action considered is the modification of the SBR. Upon the reception of the PQoS Alarms associated to a set of video sessions, the MCMS/AEM determines the best-performing combination of SBR for each degraded session that maximizes the average MOS.

■ Multi objective: Average MOS values

In this third approach the possibility of performing both adaptation actions at the same time is combined, becoming a Cross-Layer Cross-Domain adaptation system. The objective function in this case is the average of the expected MOS values for all the registered IPTV sessions, instead of taking into account only the degraded subset.

■ Multi objective: Average/minimum MOS values

The approach adopted for the CN adaptation needs also combines the two described adaptation actions. However, the objective function in this case is not only based on the average value of the expected MOS levels for all the IPTV sessions, but introduces a combination of the average and the minimum achieved values. In this way, the algorithm prioritizes those alternative configurations where the average PQoS can be lower but none of the video sessions gets ends at a degraded state. Figure 15.20 shows the obtained values for the previously described case study. For each alternative adaptation approach presented, Figure 15.20 shows the MOS values expected for every IPTV session after the adaptation decision is made by the MCMS/AEM. In addition, Figure 15.20 shows the MOS threshold established for considering an acceptable IPTV session, which is set up to 3 for this study.

As can be observed, the initial system status exhibits considerable degradations for all the IPTV sessions, being the last four sessions (corresponding to High

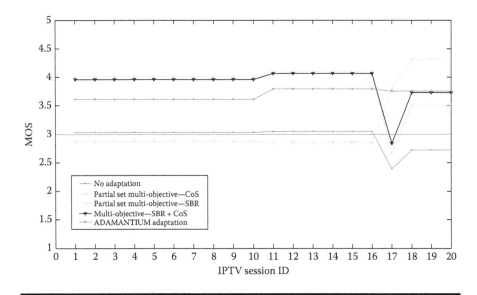

Figure 15.20 Computer MOS score evolution.

Motion video sessions) under the threshold. The second approach maximizes the PQoS of these degraded video sessions at the cost of introducing further degradation in the rest of sessions, which become under the PQoS threshold. The third approach improves the system performance in general. This approach decreases the overall traffic load supported by the CN, and thus the IPLR is decreased as well. However, the combination of SBR/IPLR is not good enough to offer an acceptable quality for one of the degraded sessions.

The multi-objective approach that considers all the registered sessions introduces a limited enhancement over the previous approach, but the final system status includes a degraded session yet. Finally, with the implementation of the objective function based on both the average and the minimum MOS values, the system reaches a point. Table 15.12 shows that, although the average of the MOS values for all the session gets a higher score for the third adaptation approach, the adopted adaptation approach is preferable for the aims of the MCMS system.

15.4.1.1.4 Adaptation Actions Driven by AN/CN Degradations

Finally, a case study where both the AN and the CN have a considerable impact on the PQoS level is presented. The aim is to show the benefits of using the GA creation and cross-over functions developed for the MCMS needs for the optimization of these general problems. As a step forward in the problem statement, the CN conditions are modified introducing additional background traffic in the BE CoS. Table 15.13 gathers the initial MOS values that can be achieved for the different

Table 15.12   Comparison of Achieved MOS Values for Different Approaches

	No Adaptation	Partial Multi-Objective − CoS	Partial Multi-Objective − SBR	Multi Objective	Proposed MCMS
Avg (degraded)	2.63725	4.19045	3.295625	3.505375	3.7559
Min (degraded)	2.3914	3.7997	2.6666	2.8363	3.7559
Avg (all)	2.95809	3.13079	3.894335	3.899005	3.69499
Min (all)	2.3914	2.852	2.6666	2.8363	3.6132

Table 15.13   MOS for Different CN Loads

Session ID	CT	SBR	BLER	MOS			
				Background 0 Mbps	Background 0.5 Mbps	Background 1 Mbps	Background 1.5 Mbps
1	1	200	0	4.5491	3.8188	2.0561	1.1075
2	1	200	10	2.6992	2.5446	1.4235	1
3	1	200	25	1	1	1	1
4	1	130	0	4.0634	3.411	1.8365	1
5	1	130	10	4.0634	3.411	1.8365	1
6	1	130	25	1	1	1	1
7	2	200	0	4.508	3.9592	1.8804	1
8	2	200	10	2.5796	2.4936	1.2327	1
9	2	200	25	1	1	1	1
10	2	130	0	3.7656	3.3071	1.5707	1
11	2	130	10	3.7656	3.3071	1.5707	1
12	2	130	25	1.6997	1.643	1	1
13	3	256	0	4.5946	3.7714	1.5236	1

Table 15.13   (continued) MOS for Different CN Loads

Session ID	CT	SBR	BLER	MOS			
				Back-ground 0 Mbps	*Back-ground 0.5 Mbps*	*Back-ground 1 Mbps*	*Back-ground 1.5 Mbps*
14	3	256	10	1	1	1	1
15	3	200	0	4.0406	3.3167	1.3399	1
16	3	200	10	3.6963	3.2224	1.3118	1
17	3	200	25	1	1	1	1
18	3	130	0	3.0739	2.5231	1.0193	1
19	3	130	10	3.0739	2.5231	1.0193	1
20	3	130	25	1	1	1	1

CN loads if no adaptation actions are performed. We compare the performance of four different approaches:

■ No adaptation

If no adaptation is performed, the additional CN load makes the IPLR of the BE CoS increase until the overall quality is totally degraded for the multimedia sessions. Table 15.14 shows the experienced values of traffic load and the associated IPLR for the different cases.

■ CN-driven adaptation

The first optimization algorithm considered for comparison is based on the logic presented in Figure 15.16. However, in this case the assumption that the ANMM is not able to provide the specific value of the AN condition for the degraded sessions is made. As a result, the first step in the creation function cannot be performed, and the algorithm must generate the individuals without any a priori knowledge of the AN conditions, modifying the SBR and the CoS in a complete random fashion.

■ Cross-domain AN/CN adaptation

In this case, the logic presented in Figure 15.16 is modified concerning the promotion of CoS. The creation function is guided for promoting degraded sessions to the immediately upper CoS, until it is full, and to the following CoS later if needed.

Table 15.14 Actual IPLR Suffered for Different CoS and Loads

	Load				IPLR			
	CoS_1	CoS_2	CoS_3	CoS_4	CoS_1	CoS_2	CoS_3	CoS_4
Back-ground 0 Mbps	3482	0	0	0	0.0562	0	0	0
Back-ground 0.5 Mbps	3982	0	0	0	0.8132	0	0	0
Back-ground 1 Mbps	4482	0	0	0	11.7571	0	0	0
Back-ground 1.5 Mbps	4982	0	0	0	29.45	0	0	0

Additionally, once the SBR is selected in the first step of the creation functions, it cannot be further modified.

■ Adaptation

This final approach is based in the complete logic of the creation function as shown in Figure 15.16. First of all, the optimal SBR is selected based on the AN conditions. After this process, the creation function generates new individuals modifying both the SBR (always with values equal or under the selected SBR) and the CoS (randomly to one of the three lower CoS) until the maximum number of individuals in the generation is reached.

All the three optimization algorithms are run with the following options:

■ Population size = 20 individuals
■ Maximum number of generations = 5

In these conditions, the optimization algorithm exits due to the maximum number of generations and returns the suboptimal solution found. Figure 15.21 gathers the MOS values for all the multimedia sessions, each subplot representing different CN loads.

The MOS statistics used for the computation of the fitness score are presented in Figure 15.22. Figure 15.22a shows the evolution of the average value of all the multimedia sessions with the increasing additional background traffic. Figure 15.22b represents the MOS value associated to the worse-performing video session.

As can be observed, the three optimization approaches considered highly outperform the no-adaptation case as soon as the additional BE load increases.

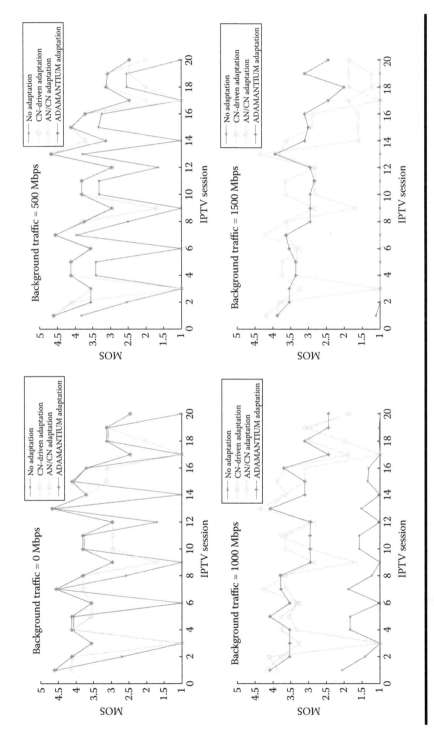

Figure 15.21   Evolution of MOS score for different CN loads.

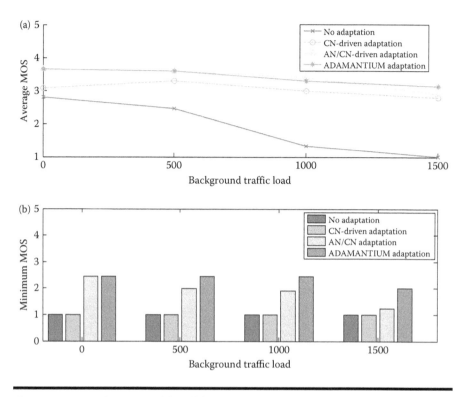

Figure 15.22 **Min MOS achieved for every adaptation scheme.**

From the three adaptation approaches, it is proven that CN-driven method entails a worse performance. Apart from achieving a bit lower average scores, there is always a video session completely degraded. This fact illustrates the relevance of the first step in the creation function, which reduces the complexity of the optimization problem and allows finding a more optimal solution in the limited execution time.

Provided that this first stage can be performed, which requires that the ANMM provides the measurements of the AN condition, the difference between the remaining two optimization approaches is examined.

The average values are quite similar for the last two optimization algorithms. However, the complete approach adopted in the user-centric IPTV system always provide a better performance with regard to the minimum value, which means that the user-centric IPTV system will prevent severe degradations more likely than the other approaches.

For comparison purposes, Table 15.15 shows the traffic splitting resulting from a run of the user-centric IPTV system optimization algorithm for the different CN conditions analyzed.

Table 15.15   Traffic Division among CoS

	Load				IPLR			
	CoS_1	*CoS_2*	*CoS_3*	*CoS_4*	*CoS_1*	*CoS_2*	*CoS_3*	*CoS_4*
Back-ground 0 Mbps	1836	940	0	0	0	0.0038	0	0
Back-ground 0.5 Mbps	2896	130	130	0	0.0007	0.1387	0.0861	0
Back-ground 1 Mbps	2500	540	340	0	0.0036	0.1349	0.0838	0
Back-ground 1.5 Mbps	2690	500	330	0	0.0205	0.2656	0.1650	0

15.5 Conclusions

This chapter has proposed and analyzed the various components that comprise a user-centric multimedia services provision system. The main component of such a system is an IMS-compatible MCMS focused on performing dynamic cross-layer adaptation for the optimization of the user experience in terms of perceptual quality for IPTV service. This multimodal management system applied in an integrated and coherent way along all the network layers and delivery-chain nodes is based on a user-/customer-centric approach rather than a typical engineering one. Toward this, the proposed management system made use of advanced IMS-compatible PQoS and NQoS monitoring and adaptation mechanisms across the network delivery chain, enhancing in this way the current IMS management functions by providing perceptual awareness to them.

Therefore, the switch-over from the legacy IMS to the novel PQoE-aware one, proved that can be performed efficiently and in an affordable way from the business aspect, without further requirements for changes to the infrastructure.

Acronyms

3GPP	3rd Generation Partnership Project
AS	Application Servers
BER	Bit Error Rate

CNR	Carrier-to-Noise ratio
CSCF	Call/session Control Factions
DiffServ	Differentiated Services
EMM	External Monitoring Module
FEC	Forward Error Correction
GGSN	Gateway GPRS Support Node
GNU	Gnu's Not Unix
HSS	Home subscriber Server
IMM	Internal Marking Module
IMS	IP Multimedia Subsystem
IP	Internet Protocol
IPTV	Internet Protocol Television
ITU	International Telecommunication Union
MCMS	Multimedia Content Management System
MOS	Mean Opinion Score
MPEG	Moving Picture Experts Group
MPLS	MultiProtocol Label Switching
MRF	Media Resource Function
MRFC	Media Resource action Controller
MRFP	Media Resource Function Processor
MSRF	Media Server Resource Function
NASS	Network Attachment Sub-system
NGN	Next-Generation Network
NQoS	Network Quality of Service
PDF	Policy Decision Function
PMB	Project Management Board
PMS	Performance Management System
PQoS	Perceived Quality of Service
QoS	Quality of Service
RACS	Resource and Admission Control Subsystem
SBLP	Service-Based Local Policy
SDP	Session Description Protocol
SGSN	Serving GPRS Support Node
SIP	Session Initiation Protocol
SSIM	Structural Similarity Index
TAM	Terminal Adaptation Module
TMF	Telemanagement Forum
UE	User Equipment
UML	User Mode Linux-Based Networks
UMTS	Universal Mobile Telecommunications System
VoD	Video on Demand
VoIP	Voice over IP

Acknowledgment

This work is supported by the European Commission in the context of the ADAMANTIUM project (ICT-2007.1.5-214751).

References

1. H. Koumaras, D. Negru, F. Liberal, J. Arauz, A. Kourtis, ADAMANTIUM Project: Enhancing IMS with a PQoS-aware multimedia content management system, *IEEE-TTTC International Conference on Automation, Quality and Testing, Robotics (AQTR 2008)*, Cluj-Napoca, Romania, May 22–25, 2008.
2. M. Poikselka, G. Mayer, H. Khartabil, A. Niemi, *The IMS: IP Multimedia Concepts and Services in the Mobile Domain*, ISBN: 978-0-470-87114-0, Wiley publications, West Sussex, England, 2004.
3. L. Boula, H. Koumaras, A. Kourtis, An enhanced IMS architecture featuring cross-layer monitoring and adaptation mechanisms, *Fifth International Conference on Autonomic and Autonomous Systems, ICAS 2009*, Valencia, Spain, April 20–25, 2009.
4. ADAMANTIUM Deliverable D2.1 Overall system architecture and specifications.
5. ADAMANTIUM Deliverable D2.3 Definition and specifications of IPTV and VoIP Services.
6. J. Arnaud, D. Negru, M. Sidibe, J. Pauty, H. Koumaras, Adaptive IPTV services based on a novel IP multimedia subsystem, *Multimedia Tools and Applications*, Springer Online First, September 22, 2010.
7. H. Koumaras, A. Kourtis, Perceptually enabled and user centric IMS architecture: The ADAMANTIUM project, *TEMU2008, Int. Conf. on Telecommunications and Multimedia Ierapetra*, Crete, Greece, July 16–18, 2008.
8. H. Koumaras, L. Sun, A. Kourtis, The ADAMANTIUM multimedia content management system for real time cross-layer adaptation of IPTV and VoIP services over IMS, *TEMU2008, Int. Conf. On Telecommunications and Multimedia Ierapetra*, Crete, Greece, July 16–18, 2008.
9. ADAMANTIUM Deliverable D4.1 Voice and Video Quality Perceptual Models.
10. ADAMANTIUM Deliverable D4.4 PQoS models, adaptation and mapping mechanisms.
11. ADAMANTIUM Deliverable D5.3 Trials and evaluation.
12. F. Liberal, A. Ferro, H. Koumaras, A. Kourtis, L. Sun, E. Ifeachor, QoE in multi-service multi-agent networks, *International Journal of Communication Networks and Distributed Systems* 4(2), 183–206, 25 January 2010, Special Issue on "Performance Assessment of New Internet Services."
13. E. Jammeh, I-H. Mkwawa, A. Khan, M. Goudarzi, L. Sun, E. Ifeachor, Quality of Experience (QoE) driven adaptation scheme for voice/video over IP, *Telecommunication Systems*, Special Issue on the "Quality of Experience issues in Multimedia Provision," Springer Online First, June 09, 2010.

Chapter 16

Immersive 3D Media

Erhan Ekmekcioglu, Varuna De Silva, Gokce Nur,
and Ahmet M. Kondoz

Contents

16.1 Introduction to 3D Media.. 348
    16.1.1 Overview of the 3D Media Research 348
    16.1.2 Overview of 3D Media Formats 350
    16.1.3 Overview of 3D Media Coding and Delivery Techniques 352
16.2 3D Content Preparation and Processing for Immersive Media
    Experience...354
    16.2.1 3D Multiview Media Capturing and Postproduction 355
    16.2.2 Content-Aware Depth Processing for Improved Coding and
         Visual Synthesis Performance 358
    16.2.3 Content and Context-Aware 3D Media Adaptation
         for Improved User Experience..361
16.3 3D Media Compression for Transmission 363
    16.3.1 A Scalable and Content-Aware 3D Multiview
         Coding Approach 363
    16.3.2 3D Immersive Multiview Media Distribution Via P2P 366
    16.3.3 Hybrid Broadband–Broadcast Approach for
         3D Multiview Delivery 368
16.4 Conclusion 371
Acknowledgments 372
References 372

16.1 Introduction to 3D Media

In this section, a brief overview of the 3D media research is presented, and then the existing 3D video formats, coding, and delivery schemes are provided.

16.1.1 Overview of the 3D Media Research

The history of stereoscopic 3D goes as far as the invention of the first photography and the equivalent motion picture. Similarly, the concept of stereoscopic cinema appeared in the early 1900s and the first stereoscopic TV was proposed in the 1920s. It is known that a mirror device was used in 1838 to deliver stereoscopic 3D images by Sir Charles Wheatstone. Soon after, it became very popular in the United States and in Europe. 3D movies became quite popular after the 1950s and 3D movie theaters spread over the world. 3D-TV broadcasting attempts date back as early as 1953, although they were only in the form of experiments. The first commercial 3D-TV broadcast took place in the United States in 1980. The overview of the technologies presented in the 3D-TV exhibitions that mostly took place between 1985 and 1996 has been provided in Ref. [1].

Besides the stereoscopic vision, other technologies, such as holographic video and integral images, have been researched on since the past. Experimental holographic videos appeared for the first time in 1989 [2]. Integral imaging, which comprises capturing a scene from many viewing angles and projecting them back optically (via a set of microlens arrays) to the geometric location of the object, is known since 1908 [2]. Both technologies differ from the stereoscopic vision in the sense that their principles are based on duplicating the physical light distribution in the viewing space, in the absence of the original scene objects. Hence, they offer the true full parallax unlike the stereoscopic video from a single viewpoint, and are seen as the ultimate candidates for the leading 3D-TV of the future. This can be valid with the deployment of media delivery-oriented future internetwork architectures, where huge loads of multiview and light field information can be transported with reasonable delays and high quality of service.

Free-viewpoint vision is a special aspect of 3D-TV technologies, which aims at enabling users to navigate through the scene space, that is, to make the user select the desired viewpoint to watch. The theoretical facts behind this concept are well known for long by the computer vision society.

However, it is just recently that the advancements have reached a certain level for the consideration of end user free-viewpoint applications. It is with the efforts by University of Nagoya, Japan, that the concept of FTV becomes popular back in the past decade. It was introduced as a generalized ray-space-based representation, rather than a pixel-based representation, but recent standardization efforts on 3D video are based on pixel-based representation.

Research institutes, universities, and leading technology companies across Europe have shown a considerable amount of interest in advancing the 3D media frameworks, to reach a state, where immersive media experience can be delivered to vastly. Many major European projects have targeted toward the 3D media framework chain. One of them, Advanced Three-dimensional Television System Technologies (ATTEST) [3], was a project that took place between 2002 and 2004 funded by the European Commission. ATTEST aimed at providing a backward compatible 3D video delivery system over Digital Video Broadcast (DVB) with manageable amendments to the existing infrastructure. The standards compatible approach presented in ATTEST lead to the development of interoperable structures for 3D-TV. 3D-TV Network of Excellence (3DTV NoE) [4], funded by European Commission and conducted between 2004 and 2008, was a follow-up of ATTEST project, which considered a very broad range of aspects of the 3D television from capturing to content representation, coding, transmission, and to display. The technological breakthroughs in optical display technologies, which led to the production of successful electronic holographic displays, were considered and exploited in this particular project with the idea of the true full parallax concept for 3D-TV services. Among other strategically targeted and larger-scale 3D technology-related projects (including delivery, content generation, visualization, mobile, and fixed platforms) are 2020 3D Media, 3D4YOU, Mobile 3DTV, 3D Life, 3D Phone, FINE, Helium 3D, SKYMEDIA, MUSCADE, and DIOMEDES. A comprehensive list of other collaborative research projects funded under EU framework programs can be found in Ref. [5]. Also, for a more recent review of ongoing and established research and development works in the field of 3D-TV in Europe, readers are referred to Ref. [6].

So far, it is evident that the 3D-TV concept is well understood by a majority of the public and is unavoidably a step change in home multimedia experience. Differently from previous breakthroughs in the area, such as color-TV and HD-TV, 3D-TV is believed to further exploit the recent networking technologies, which offer much increased throughput rates and enable the transport of very detailed and high spatial resolution media content, resulting in an immersive experience. Also, a great increase in the production rate of 3D media content, in the form of 3D movies, proves the excessive user demand on immersive video experience. In the meantime, leading consumer electronics companies are launching 3D-TV-related products. Companies such as Sony, Philips, Mitsubishi, Samsung, LG, and Panasonic are producing 3D displays suitable to view both 3D- and 2D-HD content. Broadcast companies and Internet service providers across the world also consider supplying 3D-TV services to their customers based on the excessive interest shown by audience, especially in the special occasions when the live sports events were live broadcast. Sky TV, the British broadcaster, launched the UK's first 3D channel in October 2010. British Telecom recently started to offer the Video on Demand (VoD) streaming of featured 3D HD movies to its customers. Similarly, Virgin Media's broadband Internet service provider started to deliver 3D television shows and movies to its customers having compatible set-top boxes. Vast majority of the existing 3D media delivery services are in the form of 2-view (stereoscopic) movies and mostly offline encoded and processed TV productions. Special 3D capturing rigs and 3D media encoding units have been deployed in live sports events' broadcasting. For example, Sony captured twenty-five of the football games during FIFA World Cup 2010 using eight stereoscopic cameras, and live broadcast through satellite and digital cable networks in Europe, the United States, and Korea (In Korea, also over DVB-T). However, a truly immersive 3D media experience also refers to interaction with the content (e.g., scene navigation on the fly), which is made possible with more comprehensive 3D media formats, such as multiview video that is going to be reviewed in the next section.

16.1.2 Overview of 3D Media Formats

A number of research works have aimed at implementing 3D-TV, covering all or part of its aspects, such as video acquisition, postprocessing/formatting, coding, transmission, decoding, and display [7–10]. The evolution in 3D entertainment media comprises the way from stereoscopic video to multiview auto-stereoscopic and high-resolution light field. Figure 16.1 illustrates several 3D video formats that have been established and are foreseen to realize 3D video applications [11]. There is an inherent scalability between the formats presented, that is, a lower-range format can be extracted from a higher-range format.

Stereoscopic 3D video was studied previously in the form of left-eye and right-eye channels separately and in the form of a side-by-side format, that is, the left-eye and the right-eye images are combined in a single image. Such formats suffered from limitations (e.g., bandwidth) associated with existing infrastructures designed for 2D video transmission (e.g., DVB-T, DVB-S) [12,13].

With the introduction of the color-plus-depth 3D video format, customized and flexible stereoscopic video rendering has been achieved with adjustable stereo-baseline, depending on the screen size and the viewing conditions. Furthermore, the cost of coding and transmitting this format is less compared to the cost of conventional stereoscopic video format that consists of the left-eye and the right-eye channels. This is mainly due to the texture characteristics of depth map images, which comprise large and uniform areas that can be more efficiently encoded using

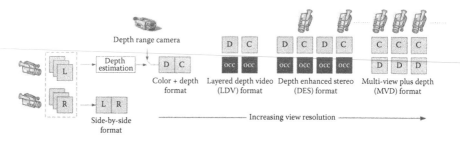

Figure 16.1   3D video formats.

predictive video coding tools. Moreover, the depth images are composed of a single luminance plane without the chrominance components, that is, gray-level images. The introduction of the color-plus-depth format did not incur significant changes on the existing video transmission infrastructure. Hence, in general, a depth-based 3D video format approach is desirable due to its advantages (e.g., transmission bandwidth, adjustable baseline).

Color-plus-depth and layered depth video (LDV) formats are similar to each other, besides the added occlusion layer information that serves to reconstruct the occlusions on the synthesized images with higher accuracy. Accordingly, better-quality views can be synthesized utilizing the additional occlusion layer information. However, the quality of the viewing range is still very restricted with the LDV format. The addition of another LDV viewpoint results in the so-called Depth Enhanced Stereo (DES) format that enhances the view range and gives premium stereo-video quality, since the base video format comprises Left and Right channels separately.

Multi-View plus Depth map (MVD) format consists of multiple (usually more than 2) viewpoints with associated depth map images to serve limitless viewing range with the ability to synthesize any view point in the range. No occlusion layer information is necessary with this format, as any occluded region in any arbitrary synthesized viewpoint can be filled using the texture information of the nearby viewpoints. With sufficiently many viewpoints, the video can also be converted into a light-field representation, which is necessary to drive most light-field displays. Nevertheless, the usage of MVD format brings an indefinite amount of bandwidth requirement depending on the total number of views in the system.

Evidently, spatial audio is essential in 3D media applications for improving the realism of the rendered scene. Spatial audio describes the audio signals that convey information about a 3D sound scene. Multichannel audio, the means to carry spatial audio information, is in many areas today, including digital cinema industry, digital broadcasting over terrestrial networks, streaming over Internet, remote collaboration services, video gaming, and teleconferencing. Several 3D spatial audio reproduction systems exist today, such as binaural [14], stereo, 5.1, 7.1, and similar multichannel audio systems [15], Ambisonics [16], and Wave Field Synthesis (WFS), where a very large number of channels (e.g., over 64) are used [17]. The fidelity of the audio scenes rendered from these source channels/audio objects, and its contribution to the overall experience of immersive media, depends on to the listening position, loudspeaker setup, and the acoustics of the listening environment. It is worth mentioning that the incurred transmission overhead due to multichannel audio coding is much lower compared to its multiview video counterpart. Nevertheless, considering more advanced reproduction systems, such as WFS, which comprises many audio channels, it is inevitable to deploy efficient coding techniques preserving spatial fidelity of rendered audio scenes.

In the next section, an overview of the 3D media coding and delivery schemes is provided.

16.1.3 Overview of 3D Media Coding and Delivery Techniques

There are various coding schemes related to the aforementioned 3D media formats, which are adopted by major international standards organizations, such as ITU and ISO. Most common format used in the market today is the stereoscopic 3D video comprising two video channels (also referred to as conventional stereoscopic video). MPEG standards provide solutions to the coding of stereoscopic video.

Simulcasting, that is, individual encoding of each view as standalone 2D video channels is possible with any video coding standard. MPEG-2 encompasses a multiview profile (MVP) suitable to encode two video channels. The provided MVP leads to temporal interleaving of Left and Right channels' frames. Its successor MPEG-4 standard provides an extension called Multiple Auxiliary Components (MAC) that is suitable to handle depth map videos, which is regarded as a nonvisual information field. Similarly, MPEG-4 Part 2 provides Stereoscopic Multimedia Application Formats (MAF), which covers storage of and basic interaction capability with stereo content, and stereo video applications in mobile devices (like exchanging). Furthermore, AVC [18] (also referred to as H.264/MPEG-4 Part 10/ AVC), the joint standardization effort of ITU-VCEG and ISO-MPEG, provides some profiles and extensions to cope with the coding of stereoscopic as well as multiview content. Indeed, stereo high profile of AVC, which is finalized in 2009, allows the exploitation of inter-view correlations (improved spatial redundancy removal) while coding and has progressive or interlaced stereo output. Figure 16.2 depicts a sample coding scenario, where the arrows represent coding dependency. Besides, AVC also provides a Scalable Enhancement Information (SEI) definition in relation with the stereo high profile, namely Frame Packing SEI that facilitates the signaling of the type of view arranging inside a coded high-definition frame (e.g., temporal interleaving, spatial side-by-side, spatial row/column, spatial top-bottom, quincunx—checkerboard) to the compatible decoders.

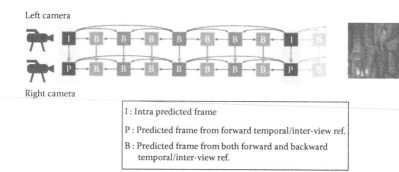

Left camera

Right camera

I : Intra predicted frame

P : Predicted frame from forward temporal/inter-view ref.

B : Predicted frame from both forward and backward temporal/inter-view ref.

Figure 16.2 Sample stereoscopic video coding representation (Group of Pictures size: 8).

For multiview applications, on the other hand, the aforementioned coding solutions cannot work straightaway. Multiview video that has usually two or more than two viewpoints accompanied with depth maps (called MVD in that case) embodies a significantly higher amount of spatially redundant information than conventional stereo. The multiview amendment of AVC standard, referred to as MVC [19], provides a bunch of coding techniques on top of the base AVC standard that helps to reduce the effective coding bit rate by 25–50% depending on the captured scene. Base MVC scheme relies on the utilization of highly efficient *hierarchical B-frame prediction* that is originally utilized in AVC in the temporal domain, on the inter-view domain additionally. Video formats that comprise depth maps (e.g., color-plus-depth and multiview depth) can be encoded using the existing standards and techniques, without the necessity to deploy dedicated encoders for depth maps. Because, depth maps exhibit similar image statistics as natural images despite some differences. Therefore, they can be compressed using block-based encoders. In most cases, it is reported that depth maps can be encoded at around 20–25% of the bit rate necessary to encode the color videos [20]. Even reduced resolution coding techniques are utilized for depth maps, compromising the accuracy in various depth map regions for more bandwidth saving [21,22]. However, depending on the complexity as well as on the rendering requirements of the application, especially in the context of free-viewpoint scenarios, more bit rate (e.g., 50%, or more) may be needed to encode depth maps. Alternative depth map encoders, which do not rely on block-based DCT encoding, have been researched. In Ref. [23], authors presented a novel concept for depth map coding based on platelets, considering the depth map frames as piecewise smooth images.

Field of view, or the viewing range of the rendered scene, is a major indicator of how immersive the 3D application can be as well as how much interaction with the scene can be achieved (e.g., navigation). This can be realized with increased number of viewpoints of that scene. As the best-performing solution, MVC is recognized and adopted commercially by 3D Blu-Ray Association. However, even the state-of-the-art solution MVC cannot maintain the bit rate at manageable limits as long as the number of viewpoints grows. Hence, it is necessary to define a generic framework with a limited number of input view channels and maximum range of viewing space that can be reconstructed. In this case, the auxiliary depth map information bears even more importance, since a lot of the views should be synthesized from the decoded depth maps and camera views. 3D Video Coding (3DV) concept has arisen from that point and a call for proposals for 3D video coding has been released by MPEG organization in April 2011 to evaluate the responses in late 2011 [24].

Hourly broadcasting of 3D television channels in the form of stereoscopic video is still not widespread. A major factor in this is the lack of 3D content acquisition and postproduction expertise and equipment to deploy for 24-h run. But next to it, other restrictions exist, such as spectrum limitation (affecting number of total channels) and minimum quality requirements (e.g., 3D video resolution and quality needed as low as that of 2D-HD service). Nevertheless, experimental 3DTV broadcasting services over terrestrial exist. In a generic framework for 3D-TV broadcasting, the live acquired and registered view channels are fed into an engine that performs parallel (or joint) encoding, transport stream multiplexing along with the audio channels and modulation for play-out. Similarly, the user side comprises a compatible set-top box to demodulate and decode the stream (or streams, depending on the coding and interleaving method) and the 3D display connected to that set-top box via a particular multimedia interface (e.g., HDMI 1.4). Service requirements and more into depth delivery mechanisms for high-quality entertainment 3D media broadcasting to home users over terrestrial, as well as to mobile users over DVB-H are reviewed in Ref. [25]. 3D video streaming over Internet is another area that has been researched on over many years. It is a well-known fact that the Internet platform that facilitates applications such as IPTV and WebTV provides a more flexible medium to transmit as many views (not restricted to stereo) as demanded by the client application and at quality levels allowed by the bandwidth of each user. However, the level of reliability can change, and therefore specific arrangements in coding and streaming (e.g., adaptive streaming) should be made by content servers. Different transport protocols, such as TCP, UDP, DCCP have been reviewed for 3D video streaming, where DCCP has shown to yield optimum reliability levels with lesser sensitivity to network congestions than TCP [26]. Adaptive streaming framework in the context of server–client architectures have been tested, where the network state is dynamically estimated to adjust streaming rate to lead to smooth playback in the client player. Besides, 3D media streaming over P2P overlays has been a hot research topic since recently. The major advantage of P2P-based 3D media streaming solutions over server–client-type streaming is that media distribution is highly scalable. The bandwidth requirement of the server can be effectively reduced by utilizing the network capacity of the multiple clients (or peers). A thorough analysis and review of P2P transport techniques regarding flexible 3D media transmission can be found in Ref. [27]. A more recent research area considers a combination of DVB-T and IP platforms to deliver multiview video, in order to provide free-viewpoint TV service that is much closer to provide a truly immersive 3D media experience [28]. In that scenario, the stereoscopic base video representation is continuously broadcast over DVB-T in a frame-compatible manner, where all auxiliary viewpoints (cameras) are broadcasted over a P2P overlay to the demanding users. The main research challenge comprises the timely synchronization of both delivery networks.

16.2 3D Content Preparation and Processing for Immersive Media Experience

In this section, state-of-the-art in multiview video capturing and postproduction (including disparity/depth map extraction) is reviewed, and then dedicated content-aware processing techniques for depth maps are explained in detail. It is reported that major coding bit rate can be saved as a result of applied processing steps, while the perceptual quality of rendered camera views improve that yields enhanced media experience for users. In addition, we review content and context aware 3D media adaptation techniques for improved user experience.

16.2.1 3D Multiview Media Capturing and Postproduction

The interest in the research community through capturing high-quality multiview content has increased, thanks to the recent advancements in multiview coding and streaming techniques paving the way of realizing immersive 3D media distribution. Nevertheless, building a multiview capture rig (usually with more than two cameras) has got its own challenges that include calibration, timely synching of cameras, rectification (as a postcapturing step), and auxiliary information extraction (e.g., accurate depth maps). Regarding the postproduction step, the real-time performance aspect is another key to consider for live systems. Most existing multiview capture rigs comprise consumer range or professional level (e.g., broadcast quality, or digital cinema quality) cameras of the same type. It is desired that the cameras are identical (e.g., same brand and model) in most cases, so that cameras' focal lengths and field of view (FOV) parameters can be adjusted to be very similar. Depending on the scene and context, variable number of cameras, difference baseline (distance between cameras) and convergence can be applied. Figure 16.3 depicts some basic multiview camera arrangement schemes.

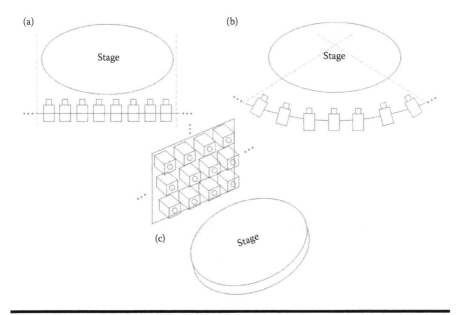

Figure 16.3   Some multiview camera arrangement models. (a) Linear camera array. (b) Camera array on arc. (c) 2D matrix type camera array.

There are also other systems, which additionally involve Time of Flight (ToF) cameras that are able to measure depth directly. In such a hybrid system, the disparity estimation step is assisted by the depth maps captured using these special range cameras, to increase the consistency and accuracy of the registered depth values. A thorough review of such a hybrid multiview acquisition system is provided in Ref. [29]. In the rest of this subsection, we will describe a recently developed multiview acquisition system. The details of another multiview rig consisting of four-cameras and a real-time calibration interface that is developed within the EU-FP7-funded project MUSCADE can be found in Ref. [30].

The first system comprises eight high-end HD professional range cameras (Thomson® Viper®) distributed on a linear steel constructed rig to capture scenes in a wide viewing space. This systems is intended for producing a wide range of formats ranging from HD Ready (720p) to Full HD (1080i/1080p at 50 fps), in order to facilitate high-fidelity capturing of any high-motion scene and hence producing immersive 3D multiview content. This rig is built and tested as part of the EU-FP7-funded project DIOMEDES [31]. Figure 16.4 gives a snapshot of that multiview camera rig.

The offline calibration stage is composed of two sessions: the first step aims at deciding the intrinsic parameters of each camera inside the multiview rig (e.g., precise focal lengths, axial distortions, and principle points), and the second step aims at finding the relative positions and rotations of all cameras with respect to each other (also referred to as extrinsic camera parameters). For individually calibrating the intrinsics of each camera, a series of images of a planar checkerboard pattern in different poses is captured. Via the detection of the features on the checkerboard in every image, initial estimation of the calibration parameters through a

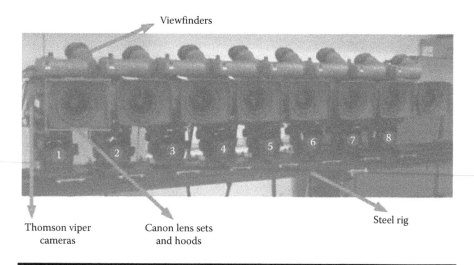

Figure 16.4   The linear multicamera set up.

closed-form solution is done and the refinement of them takes place through a maximum-likelihood estimation cycle [32]. The extrinsics are calibrated using a sequence of wand images (with two balls attached to each end; one green and the other red), where both balls can be easily detected due to color difference with the background. The method described in Ref. [33] is used that runs an automatic initialization procedure and applies a novel bundle adjustment to refine the calibration estimates. Following the calibration of the multiview rigs, multiview scene is shot using high-end multichannel video recorders that are genlocked to each other using an external synchronizer. After the capture, the three-step post processing stage takes place as: multiview rectification (to eliminate all vertical disparities in the scene), color correction (to increase the accuracy of stereo object matching and to overcome synthesis blending process), and dense depth estimation. The dense depth estimation uses color-based image segmentation and a successive block matching method, as described in Ref. [34]. To overcome temporal inconsistencies in the extracted depth maps, the initially extracted depth map videos of each viewpoint are postfiltered using the corresponding color videos. The values of the depth pixels, which were marked as background, are averaged through a set of successive time frames and the computed background depth pixels are replaced with the old background depth pixels within the successive time depth frames. As a result, the postfiltered depth map frames have shown a much improved temporal consistency.

Similarly, to accompany the visual scene reconstructed from eight different HD view channels, object-based (as well as binaural) audio is captured in synchronization with video to facilitate powerful and multiformat spatial audio. Neumann KU 100 "dummy head" makes two-channel, binaural recordings using two microphones placed inside the ear canal of the dummy head. These microphones record sounds that include the specific spectral shaping of the head and the ear, which are known as the Head-Related Transfer Functions (HRTFs). Playing these recorded tracks over headphones provides quite impressive spatial and immersive audio. Other than that, miniature DPA microphones are used for near recording of the individual significant sound sources in the captured scene. When placed close to a sound source, these omnidirectional microphones can provide clean recording of an individual audio object. A MOTU 828 MKII multichannel audio interface is utilized that works as a multichannel sound card and provides phantom power for all microphones connected to it that require such powering. This interface is connected to the recording computer via the FireWire interface. Multichannel audio recording software, such as the AudioDesk, is used for recording multiple tracks. Hence, multiformat audio tracks are automatically captured with the multiview video. For more in-depth description of the used scenario-specific audio capturing, readers are referred to Ref. [35]. Figure 16.5 depicts a representation of the combined multiview video and multichannel audio capturing suit for immersive 3D media production.

The acquired and postprocessed multiview video and extracted dense depth videos are fed in to a multiview encoder and encapsulated according to the transmission

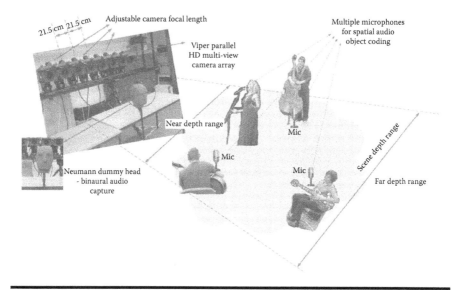

Figure 16.5   Immersive 3D media capturing suit representation.

medium along with the multiview-related metadata (e.g., media info, scene type, visual channels, audio scene format, audio channels, camera parameters, etc.).

16.2.2 Content-Aware Depth Processing for Improved Coding and Visual Synthesis Performance

In order that the delivered multiview content gives premium free-viewpoint rendering performance on the user's application platform, multiview geometry information carried along with the multiview content itself should be accurate enough. In other words, discrepancies existing in the computed and coded depth map frames affect the transformed positions of the pixels in the target view's coordinates. Hence, these result in synthesis errors and viewing discomforts. In order to minimize these, we explain in this subsection a depth map processing framework. A two-stage multiview depth map processing approach is applied at the production stage, which adaptively removes incoherencies existing between multiview depth map streams, as well as spatial and temporal incoherencies within individual depth map streams. In the next stage, the depth discontinuity regions are made consistent with the corresponding color video discontinuity regions, through the utilization of an adaptive trilateral filter that incorporates both the depth edge and corresponding color edge similarity and proximity. This first stage of multiview depth processing framework comprises three substages as view warping, adaptive median filtering and inverse view warping, as described in detail in Ref. [36]. Considering N total number of viewpoints and according depth maps, all depth viewpoints are projected to the image coordinates of the centre viewpoint. This process is depicted in Figure 16.6.

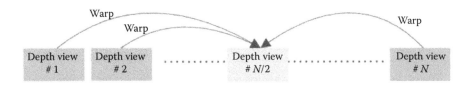

Figure 16.6 *N* **views are projected onto the image coordinates of the center view.**

During this transformation, depth map pixels of all viewpoints are first projected into 3D space coordinates. During this process, intrinsic camera parameters, rotational and translational parameters of cameras are used, and the 8-bit depth luminance values scaled within 0–255 range are converted into physical depth values. This process is reversed in the second step, and then the pixels are back-projected to the image coordinates of the target camera (center view) using the camera parameters of the centre view. The forward warping process is followed by the main processing cycle, that is, the adaptive median filtering, incorporating all the warped depth map frames. The mentioned median filter is applied on a four-dimensional window, S. The shape of the window S is adaptive to edges and also the motion in the multiview sequences. Hence, the adaptation factors are defined as: (1) the local variance of the depth values, and (2) the local mean of the absolute difference between the luminance components of the two consecutive color video frames. They are denoted as $V(x,y,t)$ and $m(x,y,t)$, respectively. Both parameters are computed in a 5×5 spatial neighborhood $N_{t,n_0}(x, y)$ taken at the time instant t and viewpoint n_0 and are centred around the spatial location (x, y). Accordingly, the first parameter is computed as

$$V(x,y,t) = \mathrm{var}\left(D'_{n_0}(N_{t,n_o}(x,y))\right) \tag{16.1}$$

And the second parameter is computed as

$$m(x,y,t) = \mathrm{mean}\left(c(N_{t,n_o}(x,y)) - c(N_{t-1,n_o}(x,y))\right) \tag{16.2}$$

The two parameters are compared to two different threshold values Thr_v and Thr_m, respectively. For instance, if the variance is lower than the threshold (i.e., not an edge), the window includes all the 5×5 neighborhoods across all warped depth representations. Otherwise, the window is restricted only to the center pixel across all warped depth representations. Moreover, if no motion is detected, that is, the computed m is lower than Thr_m, then the temporal coherence is enforced by including the locations at the previous time instant $(t-1)$ in the window. Following the construction of the adapted multidimensional window, the resulting depth values are obtained by median filtering on the same window as

$$D'_{n_o}(x,y,t,n) = \mathrm{median}\left[D'_{n_o}(N(x,y,t))\right] \tag{16.3}$$

for all cameras. It should be noted that the inter-view coherence is achieved in this case by using the same resulting real world depth value for all views $n = 1, \ldots, N$.

Following this computation, the filtered real-world depth values are mapped back to the luminance values ranging from 0 to 255, and the inverse view warping takes place to map back the filtered depth luminance values to their original viewpoint's image coordinates.

The occluded depth pixels in the image coordinates of the center viewpoint are processed in subsequent stages, by setting the target viewpoint as the cameras numbered $\lceil N/2 \rceil \pm 1, \lceil N/2 \rceil \pm 2 \dots$ instead of $\lceil N/2 \rceil$.

After the first processing stage, in order to make the resultant depth maps easier to compress, the median filtered depth values are smoothed using a joint-trilateral filter, which incorporates the closeness of depth values, as well as the similarity of both the color and depth map edges. In each depth viewpoint, for each pixel to be processed, a window of size $2w \times 2w$ is formed, which is centered at that particular depth pixel. Subsequently, in these kernels of $2w \times 2w$ (denoted by Ω), the filtered depth value is computed as

$$D_p = \sum_{q \in \Omega} \mathrm{coeff}_{pq} \cdot I_q \Big/ \sum_{q \in \Omega} \mathrm{coeff}_{pq} \qquad (16.4)$$

where q denotes a pixel within the kernel and p is the center pixel to be processed. The coeff is a multiplication of three different factors, namely the closeness in pixel, similarity in depth value, and the similarity in color texture value:

$$\mathrm{coeff}_{pq} = c(p,q) \cdot s_{\mathrm{depth}}(p,\dot{q}) \cdot S_{\mathrm{color}}(p,q) \qquad (16.5)$$

The filters related to these three factors are considered as Gaussian filters centred at point p. Accordingly, these individual factors are denoted as

$$c(p,q) = \exp\left(-\frac{1}{2}(p-q)^2 / \sigma_c^2\right)$$

$$S_{\mathrm{depth}}(p,q) = \exp\left(-\frac{1}{2}(d_p - d_q)^2 / \sigma_{S_{\mathrm{depth}}}^2\right)$$

$$S_{\mathrm{color}}(p,q) = \exp\left(-\frac{1}{2}(I_p - I_q)^2 / \sigma_{S_{\mathrm{color}}}^2\right) \qquad (16.6)$$

where d represents the depth values of pixel points p and q, and I represents the corresponding color texture luminance values. The standard deviations are calculated for each kernel, such that the group of pixels within the kernel (of size $2w \times 2w$) fall into the 95% confidence interval in the Gaussian distribution [37]. Hence, all the standard deviation parameters within these three factors are equal and calculated as

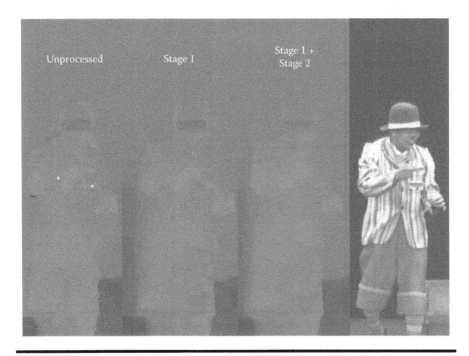

Figure 16.7 A sample initial estimated depth passing through the processing states.

$$\sigma_c = \sigma_{S_{\text{depth}}} = \sigma_{S_{\text{color}}} = w/2 \qquad (16.7)$$

Figure 16.7 depicts a visual example from a depth frame that passes through the first and the second processing stages as well as the corresponding color video frame section.

It is clearly seen that a postprocessing cycle applied on the initial computed depth map can lead to a dramatic increase in the accuracy, especially removing the discrepancies and maintaining precisely edge locations.

16.2.3 Content and Context-Aware 3D Media Adaptation for Improved User Experience

3D media adaptation technology requires in-depth investigations for enabling immersive 3D media delivery systems providing the needs of the future Internet of media. The overall target of the 3D media adaptation is to maximize users' experience in terms of 3D perceptual quality, which is affected by several factors, including content characteristics, contexts prevailing in various usage environments, and so on. The concept of 3D media adaptation is illustrated in Figure 16.8. For example, the effects of varying ambient illumination context in the environment surrounding

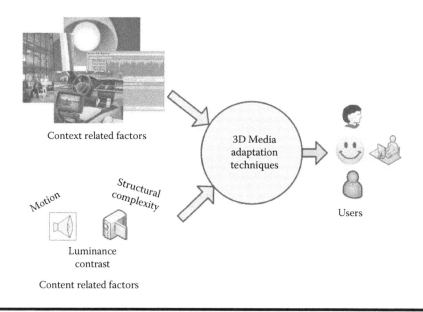

Context related factors

Structural complexity

Motion

Luminance contrast

Content related factors

3D Media adaptation techniques

Users

Figure 16.8   The concept of 3D media adaptation.

the user on 3D video quality of experience had been investigated in Ref. [38] with a group of subjective tests. Accordingly, it was reported that when the ambient illumination increases, the subjects' ratings for the perceived 3D video quality also increase. Conversely, it was found that when the ambient illumination increases, the subjects' ratings for the perceived depth decrease.

Another set of tests had been done to assess the effect of the spatial resolution of the coded depth map data on the overall image and depth perception of the 3D media [39]. The color-plus-depth format as described in Section 16.1 was used in this assessment. In order to purely observe the effect of downscaling and compressing the depth on the overall quality and depth perception, color video was not compressed or scaled. Results had showed that the higher the spatial resolution was, the higher the perceived video quality and depth perception were. When the structural complexity of the depth map increases, the perceived quality saturates and the depth perception is enhanced at higher qualities. Similarly, when the motion complexity of the depth map increases, the effects of higher spatial resolutions become more significant on the video quality and depth perception of 3D video. Based on these observations reported in refs. [38,39], a user perception model was designed to characterize the perception of users toward 3D media under a particular ambient illumination condition. Motion, structural complexity, contrast, and disparity characteristics of the 3D media are used as core content-related contextual information in this model. Using that model, an adaptation technique was developed for determining the bit rates for 3D media elements (e.g., color view and depth map streams) under different ambient illumination conditions [40]. It was observed that a dynamic bandwidth adaptation approach that depends on contextual and content-related cues helps the delivery system maintain consistent quality of 3D viewing experience at users' side.

16.3 3D Media Compression for Transmission

In this section, a brief summary of a recent, scalable 3D multiview media coding scheme is provided. Different from conventional 2D and stereoscopic media coding techniques that are agnostic to context (e.g., transmission media, application, and user requirements) and the coded media itself, the presented media coding and transmission are content-aware and allow adaptation under dynamic network conditions. Such behavior is rather important to realize future interactive and immersive (beyond simple stereoscopic) 3D media applications, as a much increased visual and aural information will be loaded over the networks. Scalability is an important aspect for wide dissemination of 3D content, for user-interactive region of interest streaming, as well as for avoiding major Quality of Experience (QoE) degradations under bandwidth throttles. Furthermore, this subsection provides a 3D multimedia transport scheme over P2P communication link to utilize the capacity of network users under continually increasing number of customers. Evidently, server-to-client type of streaming services will suffer under huge multimedia load and huge number of users, unless large-extend content distribution networks are built. Additionally, a hybrid 3D multiview entertainment transport scheme over terrestrial DVB and IP (as mentioned in Section 16.1.3) will be explained in detail.

16.3.1 A Scalable and Content-Aware 3D Multiview Coding Approach

MVC scheme is the state-of-the-art multiview coding for storage (e.g., as adopted by Blu-ray Disc Association) and multiview video streaming. Nevertheless, it is also well known that the compression rate of MVC is variable according to the multiview camera setup and furthermore, the relative MVC bit rate is indefinitely proportional to the number of viewpoints. Hence, MVC does not offer realistic savings under a large set of input viewpoints. On the other hand, due to its complicated inter-view coding dependencies, it becomes difficult to stream particular viewpoints of interest to users selectively. Additionally, since quality scalability is not a part of the MVC standard, bandwidth-adapted asymmetric streaming of stereo view-pairs is not possible unless the viewpoints are originally encoded at different quality levels. However, this option may not always result in best multiview rendering especially for customers having an MVD. Viewpoint scalability has more importance when compared with the overall compression efficiency of the multiview video content. A Scalable Video Coding (SVC) standard-based approach can provide a wide range quality, as well as maximum viewpoint scalability for streaming over IP networks. As being another important domain of scalability, quality scalability can serve to have consistent (smooth) quality variations under network congestions. It should be targeted to maintain the 3D Quality of Experience of users in every condition as far as allowed, without giving rise to noticeable viewing comfort changes during the service. In order to further increase the effectiveness of the quality scalability in the context of 3D media streaming, human visual system's adaptation capabilities can be exploited by the media encoder. Visual attention model is capable of predicting the regions of interest in a 3D audiovisual scene that can hint the media encoder how to allocate coding bit rate accordingly in between different image regions as well as between quality layers. Visual attention modelling has applications in scalable audio/video coding, where the components that are not salient or the user does not give attention to, can be coded with a lower resolution or removed completely without decreasing the perceived quality. Visually salient parts within the 3D scene can be identified and so are their according coordinates in different viewpoints, in order to prioritize them in terms of asymmetric rate allocation. In this way it can be assured that a base quality layer 3D media stream that is guaranteed to be delivered to every customer can encompass visually most significant features in the scene, and every additional quality enhancement layer coming on top would smoothly enhance the viewing experience. In other words, perceptual quality differences between different successive bandwidth adaptation levels can effectively be minimized in the favor of a higher overall quality 3D media experience.

A generic group of processing stages to extract visually salient regions involves feature extraction, feature map processing to obtain objects, object rating and classification based on different perceptual cues, and object clustering to identify different image regions' importance level. The visual system perceives the shape, size, color, brightness, location, and depth of an object. Gestalt's laws of proximity, similarity, closure, symmetry, common fate, and continuity apply to the perceptual organization of audiovisual stimuli [41]. Gestalt's theory provides effective criteria in clustering audiovisual objects. Various visual attention-based video coding schemes exist in the literature that prioritize different video regions and encode accordingly [42,43].

Previously, the effects of varying the Quantization Parameter (QP) over visually salient regions and less salient regions were observed through a set of subjective tests, where subjects were asked to assess the perceived 3D video quality of shown stereoscopic coded test videos. Figure 16.9 depicts the subjective test results for a couple of test stereoscopic video sequences (*Champagne Tower* and *Ballet*) for varying salient and nonsalient (leftover) regions QP's. It should be noted that within the MPEG video encoders' context, increasing QP represents increasing transform coefficient quantization step size, and hence decreasing coding accuracy.

It is clearly seen on Figure 16.9 that visual attention area quality has much more effect on overall perceived quality than nonvisually salient regions. Hence, a visual attention model adaptive quality enhancement layer formation scheme is applied to the SVC-based 3D video encoder. This scheme comprises adaptive QP assignment to different macroblocks within a frame to be encoded, as depicted in Figure 16.10.

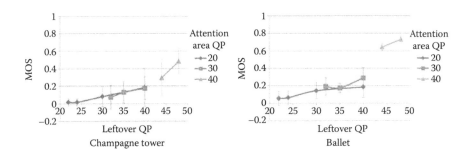

Figure 16.9 Subjective Mean Opinion Scores (MOS)-based on varying QP levels.

As seen in the base layer, salient regions (shown by micro blocks) are coded with a finer quantization parameter (QP1) than the nonsalient region that is coded with QP2. Similarly, in the enhancement layer, no extra information is encoded for the salient region macroblocks (ideally by setting the QP to the maximum value) and since interlayer prediction is performed, almost negligible bit rate is consumed for the salient region coding in the enhancement layer. On the other hand, nonsalient region is encoded with a higher accuracy in the enhancement layer, than in the base layer. This way, most visually important objects can be seen with higher accuracy, even if the enhancement layer packets have to be truncated in times of network congestion.

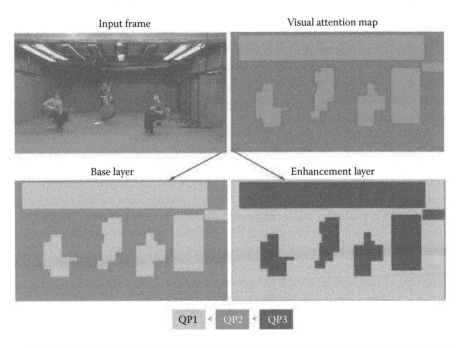

Figure 16.10 Visual attention-based QP assignment process.

The goal for the visual attention adapted encoder model is to minimize the visual perception difference between both layers. Given the saliency distribution inside a frame, QP1 and QP2 are set such that the 1:1 bit-rate ratio between both layers is maintained, provided that the difference between QP1 and QP2 is smaller than the maximum allowed difference. If the ratio is not achievable with the constrained QP settings, then the salient-classified macroblocks are gradually merged toward the most salient classified regions in the frame, according to the provided visual attention map. According to a set of comparative subjective test results (following ITU-R BT 500.11 Adjectival Categorical Judgement method), where seven expert subjects are asked to assess the overall viewing quality of a coded stereoscopic sequence of 16 s, the utilization of visual attention-based asymmetric quantization parameter assignment has shown to yield better results that using conventional SVC encoder with the same amount of quality layers under the same bit-rate conditions. Test results have shown that especially for base-layer-only streaming, the base layer generated using the visual attention model-based encoder provides significantly improved perceptual quality at the same bit rate. More details on these subjective test results can be found in Ref. [44].

16.3.2 3D Immersive Multiview Media Distribution Via P2P

In this subsection, a bandwidth-adaptive multiview video transmission scheme over P2P network, which is based on the distribution of the quality scalable content described in the previous subsection, is explained. The P2P type of network mentioned here comprises a mesh-based system. In mesh-based P2P networks, data are distributed over an unstructured network, in which each peer connects to multiple peers. Such a networking between peers alleviates the problem caused by ungracefully exiting peers from the network (by building up multiple connections among peers), despite its disadvantage of causing longer initialization intervals for peers joining the network. As an example to them, Bit-Torrent is a well-known and popular transport scheme that adopts the mash-based P2P structure. With its native rarest-first policy in forwarding data chunks, it aims at increasing the availability of each piece of information. Also, it is used to distribute equally sized pieces of data. It is not specifically designed for multimedia (e.g., video services) and furthermore yields suboptimum performance unless necessary modifications on its protocol.

As the main goal is to make the desired video information chunks available to more peers as fast as possible, the desired P2P multiview distribution scheme converts the rarest-first policy in a way to allow timely video downloading by peers, which is achieved by a windowing mechanism that restricts the random piece selection process within a particular time interval. Furthermore, since equal size data chunk partitioning is not realistic for coded multiview media stream (e.g., because chopping the media stream at fixed size bounds does not always generate independently decodable chunks due to video data—header data separation), this P2P distribution system employs Group-Of-Pictures (GOP) size media chunk generation, where such a media chunk is highly likely to be independently decoded. The reason is that each GOP starts with an Intra-coded frame and the rest of the frames within that GOP rely on decoding of that particular Intra-coded frame only, unless Bi-predictive coded frames are used. Since the media encoding is based on SVC standard that comprises base and enhancement quality layers for each access unit (encoded frame), and that base-layer Network Abstraction Layer (NAL) units do not depend on decoding the enhancement-layer NAL units, base- and enhancement-layer NAL units comprising a particular GOP unit are split into two chunks. Figure 16.11 depicts this process, where the content is encoded into two quality layers. It should be noted that enhancement layer chunks can be discarded at any point, since they do not affect the decoding process of the base quality layer chunks.

Based on the chunk generation and the deployed windowing mechanism, once a streaming session starts, peers tend to request for the highest available quality versions of the viewpoints requested and try to download all chunks within a given time window. While the downloading process continues, player consumes the downloaded content at the same time. Depending on the buffer state of the peer's player, it may be decided to discard the enhancement layer chunks (i.e. if the buffer gets overloaded) dynamically. The downloading engine keeps track of total downloading rate. If the rate is barely adequate for base layer, the enhancement-layer chunks may not be requested in the first place. Furthermore, depending on the importance of the auxiliary streams (e.g., neighbor viewpoints) with respect to the user's desired viewpoint reconstruction, downloading sessions can be initiated for them beginning from their base-layer chunks. Hence, the P2P client's operation is actively guided by an adaptation decision-taking process that tracks the buffer state of the player and ranks the importance of other streams for the rendering of a particular user-requested viewpoint. According to a set of pilot P2P over IP streaming tests with 2, 3, and 4 peers, respectively, and with different bottleneck connection rates (1, 2, and 3 Mbps), important performance characteristics of the mentioned P2P protocol are extracted. Bottleneck connection rate refers to the rate at which the

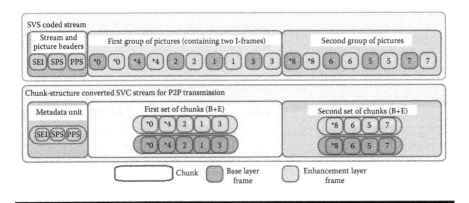

Figure 16.11 Chunk generation process.

main content server (or main seed server) serves the multiview media. Furthermore, streaming tests are carried out using different streaming window sizes, as described previously (more specifically, with window size equal to 1, 2, 4, and 8 GOP units/chunks). Hence, each peer can download parts of the media chunks either through the main connection, or via other peers (i.e., via P2P connection). Results indicate that for a very low bottleneck connection, such as 1 Mbps, the total downloading rate of peers fall down as long as the number of peers grows. However, since the percentage of the download rate via P2P connection increases, thanks to the utilization of P2P video sharing, the fall rate is below than expected. Furthermore, as the window size grows to 8, an increased percentage of the total base-layer and enhancement-layer chunks could be downloaded by peers, which results in an increased quality of viewing. Also, the percentage of P2P downloaded chunks over the total downloaded chunks grows, as the window size increases. Hence, the effectiveness of the used P2P protocol increases. For more detailed analysis of these streaming test results, readers are referred to Ref. [45].

In the next subsection, a hybrid multiview delivery approach that uses the conventional terrestrial broadcast link (DVB-T) and the described P2P delivery system together is explained in detail with the concept and system building blocks.

16.3.3 Hybrid Broadband–Broadcast Approach for 3D Multiview Delivery

In order to realize widely deployable multiview content delivery systems, a hybrid transmission scheme is considered that consists of delivery through both the DVB-T network and broadband IP with P2P communication protocol. This scheme is developed within the EU-FP7 project DIOMEDES [31], which was mentioned in Sections 16.1 and 16.2. Terrestrial DVB is a well-established multimedia broadcasting network with its underlying encapsulation, modulation and channel coding principles, and is used almost everywhere. It is reliable after all, and furthermore, the second generation of terrestrial DVB, that is, DVB-T2, offers even more transmission bit rate to carry HDTV channels and more importantly 3D-TV channels (in the form of Stereoscopic Full HD format). Broadband Internet connection offers a secondary medium to reach to broadcasted multiview material. However, since it is less reliable (nonguaranteed quality of service and usually best-effort characteristics) and highly constrained with respect to the total number of subscribers, as well as the content server's serving capacity, it currently cannot be relied on as the primary means of the media broadcast service. Nevertheless, the usage of the P2P communication protocol specialized for scalable coded multimedia delivery (as described in the previous subsection) can overcome the server side upload bottleneck up to a certain extent. Furthermore, since the coded media is scalable and content-aware (e.g., visual attention model adaptive), and the P2P transport policy is adjusted in a coded media-aware fashion (e.g., piece picking policy is adjusted to prioritize downloading of base quality layer chunks), bad fluctuations in the quality of experience can be efficiently reduced.

The 3D multiview content server is adjusted to feed part of the stream, such as a stereoscopic video pair able to be watched directly and stereo-audio, to the DVB-T2 transmission unit, and the rest of the scalable coded viewpoints and the additional audio channels for other more advanced reproduction systems to P2P-IP. The main challenge that is being actively tackled in that system is the synchronization of the delivered audio visual media over both the networks to the same user terminal (e.g., TV set). Since both broadcasting media are independent from each other, the synchronization is actively sought by the user terminal. More specifically, it is targeted to timely synchronize the viewpoints arriving through DVB-T2 and P2P connections, as well as the spatial audio with the video. For that purpose, the P2P client working on the user's terminal manages the downloading sessions of viewpoints in accordance with the time information of the arriving video signal over DVB-T2 link. In addition to that, the user's media player's buffer dynamically synchronizes the fed raw audio and video transport streams to help playback the rendered 3D video and audio in synchronization. Hence, both the synchronization between the audio and the video streams, and the synchronization between the playback times of different viewpoints' frames are handled. Figure 16.12 depicts the envisaged end-to-end hybrid multiview media transmission and consumption system. It should be noted that to increase the error resiliency of the streamed multiview video streams, an alternative approach such as multiple-description coding can be utilized as shown in Figure 16.12.

3D content server is responsible for encapsulating the coded scalable media into separate MPEG-2 transport streams and multiplexing them. Also, it creates the layer chunk units as requested by the P2P master seed (main server). However, the chunks are passed through an access control stage, where they are encrypted. In this way, only the authorized subscribers can decode and watch the 3D content. In the meantime, the copyright protection issue is also effectively addressed with such a stage. For associating the content distributed via P2P with the content distributed over DVB-T2, application signaling mechanisms are used. Depending on the application context, part of the total multiview scene can be fed to the DVB-T2 transmission module, whereas in all circumstances, all the multiview data are distributed over Internet, assuming some clients may not have DVB-T2 access. The main aim of the main seed server is to feed adequate number of peers and ensure that the immersive 3D multiview media is delivered to a maximum number of subscribers. In the user's (peer's) application terminal, a controller unit supervises the interoperation of all the remaining operational modules in terms of message and coded media flow. An adaptation decision engine module dynamically monitors the key performance index values arriving through the coded media to estimate the user's QoE, and also tracks the user's interaction with the watched media (e.g., user's selected viewing angle), to guide the P2P client module in downloading the required chunks. The synchronization

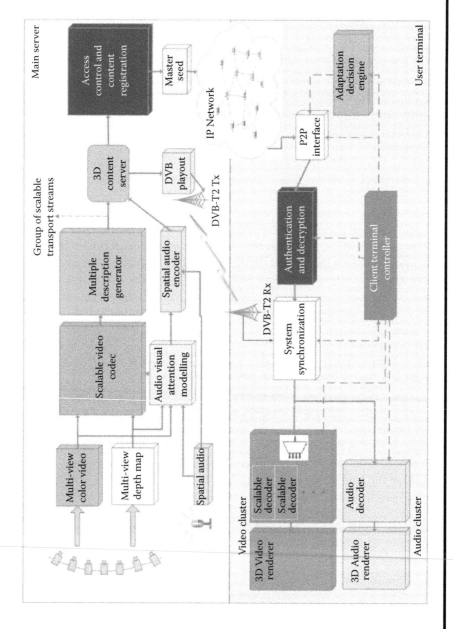

Figure 16.12 Hybrid 3D multiview delivery system overview.

module is able to receive coded media coming from both transmission media over UDP, while buffering each media stream to synchronize before sending to the video rendering and audio reproduction engines. This module regenerates the correct synchronized clock values (referred to as PCR in MPEG-2 TS terminology) for all coded streams.

Such a delivery mechanism paves the way to wide range consumption of immersive 3D multiview and multichannel entertainment content by users. Since, as the number of high-quality 3D views provided to homes increases, terrestrial broadcast system will struggle to meet the demanding requirements, P2P-IP delivery is utilized on top in a content aware fashion. Readers are referred to Ref. [28] to gain more insight on the overall system features.

16.4 Conclusion

This chapter was devoted to provide the recent advancements in 3D immersive media delivery mechanisms, providing solutions based on content aware processing, coding, and adaptation techniques. A brief introduction was provided that gives an insight into the 3D media applications, commercial deployments, and ongoing/completed research works. Commonly used 3D media formats and the standard compression approaches were presented. A recent research work on content-aware scalable media coding and P2P-based 3D scalable multiview video and spatial audio delivery was outlined. Other subjects, such as hybrid broadband-/broadcast-based solutions alongside with their impacts were also discussed.

It is a foreseeable fact that, rich 3D multimedia content in the form of multiview video with depth, as well as multichannel audio, and the use of P2P overlay systems accompanied with a variety of other transmission networks are converting the way of users' interaction with the entertainment media toward a more immersive and realistic experience. In this context, the rich interaction with content (e.g., scene navigation, free-viewpoint viewing) combined with automatic or user-interactive multimedia search across P2P network contribute toward such a vision. Similarly, dynamic adaptation to the state of the underlying transmission medium and the characteristics of diverse terminals are important and should take place in a content-aware fashion. On the other hand, collaborative entertainment media/gaming/educative applications (e.g., second life, interactive multiplayer games, virtual lectures) incur new types of demands and constraints on the utilized network architectures, such as spatial/temporal synchronization of multiple users, audiovisual communication interfaces, and so on. Together with the consideration of collaborative applications' needs and immersive 3D media creation and delivery requirements, the proceeded future Internet of media (or future Internet for media) will radically convert the entertainment media industry.

Acknowledgments

The authors of this chapter would like to thank the consortium members of DIOMEDES project, which is funded under the European Commission ICT 7th Framework Programme, for their contributions in achieving the research outcomes presented in different sections.

References

1. M. Starks, 3D for the 21st Century, The Tsukuba Expo and Beyond, 3DTV Corporation, 1996.
2. Technical Committee 4, Signal Processing Issues in Diffraction and Holography: A Survey, 3DTV NoE, 2005.
3. C. Fehn, P. Kauff, M. Op de Beeck, F. Ernst, W. Ijsselsteijn, M. Pollefeys, L. Vangool, E. Ofek, and I. Sexton, An evolutionary and optimised approach on 3D-TV, *Proc. of IBC 2002*, Amsterdam, Netherlands, September 2002.
4. 3DTV Network of Excellence, Integrated Three-Dimensional Television—Capture, Transmission and Display, http://www.3dtv-research.org/index.php.
5. European Community Research and Development Information Service, http://cordis.europa.eu/
6. O. Grau, T. Borel, P. Kauff, A. Smolic, and R. Tanger, 3D-TV R&D activities in Europe, *IEEE Transactions on Broadcasting*, 57(2), 408–420, 2011.
7. W. Matusik and H. Pfister, 3D TV: A scalable system for real-time acquisition, transmission, and autostereoscopic display of dynamic scenes, *ACM Transactions on Graphics*, 24(3), 814–824, 2004.
8. E. Kurutepe, A. Aksay, C. Bilen, C.G. Gurler, T. Sikora, G. Bozdagi Akar, and A.M. Tekalp, A standards-based, flexible, end-to-end multi-view video streaming architecture, *PV'07*, Lausanne, Switzerland, November 2007.
9. X. Cao, Y. Liu, and Q. Dai, A flexible client-driven 3DTV system for real-time acquisition, transmission, and display of dynamic scenes, *EURASIP J. Appl. Signal Processing*, 1–15, January 2009.
10. C. Fehn, Depth-image-based rendering (DIBR), compression and transmission for a new approach on 3D-TV, *Proc. SPIE Conf. Stereoscopic Displays and Virtual Reality Systems XI*, vol. 5291, CA, USA, pp. 93–104, January 2004.
11. A. Smolic, K. Mueller, P. Merkle, P. Kauff, and T. Wiegand, An overview of available and emerging 3D video formats and depth enhanced stereo as efficient generic solution, *Proceedings of the 27th Conference on Picture Coding Symposium*, Chicago, IL, USA, May 06–08, 2009.
12. I. Yuyama, and M. Okui, Stereoscopic HDTV, in *Three-Dimensional Television, Video, and Display Technologies*, B. Javidi and F. Okano, Eds. New York: Springer-Verlag, pp. 3–34, 2002.
13. N.H. Hur, C.H. Ahn, and C.T. Ahn, Experimental service of 3DTV broadcasting relay in Korea, three-dimensional TV, video and display, *Proc. SPIE* 4864, pp. 1–13, 2002.
14. D.R. Begault, *3-D Sound for Virtual Reality and Multimedia*. San Diego, CA: Academic, 1994.
15. F. Rumsey, *Spatial Audio*, 2nd ed. Oxford, UK: Focal Press, 2001.

16. M. Gerzon, Periphony: With-height sound reproduction, *J. Audio Eng. Soc.*, 21(1), 2–10, 1973.
17. A.J. Berkhout, D. de Vries, and P. Vogel, Acoustic control by wave field synthesis, *J. Acoust. Soc. Am.*, 93(5), 2764–2778, 1993.
18. P. Merkle, A. Smolic, K. Mueller, and T. Wiegand, Efficient prediction structures for multi-view video coding, *IEEE Trans. Circuits Syst. Video Technol.*, 17(11), November 2007.
19. ISO/IEC JTC1/SC29/WG11, 'Text of ISO/IEC 14496-10:200X/ FDAM 1 multiview video coding,' Doc. N9978, Hannover, Germany, July 2008.
20. M. Maitre and M.N. Do, Shape-adaptive wavelet encoding of depth maps, in *Proc. Picture Coding Symp.*, Chicago, IL, May 2009.
21. H. Abdul Karim, S. Worrall, and A.M. Kondoz, Reduced resolution depth compression for scalable 3D video coding, *Proc. of Visual Information Engineering, Workshop on Scalable Coded Media Beyond Compression*, Xian, China, July 2008.
22. E. Ekmekcioglu, S.T. Worrall, and A.M. Kondoz, Bit-rate adaptive downsampling for the coding of multi-view video with depth information, *Proc. of 3DTV Conference: The True Vision—Capture, Transmission and Display of 3D Video*, Istanbul, Turkey, 2008.
23. Y. Morvan, D. Farin, and P.H.N. de With, Depth-image compression based on an R-D optimized quadtree decomposition for the transmission of multiview images, *IEEE International Conference on Image Processing*, San Antonio, TX, USA, September 2007.
24. MPEG's call for proposals on 3D video coding technology, found in http://mpeg. chiariglione.org/working_documents/explorations/3dav/3dv-cfp.zip
25. N. Hur, H. Lee, G.S. Lee, S.J. Lee, A. Gotchev, and S. Park, 3DTV broadcasting and distribution systems, *IEEE Transactions on Broadcasting*, 57(2), 2011.
26. B. Gorkemli and A.M. Tekalp, Adaptation strategies for streaming SVC video, *Proc. IEEE Int. Conf. Image Process.*, Hong Kong, pp. 2913–2916, September 2010.
27. C.G. Gurler, B. Gorkemli, G. Saygili, and A. Tekalp, Flexible transport of 3D video over IP, *Proceedings of the IEEE*, April 2011.
28. S.T. Worrall, A.M. Kondoz, D. Driesnack, M. Tekalp, P. Kovacs, T. Adari, and H. Gokmen, DIOMEDES: Content aware delivery of 3D media using P2P and DVB-T2, *2010 NEM Summit: Towards Future Media Internet*, 13–15 October 2010, Barcelona, Spain.
29. E.K. Lee and Y.S. Ho, Generation of high-quality depth maps using hybrid camera system for 3-D video, *Journal of Visual Communication and Image Representation*, 22(1), 73–84, 2011.
30. Specification of 3D production workflow and interfaces—Phase I, Deliverable 1.2.1, MUSCADE Consortium (found in http://www.muscade.eu/deliverables/D1.2.1.pdf).
31. Distribution of Multi-view Entertainment Using Content Aware Delivery Systems (DIOMEDES) (web page in http://www.diomedes-project.eu/).
32. Z. Zhang, A flexible new technique for camera calibration, *IEEE Trans. Pattern Analysis and Machine Intelligence*, 22(11), 1330–1334, 2000.
33. J. Mitchelson and A. Hilton, Wand-based multiple-camera studio calibration, Center Vision, Speech and Signal Process., University of Surrey, Surrey, UK, Tech. Rep., vol. VSSP-TR-2/2003, 2003.
34. ISO/IEC JTC1/SC29/WG11, Depth Estimation Reference Software (DERS) with Image Segmentation and Block Matching, M16092, February 2009.

35. Report on 3D media capture and content preparation, Deliverable 3.1, DIOMEDES consortium (found in http://www.diomedes-project.eu/deliverables/3DContent/).

36. E. Ekmekcioglu, V. Velisavljević, and S. Worrall, Content adaptive enhancement of multi-view depth maps for free viewpoint video, *IEEE Journal of Selected Topics in Signal Processing*, 5(2), 352–361, 2011.

37. D.V.S. De Silva, W.A.C Fernando, H. Kodikaraarachchi, S.T. Worrall, and A.M. Kondoz, Adaptive sharpening of depth maps for 3D-TV, *IET Electronics Letters*, 46(23), 1546–1548, 2010.

38. G. Nur, S. Dogan, H. Kodikara Arachchi, and A.M. Kondoz, Assessing the effects of ambient illumination change in usage environment on 3D video perception for user centric media access and consumption" *2nd International ICST Conference on User Centric Media (UCMedia)*, Palma de Mallorca, Spain, September 1–3, 2010.

39. G. Nur, S. Dogan, H. Kodikara Arachchi, and A.M. Kondoz, Impact of depth map spatial resolution on 3D video quality and depth perception, *IEEE 3DTV Conference 2010*, Tampere, Finland, June 7–9, 2010.

40. G. Nur, H. Kodikara Arachchi, S. Dogan, and A.M. Kondoz, Advanced adaptation techniques for improved video perception, *IEEE Transactions on Circuit and Systems for Video Technology*, 22(2), 225–240, 2012.

41. A. Desolneux, L. Moisan, and J.M. Morel, *From Gestalt Theory to Image Analysis: A Probabilistic Approach*, New York: Springer Science + Business Media, 2007.

42. L. Itti, Automatic foveation for video compression using a neurobiological model of visual attention, *IEEE Trans. Image Processing*, 13(10), 1304–1318, 2004.

43. C.-W. Tang, Spatiotemporal visual considerations for video coding, *IEEE Trans. Multimedia*, 9(2), 231–238, 2007.

44. Interim report on developed audio and video codecs, Deliverable D4.3, DIOMEDES consortium (found in http://diomedes-project.eu/deliverables/default.htm).

45. S.S. Savas, C.G. Gurler, A.M. Tekalp, E. Ekmekcioglu, S. Worrall, and A. Kondoz, Adaptive streaming of multi-view video over P2P networks, *Signal Processing: Image Communications*, to appear.

Chapter 17

IPTV Services Personalization

Hassnaa Moustafa, Nicolas Bihannic,
and Songbo Song

Contents

17.1 New TV Model and Users' Consumption Model.....................................376
17.2 IPTV Services Personalization 377
    17.2.1 What is Meant by Personalization? 377
    17.2.2 Possible Enriched Services for Users and Some Use Cases.............378
    17.2.3 Users' Expectations Regarding Personalization 379
17.3 Multiscreen Approach 380
17.4 Technical Challenges for IPTV Personalization 381
    17.4.1 The Whole IPTV Service Architecture 381
    17.4.2 Context Awareness...383
    17.4.3 Profiling..384
    17.4.4 Implementing the User Profiling Component..............................385
    17.4.5 Privacy and Identity Management 389
17.5 IPTV Personalization: Current Approaches, Requirements,
    and Standardization 390
    17.5.1 Existing Personalization Approaches and their Limitations 390
    17.5.2 Personalization Requirements 391
    17.5.3 Existing Standardization Efforts..393
17.6 Possible Business Models..395
    17.6.1 Business Models Based on Content and Context Fusion
        for Improved TV Access 396

17.6.1.1 B2B2C...396
17.6.1.2 B2B...396
17.6.2 Business Models Based on Content Storage and Behavior 396
17.6.2.1 B2B2C...396
17.6.3 Business Models Based on Scalable Content Compression
and Transmission..397
17.6.3.1 B2B2C...397
17.7 Conclusion 397
Acronyms...397
Acknowledgment 398
References 399

The advances in IP TeleVision (IPTV) technologies enable a new model for service rendering, moving from traditional broadcaster-centric TV services to a new user-centric TV model. This newly proposed IPTV model is promising in allowing higher attractive services for end users with service offerings that better fit with user expectation, grant access to large interactive services (rating, sharing of recommendations within social networks) and open the legacy IPTV outline onto valuable Internet services (e.g., gaming, Internet streaming). All these service enhancements should generate higher audience, a more efficient advertising and consolidate revenues that will be shared with all the new actors involved in these new business opportunities.

This chapter aims to introduce the problem of IPTV services personalization and discuss the different technical challenges that need to be satisfied for allowing efficient and rich personalization. Some use cases for IPTV services personalization and possible business models among the different actors of the IPTV chain are also introduced.

17.1 New TV Model and Users' Consumption Model

IPTV market evolution is presenting new entertainment services: (i) Consumers are increasingly demanding instant access to digital content through various terminals; (ii) Consumers have more and more appetence for deciding by themselves which content to select at a given time also called catch-up TV or nonlinear TV; (iii) Communication devices are becoming entertainment devices as recently seen with high-end mobile handsets (also called smartphones) that embed more and more entertainment applications; (iv) Higher interaction between devices (e.g., Set Top Box (STB) and personal devices like tablets or smartphones) to receive complementary information or chatting on personal devices about a content that is displayed on the TV set; (v) Network evolutions on fixed and mobile broadband accesses are providing faster and more reliable access to digital content; (vi) Content providers are looking for new channels of distribution and revenue streams. Indeed, services personalization has an important potential in IPTV market evolution, allowing content/service providers and networks operators to create new business opportunities and to promote smart and targeting services, while enhancing users' interaction with the IPTV system and the Quality of Experience (QoE) which in turn increases users' satisfaction and services acceptability.

The existing IPTV system shows some limitations at the level of services personalization, which should be initialized by the user through for instance choosing his profile at the beginning of the service access. This approach has a number of limitations as follows: (i) it is not necessarily user friendly, where the user should always carry out the profile selection process through using the remote control, (ii) it could not distinguish between users in an appropriate manner, since the identification for services access concerns the subscription identification rather than each user's identification and it is assumed that the right user should choose his right profile each time he accesses the service, (iii) it is not a dynamic approach, in the sense of not being aware if the user having selected the profile is still being present or if there are other users instead and even the existing users profiles are static and not continuously updated, and (iv) it is not a transparent approach, where it should be repeated each time the user changes the terminal even if he is in the same domestic sphere.

17.2 IPTV Services Personalization

17.2.1 What is Meant by Personalization?

IPTV services can be defined as a set of TV services that are strongly managed by the Network Operator. Globally, IPTV services require from the network operator to closely control its network infrastructure in charge of the TV content delivery. As IPTV services put high constraints on network capabilities to handle a sufficient end-to-end Quality of Service (QoS) for both Live and Video on Demand (VoD) contents, different technical implementations may exist. Thus, IPTV services can be fully delivered through IP network connectivity and also by the use of satellite in the downlink notably when eligibility constraints occur regarding available bandwidth. As IPTV services are also embracing more and more enhanced interactivity usages with the user (e.g., rating, personalized advertising), a return channel is activated and is generally performed through IP network connectivity as with the Hybrid Broadcast Broadband TV (HbbTV) standard. The higher attractiveness of service features coupled with higher network performances in the service delivery must be guarantied by the Telcos making IPTV a key differentiating factor used by the end user when selecting the TV service provider to subscribe with. IPTV is commonly designed by the use of a dedicated device on the client premises that is called the STB embedding all features required to interface with the operator-managed network and the TV service platform. Till now, IPTV services are limited to managing household identity as the case of legacy TV broadcasting and competitive paid-offers delivered by Cablos or broadcasters. This current status of a single TV account management is moving for IPTV services opportunities built upon personalization.

Personalized TV services firstly mean the introduction of personal user identity management in complement to the unique household identity. This personal user identity that is now activated by a given user for IPTV purpose needs a unique identity distinguishing each user's access when consuming TV content on his different personal devices.

To capture a more and more significant customer base and to gain in user loyalty, Telcos had developed multiplay offers related to their fixed broadband network activities a long while ago, encompassing notably IPTV, WebTV, and Internet for content-oriented services. More recently, these multiplay offers have been enriched on several countries with the integration of mobile services in an all-in-one offer. The introduction of personalized IPTV services is a strong asset to consolidate all the possible usages around TV content consumption. Such interactions between TV streams are reinforced by a larger range of devices able to consume TV content. Tablets are for instance a key device that can foster personalized IPTV usage.

In order to achieve efficient IPTV services personalization over different IPTV systems, several enabler technologies will need to be integrated to different IPTV systems, these mainly include: (i) A context-aware module capable of monitoring and gathering the user and his environment contexts and feed them in a dynamic manner to the IPTV system, (ii) A profiling management module, capable of constructing and dynamically updating the users' profiles according to the various contexts, and (iii) A privacy management module responsible for managing the different privacy levels for each user and protecting one's personal information. Consequently, personalized services could be provided allowing content selection according to the users' preference, QoE requirements, and different contexts, while fostering trust between viewers and broadcasters through an efficient privacy management.

17.2.2 Possible Enriched Services for Users and Some Use Cases

Some examples are hereafter mentioned to illustrate the new TV experience in liaison with IPTV services.

- ■ Introduction of personalized features in the IPTV user experience:
  - Personalized Electronic Program Guide (EPG) with my favorite channels
  - Setting of reminders for my favorite IPTV programs with a display of these reminders on any personal devices or on the TV screen

- Bookmarking of VoD or catch-up programs from any personal device or TV screen and all these bookmarked content can be retrieved latter on any device (personal devices and TV set)
- Introduction of predictive viewing and associated content recommendations when the user is logged on his profile from the STB
■ Accessing from the TV set to his favorite Internet-oriented services when the user logs on his personal IPTV account

IPTV services personalization allows for some promising use cases as follows:

i. Content following the user during his mobility in his domestic sphere: allowing the user to move around within his domestic sphere while continuing accessing his IPTV service personalized according to the characteristics of the device in his proximity.
ii. Content adaptation according to each individual user and group users' preferences: allowing each user or a group of users to have personalized content matching their preferences.
iii. Content customization according to the user context and QoE: allowing each user to have personalized content matching the user context (age, gender, region, preferences, location "in the domestic sphere or outside," activity) and level of satisfaction.
iv. Content itemization: allowing each user to interact with the diffused content through providing metadata on each content item (e.g., if a movie is executed in a marvelous place, the user could interact with a specific icon on the screen to have more information on this place and the tourisms information there).
v. Content personalization during nomadism (users in nomadic situations in a hotel, in a friend's home or anywhere outside his domestic sphere): allowing the user to access his personalized IPTV content in a hotel for instance and be billed on his own bill. Content personalization during nomadism may also refer to the ability for the user to retrieve his already purchased VoDs or bookmarked contents for catch-up TV.

17.2.3 Users' Expectations Regarding Personalization

The IPTV World Forum has highlighted the operators' need to focus on the monetization of content and on making content more accessible to consumers. Indeed, operators have been increasing the amount of content they offer to their subscribers; most IPTV services now have portfolios of dozens of TV channels, supplemented by, on average, 5000 VoD assets. However, in many cases, this investment in content has failed to result in a corresponding increase in revenue, as consumers are unable to find the programs they want due to the poor user interfaces. In this perspective, recommendation engine is a key asset to help the user in selecting the right contents that match with his preferences.

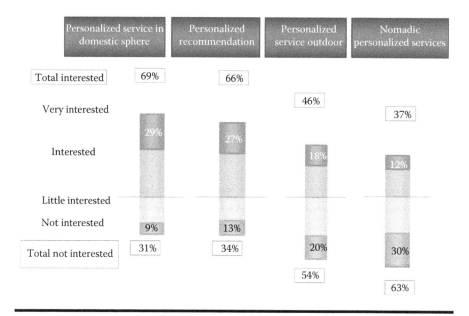

Figure 17.1 Interest rate in personalized services—Survey Lab's Orange April 29, 2011.

Figure 17.1 illustrates the results of the Survey on 1925 end users realized by lab Orange (in France Telecom) in April 2011, 29% are very interested in personalized services in the domestic sphere and 27% are very interested in advanced personal recommendation.

17.3 Multiscreen Approach

Personalized IPTV services should be fully integrated to the global TV ecosystem to which the end user has subscribed, allowing the access to personalized IPTV services must be largely opened and must no longer be limited to STB-based usage.

Personalized IPTV services is triggered and motivated by the current Ecosystem which observes:

- An explosion of audiovisual services with TV content diffusion across the three screens (TV, PC/Personal devices, Mobile phone).
- Profusion of new enhanced connected devices with advanced multimedia capabilities (codec, storage) and multiple connectivity (WiFi, Digital Living Network Alliance (DLNA) . . .).
- A crowded market, with many players providing entertainment service (network operators, content providers, and Over the Top Players "Apple, Google, . . .").

- A new IPTV user-Centric delivery model extending beyond the delivery model in today's IPTV including additional options for the distribution of IPTV tailored to the consumer (through adapting content according to users' preferences, activity, location in the domestic sphere or outside and proximity to which devices and which networks).
- New expectations and means for personalized and interactive usage enabling content sharing across many devices and with other users and seamless content access.

As personal devices are natively designed to support personal accounts like when logging-on in the welcome page, such devices have to be fully integrated to this enlarged IPTV landscape. In this multiscreen approach, personal devices can be used in the following contexts:

- To have a more and more seamless user experience with TV services now available at home on the TV screen and when on the move with the user's Smartphone for instance, while keeping a common bookmarking of favorite contents whatever the device used.
- To have a mean for scheduling Personal Video Recorder (PVR) while the user is on the move.
- To be a secondary screen to interact with the ongoing content that is displayed on the TV (e.g., the use of personal device as remote control, the selection of additional information related to the content streamed on the TV set "information on football players during an ongoing match, backstage videos linked with the ongoing Live concert," chatting on IM "Instant Messaging" applications or social networks websites about the currently displayed content on the TV set).
- Personal devices can still be used to introduce personalized IPTV services in the case where the operator is still operating STB that are limited to manage a unique household account. Indeed, the operator is able to leverage on personal devices to make some content suggestions or recommendations related to the IPTV catalogue, which would be finally rendered on the TV set.

17.4 Technical Challenges for IPTV Personalization

17.4.1 The Whole IPTV Service Architecture

All the innovating and sometimes disruptive usages related to personalized IPTV services deeply impact on the overall service architecture. Most significant impacts are:

- The ability for the STB to manage the multi-TV accounts associated to the introduction of personal identities. This evolution allows migrating from an

IPTV service limited to the household account to enriched services where the operator is able to deliver personalized services based on user preferences or habits.

■ The binding of user identities that are used in the different TV streams (IPTV, WebTV, and mobile TV). All these TV streams should be able to manage personal identities and facilitates cross-TV services (convergent bookmark, e-Sell Through for VoD) and even interactions with Internet-oriented services that are also built upon personal identities (e.g., personal account on social networks). Note that user identity formats differ between each TV stream.

– Regarding mobile TV: The mobile operator leverages on Subscriber Identity Module (SIM)-based implicit user identification and authentication and the user identity known by the service platform is the Mobile Station International Subscriber Directory Number (MSISDN).

– Regarding WebTV: The operator can typically perform a service access control from mailbox account if required. In fact, a single sign-on authentication procedure from the user mailbox account allows the operator to activate simultaneously a set of personalized Internet services once the user has been identified and authenticated.

– Regarding IPTV: The user identification and authentication method required to access to personal account directly depends on the remote control that is used. To guaranty a correct user friendly level, a Personal Identification Number (PIN) code-based solution can be generalized with basic remote control. Innovating and disruptive approaches may exist involving personal devices as Near Field Communication (NFC) or Quick Response Code (QRCode) are highly valuable technologies since the user has no longer to type his credentials like password or PIN Code. Moreover, these technologies can ensure a more secure solution against password theft.

■ The IPTV service platform must be open to a larger range of devices, shifting from an STB-centric approach to a fully integration of additional personal devices. Personal device can be variously used for IPTV concern:

– Use of the personal device as remote control
– Use of personal device to push personal recommendations by the operator
– Use of personal device to replace an STB in multiroom viewing context

■ Richer network interfaces for personal devices to participate in delivering personalized IPTV services. So far, mobile handsets were interfaced with a dedicated service infrastructure to deliver a specific mobile TV offer to cope with content adaptation related to mobile network bandwidth, device capabilities, or user mobility. Higher built-in capacities (e.g., decoding performances) and enhanced user interface clearly foster larger usage of personal devices for TV purpose and their integration in this enriched IPTV ecosystem. So, IPTV services require that personal handsets have the ability to handle new network interfaces through both a direct interface with the STB and also an interface with IPTV service platforms. It also means more complex embedded applications that have to mitigate possible heterogeneous implementations between each TV stream: multiple network connectivities with each service platform, multiple technologies suiting each TV streams, needs for network context awareness especially when being at home to directly interact with the STB as remote control.

■ Building a user profiling feature that aggregates personal information from all TV streams and are built by mixing user preferences and usage tracking. User preferences are explicitly declared by the end user from configuration forms and usage tracking is typically fed by service platforms with a consolidated view from all TV streams. Note that user knowledge can be extended with additional Internet services that become more and more coupled with the TV ecosystem. For instance, "Like/dislike" items retrieved from social networks could be inserted in the global user profiling. For privacy purpose, the user must be aware of these possible usage tracking and the operator must then allow the user to decide which information or usage can be monitored.

17.4.2 Context Awareness

The context-aware concept is promising in enhancing the personalization; however, how to use the acquired context information to realize personalized TV services is still a challenge, particularly those related to the network. Indeed, context-aware IPTV could provide services dynamically optimized taking into consideration the user contexts as well as the content and adaptation needs. First, we give a general definition of context [1]: "Context is any information that can be used to characterize the situation of an entity. An entity is a person, place, or object that is considered relevant to the interaction between a user and an application, including the user and applications themselves." This definition can include several types of context information, where the most popular ones are mostly related to users and to the surrounding environment (which can include the devices). Although context awareness can bring additional smart services and useful functionalities to IPTV systems and allow enriched personalization, how to make IPTV services context-aware and how to use the acquired context information to realize personalized TV services is still a challenge. The related contributions on context-aware IPTV mainly follow two approaches: (i) a middleware infrastructure approach, and (ii) a context server approach. Middleware infrastructures are typically built into the hosting devices or platforms on which the context-aware applications operate. However, the context server approach provides contextual information to different context-aware applications in a distributed environment.

The *Middleware infrastructure* approach, implemented at the end-user domain, can impose additional computation burden on the hosting devices, where devices with rich computing resources are needed at the end-user domain. On the other hand, the *context server* approach employs a context server (having rich computing resources instead of totally relying on the end-user devices) for providing contextual information to the different applications in a distributed manner, which can easily capture the different context information from the user, the devices, the network, and the services. The context server approach is also more adapted to contexts where several end devices are involved in personalized IPTV services. Indeed sharing and synchronizing data related to the user context are more easily performed with an interconnection of each device to a centralized context server instead of exchanging data flows between all devices.

The Middleware infrastructure approach can also suffer from higher security risks about preventing any theft of sensitive user data stored on the end device. Security mechanisms like data ciphering or access control to stored data must be implemented and controlled by the operator. Implementation of such security mechanisms is largely less constraining in a centralized context server approach.

17.4.3 Profiling

The IPTV profile contains information helping the consumption of the different types of IPTV service like broadcast, Content on Demand (CoD), and network PVR service profiles. The European Telecommunications Standards Institute/Telecommunication and Internet converged Services and Protocols for Advanced Network (ETSI/TISPAN) defines three types of IPTV profiles [2]: Content Profile, Service Profile, and User Profile. Content profile contains and maintains information about multimedia metadata and multimedia service packaging used for the provisioning of IPTV services. Service profile refers to data used to provide the service to the user (service-level user data such as user identification, numbering, addressing, security, and so on, and service-level offer data allowing the delivery of IPTV services, for example, EPG "Electronic Program Guide"/BPG "Broadcast Program Guide"). User profile refers to user customization and usage metadata containing and maintaining basic user information, service-specific information (subscription, bookmarks, activities, parental control, etc.) and user actions related to service purchase and consumption.

User profiling can globally be built by mixing two sources of information: (i) settings of preferences by the user and (ii) feeding of information related to TV usages. IPTV Preferences can be edited from account forms by the use of PC-like devices, where the user gives explicitly some parts of information related to his favorite types of films, actors, and so on. While the data collected from TV usages can be split into two subcategories: (i) explicit data from user usage. Examples are setting bookmarks, reminders, rating of content, and so on, and (ii) implicit data from user usage. Examples are scheduling of records, content ordering, and so on.

Note that data from Information System should also be taken into account for IPTV personalized services to ensure, for instance, the consistency of search results with user subscribed offers, and the time of validity for TV contents to control persistence of information in the database.

Figure 17.2 illustrates a functional architecture for the user profiling feature, where the exchanged data through the interfaces are described in Table 17.1.

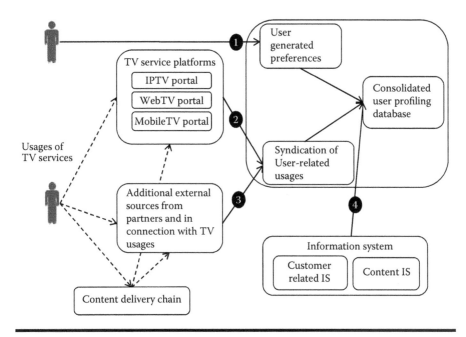

Figure 17.2   Functional architecture for user profiling.

Additional interfaces could complete this data collection like for instance the monitoring of TV-related usages between devices in the user premises. It typically corresponds to e-Sell through usage when a user has bought a VoD content and multiple viewings on different home devices are allowed. In this case, information on favorite rendering devices, preferred time of watching may be of value for the operator and its partners. Once described the way to build user profiling, it is also interesting to indicate all the TV services that can benefit from this asset for personalization purpose.

Figure 17.3 shows some features that can directly benefit from the awareness of user preferences and habits. A first list of features is given in Table 17.2 to illustrate how this user profiling becomes a key enabler for TV service personalization.

17.4.4 Implementing the User Profiling Component

The implementation of user profiling component is complex in terms of data modeling and data binding between the large amounts of information generated from usages across all TV streams. The following constraints are found:

■ Legacy constraints: Operators have largely deployed TV services from fixed broadband access and mobile access. Forthcoming new TV services on very high broadband access based on fiber technology are also under deployment. Because of specificities on service definition or on expected performances

Table 17.1 Exchanged Data

Interface	Features
1	■ Filling of the account form by the user. Some adaptation of the presentation layer may be performed following the device that is used. A complete form could be available from a PC whereas a simplified one should be designed for mobile or STB device where typing is not always convenient. ■ Ability to select which information can be tracked by the different TV service providers (through the interfaces 2 and 3 from respectively, Operator and Partners services) for Privacy purpose. It typically corresponds to explicit and implicit data related to user usage.

2	■ Tracking of TV usages per user. It includes:
	  – User activities on TV Portal like setting a bookmark or a reminder, rating a content,
	  – User requests for content, selection of a recommendation suggested by the service provider.

3	■ Tracking of additional usages for services in liaison with TV services on a per user base. It typically includes the monitoring of user-centric activities on partner services in the goal of having the most complete view of user habits. Interconnection with social service providers as partners could be performed to be aware of "like/dislike" tags set by the user for content in connection with TV purpose.

4	■ The Information System is in charge of:
	  – Providing information on which services are subscribed by a given user. Such data insures that search or recommendations results are aligned for a given user with his subscribed offers.
	  – Providing information on a given content like its metadata and exploitation duration. Such data ensure that no longer available contents are retrieved from personalized features like bookmarking.

(e.g., network QoS, encoding quality), different service infrastructures have generally been rolled out per TV stream and then user profiling data are often segmented in different databases. Moreover, any IPTV architecture evolution is at risk: (i) IPTV remains a highly differentiating service between Telcos to gain user loyalty and therefore in-depth infrastructure evolution is sensitive to maintain high QoS Key Performance Indicators (KPI), (ii) to migrate a

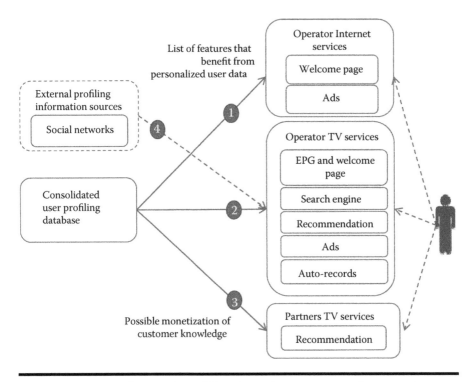

Figure 17.3 Some features benefiting from the awareness of user preferences and habits.

large installed base as already reached in numerous countries. As seen before, STB is impacted upon with the introduction of personalized IPTV services (e.g., personalized user accounts).

■ Large interfaces for ingress and egress flows on user profiling component: User profiling data are becoming more and more relevant for the operator to deliver contents aligned with user expectations (e.g., content offerings matching with user preferences) and also to monetize this user knowledge with major Partners. In this context, large interfaces will be feeding databases in charge of collecting data usages and preferences. In the same time, more and more services could request for these consolidated data. Such media services include TV, Internet, and various other universes like gaming.

■ High expected performances: Because user profiling component is initially designed for managed TV services, high requirements on data availability and response time must be achieved by the implemented solution.

■ Large volume of data to be consolidated: As a direct consequence of large interfaces for feeding TV usages from internal and also external services, the implemented solution must be efficient to build a consolidated user profiling. It consists in developing a business logic component in the front end of the

Table 17.2   User Profiles Features

Interface	Which Inputs Provided by the User Profiling Feature?
①	■ Provide data for a targeted advertising with the profile and behavior of each user; ■ Provide data on content preferences of the user (Really Simple Syndication (RSS) flows, weather preferred town ...), layout preferences for the Web page presentation.
②	On the Welcome page on TV portal: ■ User profiling component can directly provide information on Customer bookmarks (VoD and Transactional Video on Demand (TVoD)), Customer Reminders, Customer favorite channels; ■ TV Portal can use data stored by the User Profiling component to build Operator recommendations. For the Search feature: ■ TV Portal can use data stored by the User Profiling component to filter out Search results that are parts of offers or options subscribed by the user. For the recommendation feature: ■ TV Portal can use data stored by the User Profiling component to edit recommendations that fit with user preferences and habits. So, recommendations can be household or user oriented following the account that is active on the IPTV session. Regarding the Advertising feature, the same properties as for Operator Internet services can be supported. The Auto record feature consists in pushing on the STB Hard Disk all the contents that can be of interest for the end user. In order to insure that recorded content will be really displayed by the user, the auto record feature needs to take in account user preferences handled by the user profiling component.
③	The TV Operator can share with its Partners some user centric information. The partners can then expect direct benefits like higher audience thanks to customized services that match with user expectations. Partners are not limited to those delivering TV services since gaming service providers can be interfaced to this extended TV ecosystem. For instance, analyzing VoD purchases of a given user allows the gaming provider to promote to this user some recommendations with a higher success probability.

Table 17.2   (continued) User Profiles Features

Interface	Which Inputs Provided by the User Profiling Feature?
④	Opening the legacy IPTV ecosystem toward new service providers is a strengthening trend. Operator can have interest to enlarge connections with all service providers able to deliver user-centric information. In this rationale, connection with social networks services can be valuable to retrieve "like/dislike" tags in order to build future consistent Operator recommendations.

user profile database; this business logic component is in charge of managing the data consistency within the data base and also processing all data requests. A possible methodology for an efficient implemented solution can be: (i) to list all the already deployed databases that stores user-centric information, (ii) to list all the valuable user-centric data, (iii) to identify whether these user-centric data need to be shared with other services or components, (iv) to perform a feasibility study where shared user data are gathered within a common database. User data specific to a given service is not necessarily migrated to this common and central database.

17.4.5 Privacy and Identity Management

Privacy protection is a critical issue for personalized IPTV services acceptability by the users that is not yet resolved in a complete manner, although user personal information should not be a way that threatens the user privacy. In general, IPTV still raises concerns on privacy risks. With the added capability of viewing other online content and accessing Internet services, it is more urgent for IPTV service providers to incorporate mechanisms into their delivery platforms to preserve their customers' privacy. Privacy protection is also a valuable asset that IPTV services must guaranty to differentiate this premium offer from Web approach where Privacy may be weakened.

The Privacy Enhancing Technologies (PETs) have seven principles [3]: limitation in the collection of personal data, Identification/authentication/authorization, standard techniques used for privacy protection, pseudo-identity, encryption, biometrics, audit ability. More details related to the privacy requirements are given in the following section.

Beside privacy, handling personalized IPTV services required enhancement of user identity management. Unlike the legacy IPTV offers of services limited to a household account management only, personalized IPTV services must be user centric whatever the end device. Two direct requirements are:

- STB devices must be able to manage personal accounts and numerous innovating technologies that are available for a user-friendly logging on. One can mention QRCode, NFC or Radio Frequency Identification (RFID) technologies.

■ As personal devices interact more and more with IPTV stream, binding of different user identifiers and handling of heterogeneous methods for user identification and authentication must also be supported. Access to third-party services should also benefit from Single Sign On procedure once the user is logging on the IPTV home page.

17.5 IPTV Personalization: Current Approaches, Requirements, and Standardization

17.5.1 Existing Personalization Approaches and their Limitations

Personalization technologies can be used in order to recommend content and adapt services based on the profile. Existing recommendation algorithms use content-based filtering technologies [4], collaborative filtering algorithms [5,6], or hybrid solutions combining content-based and collaborative filtering techniques [7,8]. Services could be adapted to user preferences and contexts, like adapting a video-conference service by choosing the TV or the laptop as a communication device according to user context and policies (e.g., use TV when no one is using it, otherwise the laptop) [9].

Several content recommendation systems have been proposed. The proposed solution in ref. [10], is of type "content-based filtering" and considers that the users' preferences have several categories (e.g., film type, language, actor, etc.), where each category has a weight that is updated during the user's consumption of content. The description of content also follows the same categories of the user's preferences and the system recommends for the user the content having the maximum similarity with his preferences with respect to the different categories and their associated weights. Note that the matching of metadata (for content description) with user preference categories may have deep impacts on the way to edit these metadata. On the other hand, the solution proposed in ref. [11] is of type "collaborative-based filtering," in which the users are classified into groups based on the similarity between their consumptions. The system recommends for each user belonging to a certain group an aggregation of content consumed by the other users belonging to the same group.

A hybrid content recommendation solution is proposed in ref. [12], where the preferences of each user are calculated in a first-step function of the consumption frequency for each type of content. Then, in a second step, the similarities between the users' preferences are calculated based on the preferences of each user. The system recommends for each user the content that is consumed by the group of users having the same preferences as that user.

The traditional content recommendation systems consider only the user's preference and content consumption history as the recommendation criteria and the general context information is not considered (like device capacity, network states, location, time, etc.). However, the general context information takes an important role for the recommendation service because the user's preference for IPTV content could change with the change of his context and the context of his environment. Furthermore, some contents are accessible only if certain conditions are satisfied. For example, only the device which supports the High Definition (HD) could display the HD content. If a user chooses to watch TV through a terminal which does not support HD, the recommendation system would not propose the HD content for him. The content recommendation system could attain the better accuracy through considering the general context information. Another example is that content duration is directly designed following the addressed devices (e.g., short format of few minutes for rendering on mobile handsets) and it then requires to take into account the type of used device in any recommendation or search results.

17.5.2 Personalization Requirements

IPTV services personalization necessitates different investigations related to deployment strategy, targeted QoE, targeted devices, and so on. This section lists the different requirements to achieve efficient and enriched personalization for IPTV services satisfying the users' required QoE while keeping the overall network QoS guaranteed.

1. *Interoperability Requirements*:
   - Supporting different IPTV models (managed, WebTV, etc.): to allow for an open TV model not restricted to the only "wall-garden managed model" in order to match the evolution in the TV domain. Indeed IPTV is enlarging the scope of supported services with higher interaction with Internet and in this rationale, personalization features must be enriched and also be available to these Internet-oriented services.
   - Supporting an interface with social networks: to help in gathering information on users' profiles and also in sharing personalized content with other users.
   - Supporting different underlying architectures: to allow for interoperability with different underlying core network architectures.
2. *Context-Awareness Requirements*:
   - Context information flexibility, extensibility, and structuring: to allow covering the whole IPTV delivery chain through including different types of context information: (i) User Context Information: presenting the information about the user, which could be static information describing the user's subscription information (e.g., user's subscribed services, age, preferences, …) or dynamic information dynamically captured by sensors or by other services (e.g., user's location, agenda, available devices in his proximity, available network types, and network type in use), (ii) Device/Terminal Context Information: presenting the information

about the devices/terminals that could be the device activity status (on or off, volume), capacity, and its proximity to the user, (iii) Network Context Information: presenting the information about the network such as the bandwidth and congestion status, (iv) Content context Information: presenting the information about the content itself including the content description (content type, start time, and stop time, language) and the media description (codec type, image size, and resolution).

■ Dynamic context gathering and update: to allow gathering and updating in a dynamic manner the different context information on the user and his environment to be able to personalize the content accordingly.

■ Acquiring QoS parameters (as a part of the context information) from client, network operators, and service provider sides: to allow adapting the content according to the different QoS parameters, which will vary according to the underlying core and access networks, and also according to the IPTV model (e.g., managed or WebTV).

3. *Users' Profiling Requirements*:

■ Profile information structuring and extensibility: to allow a well-structured and extendable users' profile allowing to easily store user preferences, to associate them to their corresponding users' contexts, and to easily integrate new user preferences.

■ Profile accessibility from home and visited networks: to allow personalized services for users in their home networks or in nomadic situations in a transparent mean for users.

4. *Security and Privacy Requirements*:

■ Binding of the different identities used in managed and unmanaged services for a given user: to allow distinguishing each user access in a unique manner and hence, personalize the services accordingly.

■ Supporting group identity: to distinguish the access of multiple users for the same session (i.e., group of users) and personalize the services accordingly.

■ Authentication, Authorization, and Accounting (AAA) covering home and nomadic situations: to allow personalized IPTV services in the home sphere and outside the domestic sphere (nomadic access within the home network or a visited network), while defining AAA policies between different operators/service providers. It means for the service provider the ability to identify who is the user that is requesting the TV service from a given access; this user is not necessarily the line access subscriber in fixed broadband context where nomadism is activated.

■ User-controllable privacy: to allow the user to configure and adopt one's privacy any time anywhere with the ability to define different privacy levels while having full control of one's personal data.

■ Enhanced billing on Information System: to allow the service provider to charge the user requesting premium content and not the owner of the line access from which the content ordering has been executed.

5. *Nomadic Requirements:*
   - Ability to have a personalized experience in nomadic situation: to allow each user to access one's personalized TV content from any TV system not belonging to him (for instance, through the TV system in *a* friend's home) and through any access network owned by one's home or visited provider (for instance, VoD in a hotel).
   - Ability to retrieve/access users' profiles in nomadic situation: to allow the user to update one's profile anywhere.

6. *Mobility Requirements*:
   - Supporting personal mobility: to allow each user to access IPTV services at any terminal in different locations on the basis of a personal identifier, and the capability of the network to provide those services delineated in the user's service profile.
   - Supporting service mobility: to allow users to maintain access to their services while moving or changing devices and service providers.
   - Supporting terminal mobility: to allow a device to move between different networks of the same type or of different types while continuing to communicate with its corresponding peers.
   - Supporting session mobility: to allow a user to maintain a media session while changing terminals without session interruption.

17.5.3 Existing Standardization Efforts

IPTV services personalization observed lot of attention from several standardization organisms. This section illustrates some standardization efforts related to IPTV services personalization.

1. *International Telecommunication Union (ITU)* [13,14]: A big attention is given within the ITU to IPTV services personalization defining some related services as follows:

   PVR: An end-user-controlled electronic device service that records linear TV and stores it in a digital storage facility, either in standalone STB or in the network. This service can support "time shifting" and "trick modes."

   Personal Broadcast Service: is a service providing the end user with a way to advertise personal content (possibly including scheduling information) description so that other users can access such content. The service provider is responsible for relaying session information between the broadcasting and receiving end users, possibly assuming some access control functions.

   Targeted Advertising: is a commercial advertising or public promotion of goods, services, companies and ideas, usually personalized according to the end-user's preferences or centers of interest, based upon the end user's profile characteristics in order to match the audience with the campaign objective requirements.

Presence services: Managing presence information between each end user and other users or any service making use of the presence information of users. Example of services might include: personalization based on presence on top of locally stored data, targeted advertising correlated with time of day, and chatting with friends watching the same channel.

2. *Open IPTV Forum (OIPF)* [15]: OIPF considers the following aspects for IPTV service personalization:

   i. User profile management: Refers to the set of operations that allow a user to manage his profile, including the ability to create, retrieve, modify, delete, or replace the profile.

   ii. Personalized channel: A service allowing content items from scheduled content and CoD service to be lined up on a per-user basis according to the user's preferences, viewing habits or service provider recommendations.

   iii. Session transfer: Allows a user to transfer an ongoing unicast session from the device where the content is currently being streamed, to another device, where the user can resume watching the same content. Following the successful transfer of the session, the original session is terminated.

   iv. Session replication: Allows a user to replicate an ongoing unicast session from the device where the content is currently being streamed, and which will be called original device, to another device, called the target device, where the user can resume watching the same content. The original session continues to be maintained following the successful replication of the session, and indeed the original device and the target device will have completely independent sessions.

3. *Digital Video Broadcasting (DVB):* The DVB-IPTV [16] is a specification developed by the DVB [17,18] to facilitate the delivery of digital TV services over IP networks. The initial phase of DVB-IPTV work is to specify the technologies of the interface between the IP network and a Home Network End Device (HNED), for example, the STB. On the other hand, the DVB project designs an open middleware system standard called Multimedia Home Platform (DVB-MHP) for interactive digital television, where the DVB-IPTV also provides additions to the DVB-MHP middleware specifications to allow the MHP to support interactive TV application in DVB-IPTV environment.

4. *World Wide Web Consortium (W3C)* [19]: Within the W3C, the Composite Capability/Preference Profiles (CC/PP) [20] specification was developed for describing device capabilities and user preferences aiming to guide the adaptation of content for each device. CC/PP is based on the Resource Description Framework (RDF) [21], which was also designed by the W3C as metadata description model. On the other hand, the Wireless Application Protocol (WAP) Forum [3] developed the User Agent Profile (UAProf) [22] as a specification related to CC/PP, but more specified for mobile devices. UAProf defines a profile file named Capability and Preference Information (CPI)

containing the information like hardware characteristics, software characteristics, application/user preferences, and network characteristics. This information is conveyed between the mobile terminals and the servers and help content adaptation.

5. *Moving Picture Experts Group-7 (MPEG-7) and MPEG-21:* MPEG-7 [23] is a standard developed by MPEG [24], and uses a standard language called Description Definition Language to represent specific low-level features of the content, such as visual (e.g., texture, camera motion) or audio (e.g., melody) features and also metadata structures for describing and annotating the content. The metadata structures are composed by a set of related low-level features, allowing the user to find the content that interests him in an efficient way. On the other hand, MPEG-21 [25] defines an open framework to enhance the management of digital media resources exchange and consumption. It aims to achieve the functions such as digital content creation, distribution, user privacy protection, terminals, and network resource extraction, which helps in content personalization and adaptation.

6. *TV-AnyTime* [26]: Specified to support personalized TV including IPTV services over the whole value chain from the producer, through service provider to the consumer. It provides metadata to assist in the delivery of multimedia content for the user's digital video recorder (DVR) through the TV-Anytime Metadata Specification which allows describing segmented content, where Segmentation Metadata is used to edit the content for partial recording and nonlinear viewing. Another important set of metadata consists of describing user preferences, representing user consumption habits, and defining other information (e.g., demographics models) for targeting a specific audience, which is useful in content recommendation and targeted advertisement.

17.6 Possible Business Models

From a business perspective, personalized services through personalized TV are expected to create new business opportunities allowing for B2B Business-to-Business (B2B) and B2B2C Business-to-Business-to-Client (B2B2C) opportunities presenting several business benefits: (i) Users receive the content and advertisement matching their interest (more users' satisfaction), (ii) Network operators could monetize end users directly by increasing the subscription price for offering better quality during multimedia transmission, which could lead to have premium customers with increased-quality services and best-effort customers, and (iii) Opening new market for network operators based on collaboration with third parties and letting them pay for the network as a service (NaaS). The following are different examples for Business Models.

17.6.1 Business Models Based on Content and Context Fusion for Improved TV Access

17.6.1.1 B2B2C

- Service providers can sell the context-aware localization information on the user to content providers for content adaptation and targeted advertisement according to different regions (locations).
- Operator Partners like those hosting WiFi hotspots could gain advantages in collaborating with Network Operators for rendering personalized services. Indeed, to catch larger audience, Network Operators could grant access to customer knowledge to their Partners.
- Businesses can purchase air time for their advertisement from content providers for the local area or the category of clients they want to address, so that they can place their advertisements and reach the target audience.
- Users receive the content and advertisement matching their interest (more users' satisfaction). Users, in turn, pay the service providers for personalized services.
- Content items could be sold through interactive and dynamic advertisement through providing users' with the content context (for each content item) and allowing them to interact with it and even buy it.
- Content roaming agreements between operators: users are monetized by the home and visited network operators for the roaming service of the personalized home content, and network operators are monetized by the visited network operators and content providers in visited networks for the usage right of contents in different geographical areas.

17.6.1.2 B2B

- Network operators can open their networks to third parties for TV and VoD content diffusion (Youtube, Daily motion, Apple, Google, and the likes) providing them with useful information on the network context and user context and monetizing the services of providing the network.
- Operators can monetize the bandwidth offered by their infrastructures through making each actor pay for the bandwidth level used for his business (e.g., advertisers can pay not only the content providers but also the operators).

17.6.2 Business Models Based on Content Storage and Behavior

17.6.2.1 B2B2C

- Service providers owning Content Delivery Networks (CDNs) caching multimedia contents and are paid by content providers for delivering contents to clients in their local region and in nomadic situations.

- Service providers populating caches according to the local clients' consumption style and are paid by content providers for delivering personalized contents to clients in their local region.
- Clients pay the service providers for having personalized and tailored contents.

17.6.3 Business Models Based on Scalable Content Compression and Transmission

17.6.3.1 B2B2C

- Network operators could provide third parties with monetized useful information on the network context and the user context that allows these third parties to adapt the content (through choosing the most suitable compression mean) function of the network context (e.g., available bandwidth).
- The network operators could monetize end users directly by increasing the subscription price for offering better quality during multimedia transmission. It could then lead to have premium customers with increased-quality services and best-effort users.

17.7 Conclusion

Digital TV is moving from the traditional broadcast model to new user-centric model. This new model will allow users not only to access new services and functionalities from their providers, based upon their profiles and contexts, but also to become active parts in the content personalization through contributing in building their dynamic profiles and recommending content of interest to their communities for instance. IPTV allows TV services to evolve into true converged services, blending aspects of communications, social media, interactivity, nonlinear TV and search and discovery in new ways. These efforts address the growing consumer desire for personalization and customization of TV experiences. This chapter introduces the problem of IPTV services personalization and discusses the different technical challenges that need to be satisfied for allowing efficient and rich personalization. The existing IPTV personalization approaches are presented and their limitations are highlighted showing the need for a new use-centric personalization approach distinguishing each user's access in a separate manner detached from the subscription and from the used devices. The standardization efforts considering personalization are also presented as well as the technical challenges that need to be resolved. Some use cases for IPTV services personalization and possible business models among the different actors of the IPTV chain are also introduced.

Acronyms

AAA Authentication, Authorization, and Accounting
B2B Business to Business

B2B2C	Business to Business to Client
BPG	Broadcast Program Guide
CC/PP	Composite Capability/Preference Profiles
CDNs	Content Delivery Networks
CoD	Content on Demand
CPI	Capability and Preference Information
DLNA	Digital Living Network Alliance
DVB	Digital Video Broadcasting
DVR	Digital Video Recorder
EPG	Electronic Program Guide
HbbTV	Hybrid Broadcast Broadband TV
HD	High Definition
HNED	Home Network End Device
IPTV	IP TeleVision
ITU	International Telecommunication Union
KPI	Key Performance Indicators
MHP	Multimedia Home Platform
MPEG	Moving Picture Experts Group
MSISDN	Mobile Station International Subscriber Directory Number
NaaS	Network as a Service
NFC	Near Field Communication
OIPF	Open IPTV Forum
PET	Privacy Enhancing Technologies
PIN	Personal Identification Number
PVR	Personal Video Recorder
QoE	Quality of Experience
QoS	Quality of Service
QRCode	Quick Response Code
RDF	Resource Description Framework
RFID	Radio Frequency Identification
STB	Set Top Box
TVoD	Transactional Video on Demand
UAProf	User Agent Profile
VoD	Video on Demand
WAP	Wireless Application Protocol
W3C	World Wide Web Consortium

Acknowledgment

A part of this work is supported by the European Project UP-TO-US (User-centric Personalized IPTV ubiquitOus and secUre Services) under the Celtic Collaboration Framework.

References

1. A. K. Dey, Understanding and using context. *Personal and Ubiquitous Computing*, 5(1):4–7, 2001.
2. ETSI TS 182 028, NGN integrated IPTV subsystem Architecture.
3. WAP Forum Wireless Application Protocol Forum: http://www.wapforum.org/what/technical.htm
4. P. Shoval, V. Maidel, and B. Shapira, An ontology- content-based filtering method. In *I. Tech-2007—Information Research and Applications*, 2007.
5. K. Ali and W. Van Stam, TiVo: Making show recommendations using a distributed collaborative filtering architecture. *ACM SIGKDD International Conference on Knowledge Discovery and Data Mining (KDD'04)*, Seattle, Washington, USA, August 2004, pp. 394–401.
6. R. Pampapathi, B. Mirkin, and M. Levene, A review of the technologies and methods in profiling and profile classification. EPALS Technical Report, April 2005.
7. I. Cantador, A. Bellogín, and P. Castells, A multilayer ontology-based hybrid recommendation model. *AI Communications, Special issue on Recommender Systems*, IOS Press, 21(2–3):203–210, 2008.
8. S. E. Middleton, D. C. De Roure, and N. R. Shadbolt, Foxtrot recommender system: User profiling, ontologies and the world wide web. *The 11th International World Wide Web Conference (WWW2002)*, Hawaii, USA, May 2002.
9. J. Vallejos, B. Desmet, P. Costanza, and W. De Meuter, Pervasive communication: The need for distributed context adaptations. *ECOOP 2007 Workshop on Object Technology for Ambient Intelligence and Pervasive Systems (OT4AmI)*, Berlin, Germany, 2007, http://www.p-cos.net/documents/pervasive.pdf
10. Z. Yu, X. Zhou, and Z. Yang, A hybrid learning approach for TV program personalization. *International Conference on Knowledge-Based Intelligent Information & Engineering Systems (KES 2004)*, New Zealand, September 20–24, 2004, pp. 630–636.
11. Y. C. Chen, H. C. Huang, and Y. M. Huang, Community-based program recommendation for the next generation electronic program guide. *IEEE Transactions on Consumer Electronics*, 55:707–712, 2009.
12. K. Park, J. Choi, and D. Lee, A Single-scaled hybrid filtering method for IPTV program recommendation. *International Journal of Circuits, Systems and Signal Processing*, 4(4):161–168, 2010.
13. Media Content Distribution (MCD), http://www.etsi.org/WebSite/Technologies/MediaContentDistribution.aspx
14. ITU-T: International Telecommunication Union–Telecommunication http://www.itu.int/ITU-T
15. Open IPTV Forum, http://www.openiptvforum.org
16. DVB-IPTV, http://www.dvb.org/technology/standards/
17. DVB Digital Vides Broadcast http://www.dvb.org/
18. ETSI TS 201 812 V1.1.1 "Digital Video Broadcasting (DVB); Multimedia Home Platform (MHP) Specification 1.0.3."
19. W3C World Wide Web Consortium, http://www.w3.org
20. C. Kiss, Composite capability/preference profiles (CC/PP): Structure and vocabularies 2.0 April 2007. http://www.w3.org/TR/2007/WD-CCPP-struct-vocab2-20070430/
21. G. Klyne, J. J. Carroll, and B. McBride, Resource description framework (RDF): Concepts and abstract syntax. 2004. http://www.w3.org/TR/rdf-concepts/

22. WAG User Agent Profile 2001. http://www.openmobilealliance.org/tech/affiliates/wap/wap-248-uaprof-20011020-a.pdf
23. J. M. Martinez. ISO/IECJTC1/SC29/WG11 N6828 MPEG-7 Overview 2004. http://www.chiariglione.org/mpeg/ standards/mpeg-7/mpeg-7.htm
24. MPEG http://www.chiariglione.org/mpeg/
25. J. Bormans and K. Hill, ISO/IEC JTC1/SC29/WG11/N5231 MPEG-21 Overview v5. 2002 http://www.chiariglione.org/mpeg/standards/mpeg-21/mpeg-21.htm
26. TV-Anytime http://tv-anytime.org/

Chapter 18

Context-Awareness for IPTV Services Personalization

Radim Zemek, Songbo Song, and
Hassnaa Moustafa

Contents

18.1 Introduction 402
18.2 Context in IPTV Services 403
    18.2.1 Context Definition 403
    18.2.2 Obtaining Contextual Information Related
        to IPTV Services.. 404
    18.2.3 Required High-Level Contextual Information
        for Enabling IPTV Services Personalization 404
18.3 Context Modeling...413
18.4 Existing Personalization Approaches and Their Limitations.....................415
18.5 General Context-Aware IPTV Architecture 417
18.6 Conclusion 418
Acronyms...419
Acknowledgment 419
References 420

The new digital TV allows users not only to access new services and functionalities from their providers, based upon their profiles and contexts, but also to become active parts in the content creation and distribution process. With the consolidation of services, such as network Time Shifting (nTS) and network Personal Video Recorder (nPVR), users are allowed to record their own content and could also make them available to other users. In this new TV model, context-awareness is promising in monitoring the user's environment (including networks and terminals), interpreting the user's requirements and making the user's interaction with the TV dynamic and transparent. Consequently, content personalization is achieved matching the user's needs and one's surrounding environment (networks and devices). Nowadays, there is an increasing trend for context-awareness TV systems, in which promising applications and services enhancements are expected.

This chapter focuses on context-awareness for IPTV services personalization. First, a definition of the different context information types for IPTV services is given as well as the means for gathering and modeling the context information. Then the contextual information requirements for enabling different scenarios for personalized IPTV services are given, and several existing approaches for context-aware IPTV are presented showing their limitations. Finally, the chapter ends by providing a general context-aware IPTV architecture enabling enhanced services personalization.

18.1 Introduction

Users are accustomed to access multimedia content anytime and anywhere. They copy video and movie files to their devices that allow consuming the content while being away from home, they actively search video content on Internet, they access and watch movies, thanks to their Video on Demand (VoD) subscription, and they are also capable of receiving digital terrestrial broadcasting on their devices, thanks to advanced broadcasting systems (as oneSeg broadcasting system) [1].

However, such content management requires extra interaction, for example, in situations when a user wants to utilize resources of nearby devices, or when the user wants to receive content recommendation considering all content providers while taking user situation into consideration. Context-aware systems automatically adopts to user situation, for example, a context-aware IPTV automatically enables time shifting function whenever a user temporarily leaves the room or when the user is in an environment that does not support the content to follow the user from screen to screen.

Furthermore, personalized IPTV does not only mean providing meaningful content recommendation to the user, but also actively supporting user's lifestyle, for example, customizing any available screen to settings user is accustomed to. Context-aware and personalized IPTV understands user and adapts according to the user's situation or needs. For example, observation user's daily routines provides for anticipating user's actions—switching on television with news channel when the user arrives home.

In order to enable such high-level automation on behalf of user, variety of information expressing current situation of user, device, network, content and service needs to be collected and processed. Such information is called contextual information and two basic forms exists—low-level contextual information and high-level contextual information. Low-level information is often raw data and features. High-level information is often derived from one or more low-level contextual information.

18.2 Context in IPTV Services

18.2.1 Context Definition

The idea of utilizing environment information in computing systems was advocated by Mark Weiser [2]. He first introduced the term "pervasive" which refers to the seamless integration of devices into the users' everyday life. However, the term "context-aware computing" was not defined until 1994 by Schilit and Theimer [3], who described context as "location of use, the collection of nearby people and objects, as well as the changes to those objects over time." Much of the early work on context-aware systems used similar extensional definitions which define context by enumerating the constituting parameters. Brown et al. [4] enumerate "location, time of day, season of the year, and temperature." These definitions are very special and only reflect the types of information that have been used by the researchers in their context-aware applications.

A general definition of context was proposed by Chen and Kotz [5]: "Context is the set of environmental states and settings that either determines an application's behavior or in which an application event occurs and is interesting to the user." Considering the IPTV service, context can be considered as any information that can be used to characterize the situation of an entity related to the IPTV service. An entity could be the user, device, network, and service itself. Four types of context information are defined for IPTV services.

i. *User Context*: Includes information about the user, which could be static information, dynamic information, and inferred information. Static information describes the user's personal information which is stored in the database (e.g., name, age, sex, and input preference). Dynamic information is captured by sensors or by other services (e.g., user's location, agenda, and usage history). Inferred information is high-level information, which is inferred by other information (e.g., user's action "user is going to the bed' is inferred by the changed location").

ii. *Device/Terminal Context*: Includes information about the devices (terminals), which could be the device identity, status (turn on or off, volume), device capacity, and the device proximity with respect to the user.

iii. *Network Context*: Includes information about the network as the bandwidth, Quality of Service (QoS), the network load, and the user subscription type with respect to the access network.

iv. *Service Context*: Includes information about the service, which could be the service description and requirements. For example, the information about the broadcast service includes the content, language, and format.

18.2.2 Obtaining Contextual Information Related to IPTV Services

A context-aware service utilizes low-level (features) and high-level contextual information that are acquired from various sources and often processed in order to provide new high-level contextual information. The process of creating high-level contextual information from low-level one is called reasoning or inferring. For example, a camera that captures audience in front of television produces low-level contextual information—sequence of two-dimensional images. The sequence can be processed in order to provide high-level contextual information such as genre, age, or even user identification.

18.2.3 Required High-Level Contextual Information for Enabling IPTV Services Personalization

1. *User's profile:* User profile includes basic information that is often explicitly indicated by the user and is often static or changes only slowly over time. The profile includes the following almost static information: (i) unique ID—identification that uniquely identifies the user for personalizing his IPTV services, (ii) name, surname and nickname—provided by the user if the user wishes to do so, these details could be utilized by other content-based services, for example, participation in a discussion related to particular content, (iii) address—user can provide this information that could be used in interactive shopping channels, (iv) birthday, age or age group, gender, and (v) identification checksums/hashes/private–public encryption keys.

2. *User identification:* User's identification is an important part of any context-aware and personalized IPTV. It should be nonintrusive while requiring minimal interaction from the user who is being identified. Effective identification is also important for a user's authentication, authorization, and accounting (AAA). The checksums or hashes of any user's identification should be stored in user's profile information. A user can be identified using the following techniques:

   a. Identification based on login credentials: This identification method is ubiquitous in many systems; however, it requires tedious interaction from the user—entering login identification and password. Although the method is familiar to many users, it is not a preferred method when the user constantly needs to enter these credentials. It is also not a preferred method for any context-aware system.

b. Identification based on face recognition [6]: Although this method requires available cameras, many new television models are equipped with built-in cameras that are used for video calls, conferences or as a user interface for controlling games or TV (e.g., Microsoft's Kinect). Some 3D TV are also equipped with cameras in order to track face and eyes of the user in order to provide 3D TV experience without wearing glasses. Users are also getting accustomed of consuming multimedia content on tablets that are often equipped with a camera. The low-level contextual information required for this identification process mainly concern video stream from a camera (or infrared camera) and information that represent unique face features.

c. Identification based on biometric data: Users are well aware of the possibility to be identified based on biometric data such as fingerprints or eye's iris. Although such methods are very precise and suitable for authorization (e.g., transaction authorization), their acceptability is generally low [6]. The low-level contextual information required for this identification process mainly concerns information collected by biometric sensors or processed features.

d. Identification based on speaker voice: Speaker recognition and identification are based on one's voice characteristics. However, A.K. Jain et al. [6] argue that voice is not very distinctive and thus this identification method should be avoided in large-scale identification systems such as IPTV services. The low-level contextual information required for this identification process mainly concerns recorded voice or distinctive features.

e. Identification based on Near Field Communication (NFC), Radio-Frequency Identification (RFID), Personal Area Network (PAN), BAN: The main difference of this method to all the previous ones is that this method requires users to carry a device that enables their unique identification. Such device could be either a card with NFC or RFID embedded chip or a mobile phone supporting NFC either directly by the mobile phone or through NFC-enabled SIM-card. However, the use of NFC dictates that any equipment requiring user identification is equipped with NFC reader. On the other hand, it could be expected that ubiquitous screens will be equipped with wireless communication modules (e.g., digital signage) supporting PAN. Consequently, it can be envisioned that a user's mobile device communicate to such screens directly (e.g., Wi-Fi Direct, Bluetooth), and an inherent part of this communication such as mobile phone's Media Access Control (MAC) address can be used for user identification. The low-level contextual information required for this identification process mainly concerns unique identification number (e.g., mobile phone's MAC address) registered in the user's profile or device profile associated with the user.

3. *User's localization:* The knowledge of a user's both indoor and outdoor location is important for services personalization. The location accuracy varies depending on the used localization technology and the proper translation of the absolute location coordinates or low-level information to relative or annotated locations (e.g., home, home/living_room).

   Indoor localization could be achieved through using the following technologies:

   a. RF radio signal-based localization: Location estimation technique based on RF signal (e.g., Wi-Fi signal) utilizes either Radio Signal Strength Indicator (RSSI) or Time-of-Arrival (TOA) parameters [7]. The accuracy of TOA-based techniques is superior to RSSI-based ones; however, RSSI is often readily available on any device that communicates wirelessly and is sufficient for room accuracy [8]. The drawback of such a method is that precise location of Wi-Fi Access Point (AP), or map of radio signal in an environment is required. Furthermore, in order to utilize this method, the user needs to carry a device that periodically receives or transmits radio signal. The low-level contextual information required mainly concerns: Wi-Fi AP's SSID or MAC address and RSSI.

   b. RF radio signal localization based on proximity: Other option for indoor localization is based on devices proximity. In such case, neither absolute nor relative location information is used, only the knowledge of two or more devices being in proximity to each other is utilized. Bluetooth network is a good candidate for such localization—the range of Bluetooth network is up to 10 m and thus the knowledge of devices being in proximity is readily available. Other potential option is based on MAC address and processing RSSI variation (e.g., Wi-Fi Direct) in order to reason whether devices are in proximity. Provided that all considered devices utilize the same wireless communication technology and protocols, two possibilities exist: a static device can control and manage the session, and a mobile terminal can control and manage the session. In the former case, the static device monitors the MAC addresses and RSSI signal of a new device. When a new device (i.e., mobile device) is detected and reasoned as being in proximity utilizing RSSI, the static device queries context-aware and personalized IPTV platform for information about the content that is currently being accessed on the mobile device and either automatically start using own resources to provide the content or offers user to use its resources. In the latter case, mobile device senses MAC addresses and RSSI of static devices surrounding the mobile device. When a device whose resource can be utilized is detected, the mobile device negotiates with the static device to use its resources and transfers the ongoing session to the device. This process is either automatic without user intervention, based on other contextual information and past user habits or is initialized after the user's confirmation.

Basic localization based on this technique does not provide information on absolute or relative location; however, such drawback can be overcome provided that a stationary device knows its relative or absolute location and consequently communicates it to devices in proximity that express interest in the information. If stationary devices do not posses their relative or absolute location, their Bluetooth device profile could be utilized to reason the nature of the environment. The low-level contextual information required mainly concerns: MAC address (either mobile or static device, depending on the implementation), RSSI (either mobile or static device, depending on the implementation), and Bluetooth device profile (both devices, e.g., TV can suggest using nearby Bluetooth headphones).

c. Vision-based indoor localization: This technique is based on cameras that are fixed to the room's ceiling. A user is then located through image recognition and processing. Because such system setup is perceived invading privacy and thus lacks of consumer acceptance, the cameras often record in infrared spectrum [9]. Such arrangement allows sensing changes of thermal spectrum and at the same time avoids privacy issues. The advantage of vision-based localization method is that users are not required to carry any device in order to felicitate user's localization. The low-level contextual information required mainly concerns video stream from cameras.

d. Dead reckoning module [10]: Dead reckoning modules provide accurate position information for places without GPS signal such as indoor environments. The positioning is based on analyzing data from gyroscope, accelerometer, digital compass, and barometric altimeter sensors in order to obtain position information. The claimed precision is within 2% of the actual distance traveled by the user in environments without GPS signal or any other reference nodes. Although the method is precise and sensors in a mobile device could be utilized for such localization, the necessary data analysis and processing poses drawback on the method. The low-level contextual information required mainly concerns accelerometer data, gyroscope data, digital compass data, and barometric altimeter data.

The outdoor localization could be achieved through the following technologies:

e. GPS: Currently all Smartphones at the market are capable of receiving GPS signal and providing coordinate information together with estimated velocity and heading direction. The accuracy of GPS depends on the environment and direct visibility of GPS satellites; however, the accuracy is often between 2 and 10 m outdoor [8]. The low-level contextual information required mainly concerns longitude, latitude and altitude coordinates, estimated velocity, and estimated direction.

f. Cell-ID: Cell-ID localization is based on the knowledge of base stations locations and coverage. The accuracy ranges from hundreds of meters in urban environment to kilometers in rural areas. Nevertheless, the accuracy is sufficient for providing personalized and location-based news as presented in NA#3: "Personalized news/day summary" UP-TO-US use case. The accuracy is also sufficient for recommending local content to the user. The low-level contextual information required mainly concerns cell-ID (either directly from mobile phone of from operator's Presence Enabler).

g. Wi-Fi: Outdoor localization based on Wi-Fi is often based on look-up (fingerprinting) technique. The accuracy of this method ranges between 10 and 20 m [11]. The low-level contextual information required mainly concerns SSID and RSSI.

4. *Detecting user's mood:* Although the selection of multimedia content is often influenced by the user's mood, it is one of the hardest high-level contextual information to reason. Reasoning user's mood is based either on face expression [11] or voice analysis. IETF's RFC 4480 Rich Presence Extension (RPID) to the Presence Information Data Format (PIDF) discusses presence information. One of the RPID element indicate various user's mood—8 different moods are considered in total [12]. Current mood detection research approaches utilize voice [13], face analysis [14], and vital sign sensor analysis. Potential low-level contextual information includes:

a. Voice/speech streams from a microphone—mood detection/analysis based on voice

b. Video stream from a camera—analysis of a user's expressions

c. Readings from vital sign sensors (heart rate, blood pressure, skin conductance)

5. *Detecting user's schedule/activity:* A user's personal schedule or calendar can provide information about the users' activity, availability, status, and also location. Many Web-based calendars and Android-based devices calendar that allows synchronizing with Web-based service provide APIs to access such information. The low-level contextual information required mainly concerns login IDs and passwords, event name, event time and duration, event location, and event description.

6. *User's social network credentials:* Information from user's social network can provide valuable information, for example, content recommendation based on social networks applications/services (as Facebook and Twitter) status. In order to access the information, the user needs to provide login credential to those services. Optionally, access to those services can allow automatic update of user status related to currently watching content. The low-level contextual information required mainly concerns login IDs and passwords.

7. *User's preferences:* Preferences are either explicitly indicated by the user or learnt implicitly from the user's behavior, interaction with the content, and consumption history. Part of user preferences are also details about the user's

favorite content, genre, actors and actresses, and other details important for content recommendation. User's preference includes the following details:

a. Preferred audio language: Preferred content audio language (e.g., original, English, French, Spanish) is either explicitly specified by the user or reasoned automatically over time from content consumption history. Default option can be reasoned from a user's profile address or set to language of country where user is located most of the time. The low-level contextual information required mainly concerns selected audio language, and available audio language (e.g., content metadata).

b. Preferred subtitle language: Preferred subtitle language settings (e.g., none, hearing impaired, German, Polish) is either explicitly specified by the user or reasoned from content metadata and observing user's choices over time. Optionally the subtitle setting also depends on other people presented in a room—for example, one user prefers original audio language (e.g., English) and another user likes having subtitles in mother tongue (e.g., French). The low-level contextual information required mainly concerns the selected subtitle language, and available subtitle language (e.g., content metadata).

c. Favorite content genres: Information about favorite user's genre is often reasoned from content consumption history and content's metadata. This information is often used for content recommendation. However, other parameters such as time, location and nearby people should be considered for content recommendation. The low-level contextual information required mainly concerns the genre.

d. Favorite leading and supporting actors and actresses: The knowledge of user's favorite actors and actresses is useful information for content recommendation. Such knowledge can be reasoned from content consumption history. The low-level contextual information required mainly concerns the leading actors and actresses, and supporting actors and actresses.

e. Favorite director: User's content selection is often influenced by favorite director and thus this metric provides useful information for content recommendation. The low-level contextual information required mainly concerns the director.

8. *User's watching habits and consumption history:* Watching habits indicate a user's repetitive actions such as watching news every evening and watching a movie every Friday night. Consumption history records enable data mining and consequently content recommendation, automatic recording of favorite programs, or anticipating the user's actions. It also enables avoiding recommending content that the user (or users in case of group recommendation) has already seen. Since a user consumes content across many devices, it is important to record both watching habits and consumption history regardless of device being used. The low-level contextual information required mainly concerns unique content identifier (e.g., obtained by a hash function),

time, part of day, day of week, holiday flag, Electronic Program Guide (EPG), content metadata, unique ID of people nearby, location (i.e., location where user consumes the content), and device ID—particular content consumed on a specific device.

9. *Used Device:* Multimedia content is accessed from various devices and thus the device's screens size, capability, and communication means are different and correspond to the current user's preferences. The knowledge of device that is being used to access or present content and the device's parameters enables creating context-aware system that adapts content to the device and also benefits from the knowledge. For example, knowing the primary device's location and content most often accessed at such location can provide framework through collaborative filtering for content recommendation based on particular location. Such knowledge can enable location-based recommendation—various users access the same or similar content at certain locations and consequently such content can be considered as relevant to the particular location and recommended to other users.

Information related to device encompasses:

a. Device identification: Unique device identification is important for several reasons. First, it enables one to combine devices and consumption history in order to provide personalization, for example, knowing that user consumes news highlight on a particular device (e.g., portable device) and movies on another (e.g., television) will enable content recommendation not based only on the user but also on the device being used to access the content. Second, it enables one to associate devices to particular users (e.g., for identification purpose). The low-level contextual information required mainly concerns the unique device ID (e.g., set of MAC addresses).

b. ID of users utilizing the device: Whenever a device is utilized by a user, it should be able to obtain the identification of a user as has been discussed in the User Identification section. The low-level contextual information required mainly concerns the users' ID.

c. Capability: Capability identifies device's features such as whether the screen is a touch screen, capable of gesture recognition and control, or if any control devices is available (e.g., keyboard). The low-level contextual information required mainly concerns available microphone, available audio system (e.g., mono, stereo, surround system), touch screen capable, gesture recognition capable, and available peripherals (e.g., headphone, keyboard).

d. Screen size/resolution: Screen size and resolution affect several aspects with main ones being User Interface (UI), content presentation and required resolution of the content. The low-level contextual information required mainly concerns screen size, screen resolution, selected contrast, brightness, and color saturation levels.

e. Ambient noise, volume level: Information about ambient noise level enables automatically adjusting volume level. The low-level contextual information required mainly concerns device's microphone output and volume level.

f. Supported access networks, network availability, supported protocols: Information about supported network interfaces and availability of corresponding access networks provides details about nominal bitrates and connections. Protocols such as LLTD [15] developed by Microsoft and LLDP [16] provides network topology discovery and quality of service diagnostics capabilities. The knowledge of supported protocols such as UPnP [17] and DLNA [18] enables device discoveries and standardized utilization of nearby devices resources. The low-level contextual information required mainly concerns support of Body Area Network (BAN), PAN, LAN, WLAN, cellular network, and network availability.

g. Available energy resources: When a user accesses content on devices that are battery powered, the available battery resources dictates the prioritization of services and content distribution. For example, if the device is used as a primary authentication device, then it is necessary to maintain operability as long as possible for a system to provide services on nearby devices that might have independent power source. The low-level contextual information required mainly concerns battery level (e.g., unlimited when device is being charged or plugged into the power socket), remaining operation time, and last charging time.

h. Location: Location information provides not only absolute or relative location details both in indoor and outdoor, but also whether the device is in private or public space, for example, public kiosk. Such information enables displaying only content that is appropriate for a given screen. Furthermore, a device knowing its location can provide such information to devices that are not capable of locating themselves, for example, lacking a GPS receiver. The low-level contextual information required mainly concerns absolute or relative location information, and information on private or public location.

i. Nearby devices: Information on devices in each other's proximity enables utilizing those devices resources, for example, displaying content on nearby screen. Such information can be obtained through LLTD, LLDP, UPnP, and DLNA protocols [15–18]. The low-level contextual information required mainly concerns: nearby device ID, and nearby device's primary network.

Other useful information is about device portability, for example, mobile phone, tablet, and television have different portability. Furthermore, knowing whether certain device is currently being used by a user can provide seamless session continuity or adaptation. For example, when a user starts using tablet while watching TV, the displayed information adapts to the current program, and when a user leaves room with the tablet, the session from TV continues on the tablet after the user's confirmation.

10. *Network States:* Content is accessed over multiple access and transport networks and their states directly affect means to access content, content presentation (especially quality), and also user experience, for example, for an interactive content where low latency is an important prerequisite. To this respect, this section discusses the main parameters (context information) reflecting the network states. Some protocols such as LLTD and LLDP aim at standardizing the ways to obtain these parameters.

  a. Access network: Information about available access networks enables selecting the most appropriate network while considering other parameters such as transmission speed, latency, and cost.

  b. SSID (i.e., Wi-Fi) and IP address.

  c. Nominal and effective transmission rate, latency, and QoS.

  d. Connection setup time: Knowledge of precise setup time enables selecting the most appropriate connection for session continuity.

  e. Security parameters and supported protocols.

  f. Received Signal Strength Indicator (RSSI): RSSI indicates the strength of wireless connection. The value is inversely proportional to distance between devices. Trend of the value can also indicate whether devices are approaching toward or departing from each other. Such information can be used for initializing session continuity across devices. The low-level contextual information required mainly concerns RSSI, and current data transmission rate.

  g. Location: Information on network availability as a function of geographical location can be used to appropriately manage network resources and content. For example, such knowledge together with prediction of user location can help schedule content delivery or caching. The low-level contextual information required mainly concerns available networks at particular locations.

  h. Status: Status indicates the states of any network connection such as active, disabled, not available.

11. *Service/Content:* Considering that users' accessed content provided by various services, those services will be competing for users' attention. Therefore, it is in their best interest to provide comprehensive information about the offered content to increase content discovery and customers' retention. Consequently, availability of detailed information can be further exploited by content recommendation services, services that automatically record contents that user might be interested in or by services providing real-time discussion on content. This information includes:

  a. Content identifier: Unique identifier that enables exclusive identification of content and content provider. Such information is used to prioritize subscribed content providers, for billing, for recording consumption history and preferences. Optionally, users have means to review and rate content providers, and such information assists users with content provider selection. The low-level contextual information required mainly concerns unique content provider identifier.

b. Content information: In order to navigate and discover content, the following information is provided either directly by the content creator/production or by service providers.

    i. Unique content identifier

    ii. Description

    iii. Keywords

    iv. Genre

    v. Content duration—helps to recommend content based on user's availability

    vi. Rating

    vii. Reviews and discussions

c. Content price: Indicates actual content or subscription price and special offers.

d. Required or supported QoS: Information about required QoS and related parameters such as minimal supported bit rate, minimal and maximal supported resolution and required latency facilitates content presentation and session continuity. Such information also enables one to select the most appropriate device. The low-level contextual information required mainly concerns minimal bit rate required, minimal and maximal supported resolution, and required latency.

e. EPG information: EPG enables a user to conveniently navigate and select scheduled content, it also assists the user to schedule recording.

f. Digital right management (DRM): Copyright, content management, and user rights are periodically discussed across the entertainment industry. In order to evaluate whether session transfer between devices is possible, the following low-level contextual information is considered in order to evaluate such possibility: recording limitations—indicating whether content can be recoded and how many copies can be created, content availability with respect to location and content availability with respect to device.

18.3 Context Modeling

Effective deployment and easy utilization of context-aware system is partially dictated by models that are used for contextual information acquisition, storage, and retrieval. Furthermore, a good model not only provides ways for effective contextual information retrieval but also effective processing of such information. A model should be also easily extensible since new contextual information often becomes available during service operation. Strang and Linnhoff-Popien [19] summarized the most widely used data structures:

- Key-Value model: This model is the simplest data structure for context modeling, and they are frequently used in various service frameworks. Schilit et al. [20] models the contextual information in a key-value pair, with an environmental variable acting as the key and the value of the variable holding the actual context data. This model is powerful enough to allow pattern-matching queries, but is not efficient for structuring purpose.

- Markup scheme models: Markup scheme modeling uses standard markup languages or their extensions to represent context data. These models use a hierarchical data structure. Contexts are modeled as tags and corresponding fields. In particular, the content of markup tags is usually recursively defined by other markup tags. Because of the scheme representation and the recursive definition, these models are strong in context sharing and partial validation.

- Graphical models: The Unified Modeling Language (UML) is frequently used to model contextual aspects. The strengths of graphical models are on the structure level. They are mainly used to describe the structure of contextual knowledge (the objects and their relationships) and derive some code from the model. But they are usually not used at instance level.

- Object-oriented models: These models possess the main benefits of any object-oriented approach. They encapsulate all the details of data collection, data fusing, and context processing within the active objects. The context information is accessed through well-defined interfaces, and can be reused. These characters make object-oriented models strong regarding distributed composition. And the propriety inheritance makes applications easier to define the objects and theirs relationships. The key problem with this approach is that the represented information lacks expressiveness and extensibility. And these models usually have strong additional requirements on the resources of the computing devices which often cannot be fulfilled in ubiquitous computing systems.

- Logic-based models: Logic-based models have a high degree of formality. Typically, facts, expressions and rules are used to define a context model. The contextual information which is expressed as facts is added to, updated in, or removed from a logic-based system. Using the inference process can obtain new facts based on existing rules and facts in the systems.

- Ontology-based models: Ontology represents a description of the concepts and their relationships. Therefore, ontology is a very promising instrument for modeling contextual information due to their high and formal expressiveness and the possibilities for applying ontology reasoning techniques. Various context-aware frameworks use ontology as underlying context models.

18.4 Existing Personalization Approaches and Their Limitations

Although context-awareness can bring additional smart services and useful functionalities to IPTV systems and allow enriched personalization, how to make IPTV services context-aware and how to use the acquired context information to realize personalized TV services is still a challenge.

Several solutions for context-aware TV systems have been proposed employing either a distributed or a centralized approach. In the former, several entities have the capacity to acquire, process, and store the context information, while cooperating with each other to provide context-aware services. While the latter approach treats and stores the collected context information in a centralized server that could be located in the domestic sphere (for instance, server attached to the STB "Set-Top-Box"—"partially centralized" or in the operator network "fully-centralized").

A distributed context-aware interactive TV (ITV) solution is proposed in Ref. [21], implementing software agents on top of physical devices (STBs, mobile phones, and PDAs) for context acquisition, treatment, and storage, where each device agent discovers the other devices agents for exchanging/analyzing of the context information acquired. In this solution the user locations are acquired by sensors while other context information of terminals (e.g., screen size, supported content format) is provided by the terminals themselves. The network-related context is acquired through the "active network" principle, used to route the content according to the context (e.g., traffic condition, user's location, etc.) through active nodes that process the contextual information and make a decision on the best network path to deliver the content. Finally, the content context information is provided by the content provider and is stored in the content server at the service provider or operator side. In this solution, the computing capacity of the distributed devices limits the service performance, and the context information is simple and falls short to reflect the user preference, which in turn limits the service personalization.

In Ref. [22], a partially centralized context-aware TV architecture is proposed for the selection and insertion of personal advertisement in the broadcast content, based on the aggregation of past sequence of individual contexts (i.e., past viewing) and the association of the current user context to those past contexts in order to determine the most appropriate advertisement. The context is described as: (i) location: user different locations at home, (ii) identity: user identity (name, age, occupation, favorite channel, or movie), device identity (screen size, required resolution), content identity (EPG "Electronic Program Guide" and other related information such as category, channel type), and event identity (information that might have an impact on the user's watching behavior), (iii) user's activity (e.g., user's control of the program), and (iv) time. This solution is mainly targeting home networks and does not consider the network contexts and the service personalization is limited to personal advertisement insertion.

A client–server approach-based TV system is proposed in Ref. [23] aiming to realize the TV Set automatic control and personalized content recommendation through presenting a personalized EPG "Electronic Program Guide." In this framework, Service Agent Managers (SAM) present the client-side and sends the acquired context information to a server named CAMUS (Context-Aware Middleware for Ubiquitous System), which in turn manages this information. The CAMUS server includes: (i) a context manager that is responsible for context representation, inference and storage as well as discovering the ongoing application through receiving the service request initiated by the SAM, and (ii) a task manager having a rule-based inference engine which monitors the context manager and generates tasks according to the stored rules and ongoing applications. On the other hand, the SAM contains Service Agents that communicate with physical devices (sensors or equipment like TV Set), and notify the CAMUS when each device detects some noticeable changes and receive commands from the CAMUS to control the devices. The context information used in this solution does not consider the network context. In addition, no privacy protection exists for the user context information.

Another client–server approach-based personalized recommendation architecture for Digital TV is presented in Ref. [24]. In this work, the used context information is divided into five dimensions who (identity), when (time), where (location), what (activity, the content information), and how (how can the user receive the service, through a mobile, portable, or fixed device?). The user-side subsystem is implemented in the end-user terminal such as set-top-box (STB), computer or a mobile phone and includes the following modules: (i) User profile manager, responsible for the acquisition of the user profile information including personal data (name, age, occupation) and user's explicit preferences (which are input manually by the user including his preferred content type, his favorite author's name, etc.); (ii) User context manager, responsible for the acquisition of the user's current context information like location, activity, etc.; (iii) User context interpreter, responsible for the inference of the implicit preference through analyzing the user's current and past context information using the rule-based reasoning method. The implicit preference indicates the user's current preference which may be different with the explicit preference; and (iv) Recommendation manager, coordinating the other modules for the personalized content recommendation and analyzes the feedback of the user. On the other hand, the service-provider subsystem includes the following modules: (i) Context-based content filer, filtering TV programs through considering users' explicit and implicit preference, and the content description of the TV programs; (ii) TV Program Description, responsible for consulting the TV programs information captured by the TV programs collector and storing it in a database; and (iii) TV programs collector, collecting the information relative to TV programs form outside sources such as the WEB. The privacy protection in this work is carried out through storing the user's personal context in his devices, which is not an always practical and efficient approach, and it limits the nomadic service access through passing by other's devices.

18.5 General Context-Aware IPTV Architecture

It is noticed that the existing contributions could not satisfy TV services personalization in a complete and appropriate manner, and most of the existing context-aware-based TV services are limited to the home sphere with no focus on the whole IPTV architecture.

Indeed, efficient IPTV service personalization require efficient context-awareness integration along the IPTV architecture through including the components related to context-awareness following a hybrid approach (employing distributed and centralized components) [25] as illustrated in Figure 18.1.

The following is a description of these components (functional entities):

i. Context information gathering: to dynamically acquire context information for different context types (user, device, network, content) and is then distributed among the user domestic sphere, network domain, and content provider domain.

ii. Context information management: is a centralized entity for organizing the acquired context information and inferring high-level information. The gathered context information can be divided into two classes: common context information like content and network information and personal context information like user's profile and activity.

iii. Context Database: is a centralized entity that stores the gathered and inferred context information.

iv. Service control and adaptation: is a centralized entity that presents a bridge between context-aware management and service provider functions (like service selection and content delivery functions) and should be responsible for configuring the services according to the context information gathered and inferred.

v. Local service adaptation: is a distributed entity in the user domestic sphere, for locally adapting the service (after content reception through the IPTV chain) according to the gathered context information from the user and terminals.

vi. Privacy protection: is a centralized entity to control the user context information publishing and manipulation by only authorized entities while guaranteeing different privacy levels.

Figure 18.1 shows the distribution of these components along the different IPTV domain as well as the interaction between them which could be initiated by the user himself whenever he wants to access the service or by the system itself in a continuous manner during the user session.

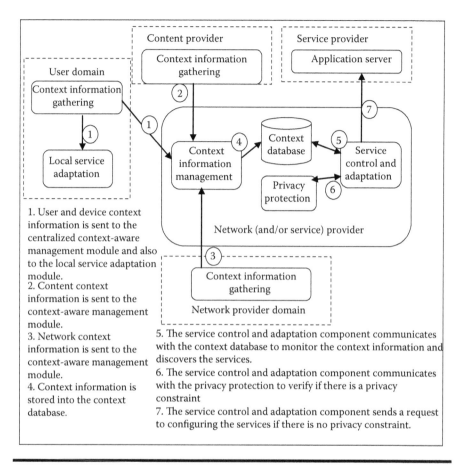

1. User and device context information is sent to the centralized context-aware management module and also to the local service adaptation module.
2. Content context information is sent to the context-aware management module.
3. Network context information is sent to the context-aware management module.
4. Context information is stored into the context database.

5. The service control and adaptation component communicates with the context database to monitor the context information and discovers the services.
6. The service control and adaptation component communicates with the privacy protection to verify if there is a privacy constraint
7. The service control and adaptation component sends a request to configuring the services if there is no privacy constraint.

Figure 18.1 Overview of the functional entities for an IPTV service personalization system.

18.6 Conclusion

This chapter focuses on IPTV services personalization through applying context-awareness. Context-awareness is promising in enhancing the general IPTV system through providing smart and advanced services with appropriate content recommendation and adaptation and hence allowing services interactivity and personalization. The existing context-aware IPTV solutions could not satisfy enriched IPTV services personalization in a complete and adequate manner and are limited to the home sphere without focusing on the architecture of the context-aware IPTV from the deployment point of view. This chapter presents the different context information types, gathering means and modeling approaches suitable for IPTV services. A presentation of the existing context-aware IPTV solutions and their limitation is also given. Finally, the chapter ends by illustrating candidate architecture for context-aware IPTV deployment.

Acronyms

AAA	Authentication, Authorization, and Accounting
AP	Access Point
BAN	Body Area Network
CA	Context Awareness
DLNA	Digital Living Network Alliance
EPG	Electronic Program Guide
GPS	Global Positioning System
IETF	Internet Engineering Task Force
ITV	Interactive TV
LLDP	Link Layer Discovery Protocol
LLTD	Link Layer Topology Discovery
MAC	Media Access Control
NFC	Near Field Communication
nPVR	network Personal Video Recorder
nTS	network Time Shifting (nTS)
PAN	Personal Area Network
PIDF	Presence Information Data Format
RF	Radio-Frequency
RFC	Request for Comments
RFID	Radio-Frequency Identification
RPID	Rich Presence Extension to PIDF
RSSI	Received Signal Strength Indicator
STB	Set Top Box
TOA	Time of Arrival
UI	User Interface
UML	Unified Modeling Language
UPnP	Universal Plug and Play
VoD	Video on Demand

Acknowledgment

This work is supported by the European Project UP-TO-US (User-centric Personalized IPTV ubiquitOus and secUre Services) under the Celtic Collaboration Framework.

References

1. Kenichi Tsuchida, Research related to "One-Seg" transmission—Aiming at a "One-Seg" service that can be received anywhere, *Bradcast Technology*, no. 26, pp. 19, Sept. 2006 (http://www.nhk.or.jp/strl/publica/bt/en/rd0026.pdf).
2. Weiser, M. 1991. The computer for the 21st century. *Scientific American*, 265(3):66–75.
3. Schilit, B. N. and Theimer, M. 1994. Disseminating active map information to mobile hosts. *IEEE Network*, 8:22–32.
4. Brown, P., Bovey, J., and Chen, X. 1997. Context-aware applications: From the laboratory to the marketplace. *IEEE Personal Communications*, 4(5):58–64.
5. Chen, G. and Kotz, D. 2000. A survey of context-aware mobile computing research. Technical Report TR2000-381, Department of Computer Science, Dartmouth College.
6. Jain, A. K., Ross, A., and Prabhakar, S. 2004. An introduction to biometric recognition, *IEEE Trans. on Circuits and Systems for Video Technology*, 14(1), 4–20.
7. Patwari, N., Ash, J. N., Kyperountas, S., Hero III, A. O., and Moses, R. L. 2005. Locating the node: Cooperative localization in wireless sensor networks, *IEEE Signal Processing Magazine*, 22, 54–69.
8. Hightower, J. and Borriello, G. 2001. Location systems for ubiquitous computing, *IEEE Computer*, 34, 57–66.
9. Kemper, J. and Linde, H. 2008. Challenges of passive infrared indoor localization, *Proc. of the 5th Workshop on Positioning, Navigation and Communication 2008 (WPNC 2008)*, pp. 63–70, March 2008, Hannover, Germany.
10. Honeywell, Dead Reckoning Module, http://www.magneticsensors.com/dead-reckoning-module.php
11. Skyhook, Inc., http://www.skyhookwireless.com
12. RFC 4480, http://tools.ietf.org/html/rfc4480
13. Polzin, T. S. and Waibel, A. 2000. Emotion-sensitive human–computer interface. *Proc. of the ISCA Tutorial and Research Workshop on Speech and Emotion*, September 2000, Newcastle, UK.
14. Sun, Y., Sebe, N., Lew, M. S., and Gevers, T. 2004. Authentic emotion detection in real-time video. *Computer Vision in Human-Computer Interaction, ECCV 2004 Workshop on HCI*, pp. 94–104, May 2004, Prague, Czech Republic.
15. Link Layer Topology Discovery Protocol Specification, http://msdn.microsoft.com/en-us/windows/hardware/gg463024
16. IEEE 802.1AB Standard, Station and Media Access Control Connectivity Discovery http://standards.ieee.org/getieee802/download/802.1AB-2005.pdf
17. Universal Plug and Play, http://www.upnp.org/
18. Digital Living Network Alliance, http://www.dlna.org/home
19. Tomas Strong and Claudia Linnhoff-Popien A Context Modeling Survey.
20. Schilit, B. N., Adams, N. L., and Want, R. Context-aware computing applications. In *IEEE Workshop on Mobile Computing Systems and Applications (WMCSA '94)*, pp. 89–101, Santa Cruz, CA, USA.
21. Santos, J. B. D., Goularte, R., Faria, G. B., and Moreira, E. D. S. 2001. Modeling of user interaction in context-aware interactive television application on distributed environments, *1st Workshop on Personalization in Future TV*, Sonthofen, Germany.

22. Thawani, A., Gopalan, S., and Sridhar, V. 2004. Context aware personalized ad insertion in an interactive TV environment. *4th Workshop on Personalization in Future TV*, Eindhoven, the Netherlands.

23. Moon, A., Kim, Hs., Lee, K., and Kim, H. 2006. Designing CAMUS based context-awareness for pervasive home environments. *International Conference on Hybrid Information Technology*, pp. 666–672, Cheju Island.

24. Santos da Silva, F., Alves, L. G. P., and Bressan, G. 2009. PersonalTVware: A proposal of architecture to support the context-aware personalized recommendation of TV programs. *7th European Conference on Interactive TV and Video,* Leuven, Belgium.

25. Song, S., Moustafa, H., and Afifi, H. 2012. A survey on personalized TV & NGN services through context-awareness. *Journal of ACM Computing Surveys.*

Chapter 19

Metadata Creation and Exploitation for Future Media Networks

Thomas Labbé

Contents

19.1 A Moving Audiovisual Experience 424
    19.1.1 Social Networks..424
    19.1.2 Convergence 424
    19.1.3 Recommendation..425
    19.1.4 Interactivity 425
19.2 Metadata: A Big and Heterogeneous Family 426
    19.2.1 Definition 426
    19.2.2 Audiovisual Content Chain and Associated Metadata 426
    19.2.3 Metadata Types 427
    19.2.4 Standards..430
19.3 Key Players...432
    19.3.1 Pure Media Players 432
    19.3.2 Metadata Creation Players 432
    19.3.3 Metadata Aggregation Players...433
    19.3.4 Metadata Exploitation Players...433
    19.3.5 Key Players and Value Chains Positioning 434
19.4 Metadata Processing...434
    19.4.1 Technologies Involved...435

    19.4.1.1 Backend Services 435
    19.4.1.2 Frontend Services..436
  19.4.2 Metadata Involved 437
19.5 Metadata Systems within Media Networks 438
  19.5.1 Content Management Systems...438
  19.5.2 Data Mediation 439
    19.5.2.1 Semantics 439
  19.5.3 Context-Awareness..441
19.6 Conclusion 443
Acronyms... 443

19.1 A Moving Audiovisual Experience

Traditional audiovisual contents services are quickly moving to new content consumption experiences. Indeed, the Web culture is overflowing its original frame to influence all audiovisual services whatever the target support is (TV, PC, mobile): ergonomics navigation, open Application Programming Interface (API), personalization, consumption, and so on. For instance, the user browses on TV like on PC, with the catch-up TV he can consume the content when he wants or come back to it easily. This is no more than the reflection of the societal evolution.

The new generations, who have grown up with Internet and global Information and Communication technologies, experience them differently. They cross media easily, they are multitasking, and dematerialization is not a blocking point. In that context, the content accessibility and the way to experience it are more important than owning it: the mutation of the audiovisual experience has already begun, and so on.

We can illustrate this mutation in four main services families that are: social networks, convergence, recommendation, and interactivity.

19.1.1 Social Networks

Social networking is becoming one of the most trusted forms of advertising together with customers' opinions.

Exploring the possibilities of metadata like social information (social graph, profile, and eventually usages, etc.) and opinions, comments, votes linked to a content issuing from social network can be a good approach to improve contents recommendation tools.

19.1.2 Convergence

There are several levels to convergence.

A first level is the internal convergence within a provider environment with multiplay offers (providing services and/or devices for different access networks). This is typically an important issue for an operator or manufacturers.

The second level is the convergence between these internal services/devices and external services/devices linked to audiovisual contents (news, databases, social networks, user generated contents (UGC), recommendation, game consoles, boxes, etc.) which may be provided by a lot of various players. This kind of convergence can only be reached with the exchange and exploitation of contents metadata.

If network operators had given priority in the last few years to the first level of convergence, one has to notice that the second one has already been explored by audiovisual market players. TV set manufacturers, for instance, have created links with Video on Demand (VOD) suppliers or with some Web actors (like games actors, news, social networks, etc.) or have facilitated interfaces with different types of boxes (Private Video Recording (PVR), games, etc.). This, in order to include services in their TV set. Internet content services are coming to the TV screen with adapted interface, better image quality, and professional content.

Having a convergent approach of metadata is obviously a necessity to perform this kind of service.

19.1.3 Recommendation

We can organize recommendation into four types:

- Editorial recommendation which highlights contents for all users: blockbusters, best sellers, and so on.
- Contextual recommendation suggests new contents by similarity: movies from the same producer, next part, those who have bought this video, also buy this one, and so on.
- Personalized recommendation which aim to suggest new contents according to the profile: favorite actors, type of movies, and so on.
- Social recommendation which allow the user to recommend content to his social network but also to share its universe: to be fan of, and so on.

In these cases, we find metadata about user identification and his devices, some very precise information about contents such as the type of the movie, the mood, the scenario, associated awards, type of publics, the speed of the movie, and so on.

19.1.4 Interactivity

Interactivity is the possibility for the user to interact with content. Different types of interaction can be considered:

- Click interaction: possibility for a given moment to click in a defining part of a main content to access to other information.
- Browsing interaction: chapter browsing, search inside the content (e.g., this can be linked to image or vocal recognition).

- Comments and votes interaction: adding comments or votes information to content.
- Chat between people consuming the same content.

This kind of interactivity services are based on different levers: defining clickable spaces in the content and give the possibility to browse; having "time stamp" metadata to identify different instants within a content (these metadata could feed new services like clickable video, navigation through the content, etc.); and sometimes enrich offer with additional contents.

In that cases we are dealing with "temporal" metadata—time stamp information or time code linked to an area of the video—image or vocal recognition metadata.

As a conclusion of this chapter we can observe that all these services evolution will allow content consumers to experience differently contents consumption and enriched their usages.

These services are widely based on "audiovisual metadata," whose good mastery opens the door of new users' experiences and allow us to consider new ways of consuming audiovisual contents.

19.2 Metadata: A Big and Heterogeneous Family

19.2.1 Definition

Metadata (metadata, or sometimes meta information) is "data about data," of any sort in any media. Metadata is structured, encoded data that describe characteristics of information-bearing entities (*in our case audiovisual contents*) to aid in the identification, discovery, assessment, and management of the described entities.[*] Metadata may be recorded with high or low granularity: network information, contents characteristics, user profile, and so on.

In this chapter, we will focus on "multimedia metadata": metadata linked (directly or not) to a multimedia content (also called "audiovisual" content by a voluntary misuse of language) and which is not the video nor the audio stream. This definition is voluntarily large, including data sometimes identified as "additional content" (e.g., users' comments on a given movie).

The audiovisual content exploitation chain will be presented in this part, and then the different corresponding metadata types will be described.

19.2.2 Audiovisual Content Chain and Associated Metadata

Historically, the audiovisual content exploitation chain can be divided in different steps (see Figure 19.1).

---

[*] American Library Association, Task Force on Metadata *Summary Report*, June 1999.

Figure 19.1 Traditional audiovisual content chain.

Each step creates and/or uses information linked to a given content, like: title (writing), storyboard (preparation), timecode (shooting), subtitles (postproduction), poster (edition), id (aggregation), exploitation clauses (distribution), usage (watching), and so on. This nonexhaustive list shows that these data, we called *audiovisual metadata*, are the links required for the continuity and exploitation of the audiovisual chain.

These basic metadata are then widely exploited all along the distribution chain, and may be enriched all the way through the different steps. The enrichment can be for instance the completion of missing information that have to be created or retrieved *a posteriori* (manually, or using a mediator connected to external sources, or making the end-user participate to data updates, etc.).

However, new audiovisual metadata are created outside this traditional distribution chain (new Web players, uploaded information, content scenes descriptive metadata, interactivity services data, etc.), that will be reintegrated into the chain to feed new services allowing new user experiences.

Thus, in order to answer to service needs, we can see that there is an evolution of audiovisual metadata along and outside the traditional multimedia content distribution chain, as well as an evolution of its temporal exploitation (which is no more chronological-only).

In spite of this heterogeneity of audiovisual metadata, we can classify them within different types.

19.2.3 Metadata Types

Figure 19.2 shows the different types of metadata involved in the audiovisual content distribution chain.

Metadata can be parsed into five main families (enhanced with new types) to which two connate ones can be added (Table 19.1).

Once generated, these basic metadata are then widely exploited all along the distribution chain, and may be enriched all the way through the different steps.

However, new audiovisual metadata are created outside this traditional distribution chain (new Web players, uploaded information, content scenes descriptive metadata, interactivity services data, etc.), that will be reintegrated into the chain to feed new services allowing new user experiences (Figure 19.3).

Thus, in order to answer to service needs, there is an evolution of audiovisual metadata along and outside the traditional content distribution chain, as well as an evolution of its temporal exploitation (which is no more chronological-only).

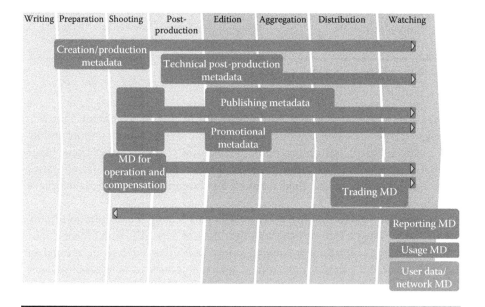

Figure 19.2 Audiovisual metadata type.

Table 19.1 Metadata Families

Metadata Families	Types
	Identifier, title
	Facts about shooting (date, location, team, etc.)
	Rights (director, producer, etc.)
	Scene description: information about what is seen and heard within a content
	Fine Genomic characteristics: using the same ideas as for the Music Genome Project where a song is represented by around 150 "genes" (musical characteristics), the movie is postanalyzed through a palette of information (up to more than thousand "genes"):
Creation/production metadata	Genre
	Time period
	Pace
	Mood (Calm, Contemplative, Violent, Fast, Suspense, Humor, etc.)

Table 19.1   (continued) Metadata Families

Metadata Families	Types
	Audience
	Character depth
	Plot type
	Genre
	Temporal Metadata used for time stamping other creation metadata
Technical, postproduction metadata	Film editing and special effects data integration
	Codec, frame rate, bit rate, and so on
Publishing metadata	Metadata for edition and reproduction
	Descriptive editorial metadata
	Metadata for advertising
	Metadata for operation and compensation (settlements, contract, operating and trading clauses, actors specific conditions, exploitation, etc.)
Distribution metadata	Property protection and Rights Management
	Metadata for trading (price, channel offer, price cut, availability, etc.)
Feedback metadata	Usage data (rate of viewing, votes, comments, appreciations)
	Reporting data (reference, entitled, views counters, contract and amount of revenue, etc.)
	Web identification
	Consumed content
User data	Social networks data, and so on
	Localization of content
Network metadata	Network condition for delivery (available bandwidth, delay, terminal type, etc.)

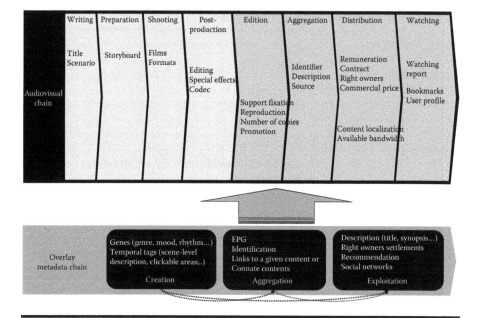

Figure 19.3 The overlay metadata chain.

The classical audiovisual chain produces and uses different types of metadata, which are nowadays enriched to provide new usages. So mastering metadata is the key to provide new services and user experiences.

19.2.4 Standards

In the objective to facilitate interworking between the various actors of the content production and delivery chain, standardization seems a natural stake.

Thus, we find several initiatives or standards which are difficult to address the entirety. Nevertheless, we can point out significant initiative within the multimedia industry:

■ Moving Picture Experts Group-7 (*MPEG-7*) is a standard of description of which goal is to facilitate the indexing and retrieval of multimedia documents. MPEG-7 format includes several elements: Visual descriptors, Audio descriptors, segment description, author information description, Media physical information description. TV-anytime standard, developed for the description of television and VOD contents, uses MPEG-7 descriptors.

■ The metadata of *TV Anytime* are intended for the receivers. They make it possible to describe and locate the contents in a context of diffusion of television channels and in a context of on demand contents distribution. TV-Anytime specifies an Electronic Program Guide (EPG) format, Extensible Markup Language (XML) based, interoperable and directly readable by a Set-Top Box (STB) or a PVR. TV-Anytime does not define how this EPG arrives in the receiver: Digital Video Broadcasting (DVB), Association of Radio Industries and Business (ARIB), Advanced Television Systems Committee (ATSC), Open Internet Protocol TeleVision (IPTV) Forum, Digital Living Network Alliance (DLNA) are in charge of defining the metadata distribution protocols.

■ The *Open IPTV Forum* specifies a platform that will allow any consumer end-device to access enriched and personalized IPTV services either in a managed or a nonmanaged network. A number of standard bodies are already addressing specific elements of IPTV, but the Open IPTV Forum says it will work to aggregate these diverse standards into a complete delivery solution, with the goal of accelerating the full standardization of IPTV-related technologies. TV-Anytime has been selected by the Open IPTV Forum as metadata format.

■ The *CableLabs* Asset Distribution Interface *(ADI) VOD Metadata* specification is the result of a cooperative effort undertaken at the direction of Cable Television Laboratories, Inc. for the benefit of the cable industry and its customers. It specifies metadata for the distribution of a VOD show from a Provider to one or more cable operators. The ADI metadata format is similar to TV-Anytime format.

■ The *Digital Living Network Alliance*, shortened under initials DLNA, is an alliance of production companies of electronic devices, mobile peripherals and personal computers. Its goal is to define a multimedia standard of service of files, with objectives of interworking between marks and convergence of the very varied electronic devices. Inter alia, the mapping of TV-Anytime toward Upnp was defined in DLNA.

■ The building block proposed by *MPEG-21* is the DIGITAL Item (DI) which is an object structured with a standardized representation. A DI understands multimedia contents, metadata associated with this file, the whole structured by a set of elements of syntax which describe the relations between these files and their metadata.

Other standards cover others universes:

■ *Material eXchange Format* or MXF is a container used by the professionals for the audio and video numerical data. It is a content production exchange standard.

■ *Digital Cinema Initiatives* (DCI) has completed an overall system requirements and specifications to help theatrical projector and equipment manufacturers create uniform and compatible digital cinema equipment.

Although the standards are multiple and cover the different distribution channels, one notes a weak placement of these standards such as

- The final standards arrived after the first deployments
- The standards do not meet the definite needs and are somewhat too comlex/complete

That is the reason why the different ecosystem players have defined their own format (nevertheless often based on standard ones). Indeed, the value added of metadata is less on the format itself than on the exploitation of these metadata to provide content services (search, recommendation, interactivity, etc.). This led to the necessity for players to deal with multiple formats, bringing back the metadata unification issue (essential to facilitate interoperability and convergence) to an internal management.

19.3 Key Players

This part presents the different kind of players dealing with content metadata.

19.3.1 Pure Media Players

Pure media players are players historically involved in the traditional audiovisual chain: movie studios, laboratories, cinema distributors, broadcast companies, and dematerialized contents providers. All of these players are spark off the traditional audiovisual chain introduced previously.

Some pure media players are: Warner, Disney, INA (Audiovisual National Institute), Sony, Globecast, and so on.

As said before, a new audiovisual metadata value chain has emerged in overlay of this traditional chain, involving new players. Hereafter is a proposed classification of these metadata players.

19.3.2 Metadata Creation Players

Definition: Players creating original metadata linked to a given audiovisual content.

These players can be divided in four different types: Web actors dealing with advanced discovery and navigation, traditional producers, laboratories, and institutions.

- *Advanced VOD discovery/recommendation and navigation Web players*: Generation of new metadata characterizing audiovisual contents (video genome, etc.) in order to propose a new experience to the user. These players (Jinni, Nanocrowds, Clerkdogs) are building revenues from online video distributors, who share a portion of the transactional and advertising revenues (Amazon, Netflix, Blockbuster, Movielink, Hulu, iTunes).

- *Producers*: Creation of metadata from the shooting step. These players are not new comers, but they are incontrovertible for some metadata (time-code, title, crew cast, etc.): Warner, Disney, 20th Century Fox, EuropaCorp, Gaumont, and so on.
- *Laboratories*: Creation of new contents from an original one (subtitles, different encoded versions, etc.). As for the producers, these players are not new comers within the metadata value chain: Sony, Cognac-Jay image, and so on.
- *Institutions*: Institutional bodies focusing on audiovisual content (INA, International Standard Audiovisual Number (ISAN)). They usually create metadata (including identifier which is an important problematic).

19.3.3 Metadata Aggregation Players

Definition: Players aggregating multiple contents in order to provide/sell contents or information linked to contents (gross metadata or EPG). Sold metadata can be created by the metadata aggregation player itself or by external ones.

Data Mediation and Content Management System (CMS) players can be part of metadata aggregation players, even if the technical skills and the business models are quite different as they sell a product managing metadata and not the metadata themselves.

These players usually generate a specific identifier for each content. They answer to the problematic of content sources and types heterogeneity which became more and more complex with the increasing number of contents on different devices.

19.3.4 Metadata Exploitation Players

Definition: Players exploiting audiovisual metadata to provide a service. These players can be the traditional ones (Samsung, Nokia, Apple, etc.), operators (Vodafone, Telefonica, Orange, etc.), over-the-top players (Youtube, Dailymotion, etc.), or new comers (in the audiovisual ecosystem) proposing new services based on new or aggregated audiovisual metadata (Netflix, Jinni, Amazon, etc.). Thus, assuming the different positioning of these players, metadata are exploited in all the steps of the audiovisual chain.

Moreover, depending on the regulation context, the leverages of the different players are not the same. In some countries, the possibility to manage both network and services allows an operator to fully control the end-to-end distribution chain, and consequently to provide potentially most quality and enriched user experiences. This is not the case in other countries where network and services are clearly separated by the regulation body.

One has to note that this metadata exploitation players category includes an increasing numbers of connected boxes providers in the home network (Roku, Boxee, etc.).

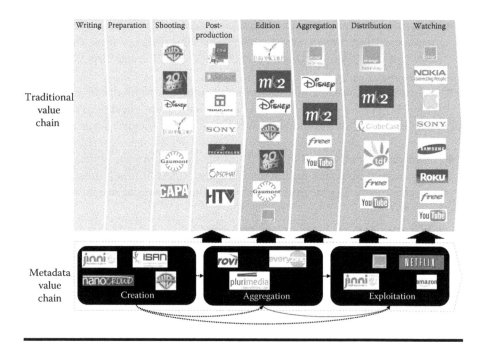

Figure 19.4 Key players and positioning.

19.3.5 Key Players and Value Chains Positioning

Some of the players previously described and their position within the distribution chains (the traditional audiovisual one and the overlay metadata value chain) can be shown in Figure 19.4.

Thus, we observe that **the audiovisual content distribution ecosystem is evolving** under the pressure of new players developing their activities on a metadata-basis to propose new experiences of content consumption.

Indeed, the marketplace around audiovisual metadata is growing along of two axes: a **metadata-centric business** on the one hand (creation, aggregation, and resale), and a **new value-added services business** based on the exploitation of these metadata (navigation, recommendation, etc.) on the other hand.

One has to note that more and more connate players are investigating this field (Atos, Thomson, NDS, BT, etc.) confirming the interest of an innovative exploitation of audiovisual metadata for new content services.

19.4 Metadata Processing

As seen previously, a lot of players are dealing with metadata which are passing through many processes depending on the complexity of the goals to reach. We will see in this paragraph what are the technologies involved with the different kind of metadata, and some existing built solutions that have emerged as central tools.

19.4.1 Technologies Involved

A lot of technologies generating, processing, and using metadata exist. We can classify the technologies involved according to two categories of services: Backend Services and Frontend Services (Figure 19.5).

19.4.1.1 Backend Services

- Delinearization/capture: Uses engines of contents analysis (logo recognition, audio signatures) and the already-existing metadata in an EPG to separate the contents recovered from a Live stream. This feature is used in particular for Television On Demand (TVOD) services.
- Contents Analyses:
  - Speech-to-text: Transcription of audio data in textual data, often associated with the other technologies such as speaker's change. Gives temporal information. Feeds Search, Recommendation, navigation through the content.
  - Speaker's recognition/speaker's change: Temporal metadata and identification by audio signatures: new interface navigation through the content (interactivity) spatial presentation of the audio data.
  - Face recognition/objects recognition/watermarking: Requires images signatures, and generates metadata associated with temporal segments and two dimensions spatial coordinates that are necessary for the services of clickable videos.
  - Word Spotting: Detection of words in an audio sequence from a lexical base: adapted to the real time (the speech to text being a much longer process according to the used technology), in association with technologies of named entities detection. Generation of metadata for the search and the recommendation.

Figure 19.5 **Metadata processing technologies.**

- Mediation: Used to interface different (and possibly heterogeneous) databases. Case of applications: automatic enrichment of metadata from external videos sources, researches for elements on not fixed bases.
- Reconciliation/Aggregation at two levels:
  - Combining the various metadata of the same item resulting from different sources
  - Combining the results from various search engines and/or recommendation engines

    The reconciliation has to allow covering the problems of information redundancy (repeat broadcasting and multiple sources cases). In this purpose some technologies of "doubloons" detection can be used. The metadata aggregation coming from different sources has to eliminate certain incoherencies (by crossing sources) and to obtain better qualities of metadata.
- Network Metadata tools: The network and devices metadata came from technologies which allow obtaining information on the network use (Local-Area Network (LAN) and Wide-Area Network (WAN)) to adapt the contents diffusion to the consumer and to permit communication between different devices.

19.4.1.2 Frontend Services

These services use the metadata generated by the backend services.

- Recommendation: Aims to suggest contents according to users' profiles, social network, editorial metadata, and similarity with the searched contents.
- Search: Includes ranking and indexation technologies. In theory, all the metadata can be subject to the ranking but in practice only a part is useful (no need to index and to rank the URL or broadcasting metadata). The search returns a list of items matching the user query and has to possess a very high-quality ranking technology to increase the answer's relevance.
  - Search through the content: It is similar to Search but there is less importance on the ranking, and a dominance of used metadata like the temporal segments.
  - Clickable videos: There are several modes of implementation.
- One solution is to use a particular player to localize a part of a picture in each video frame knowing some given metadata (e.g., from the position of a face recognized in a particular frame).
- Another solution is to analyze all the frames of a video in order to fill-in metadata on every frame. This solution leads to a very heavy XML document.
- User Profile/personalization: Include all the technologies allowing of follow-ups of the consumptions of the user and the services of profiling (with a questionnaire or bases of information, social networks).

19.4.2 Metadata Involved

Table 19.2 presents the specific metadata the different technologies use to render the four typical services trends.

To answer **societal evolution in audiovisual contents experiences**, players using media networks has to face different technical challenges (metadata format heterogeneity, various metadata sources, etc.) involving many technologies. But how these technologies have to be bundled and integrated to media networks? The last coming part of this chapter intends to give some answers.

Table 19.2 Metadata and Technologies

Technologies/ Services Trends	Social Network	Convergence	Recommendation	Interactivity
Search tools			X	X
Speech-to-text			All the word	Temporal segment
Speaker's recognition/ speaker's change			Speaker	Temporal segment
Face recognition/ objects recognition/ watermarking			Person, object, and so on	Spatial metadata
Word spotting			Detected words	Temporal segment
Mediation		X	X	
Reconciliation/ Aggregation		X	X	X
User profile/ personalization	Membership in a group, contacts		Comments, direct recommendation, sharing, social graph	
Network metadata		Profile of device Device in use		

Note: X: These technologies do not use specific metadata, but all types of metadata.

19.5 Metadata Systems within Media Networks

The previous paragraphs of this chapter have highlighted the interest of the metadata as well as technical issues that should be solved: format heterogeneity, distributed data sources, content preparation to fit different access networks and so on. In order to solve these issues and interface together many technological components, different architectures have been defined, giving rise to dedicated systems. We present below three of these systems we considered representative for the metadata processing within present and future media networks.

19.5.1 Content Management Systems

In the Web environment, CMS is a family of software intended for the design and the dynamic update of website or multimedia application.

In the TV and Video business, CMS are systems that aim to aggregate contents and metadata, to prepare contents (coding, linearization, etc.) and update metadata and to deliver contents and metadata to the distribution chain (delivery, portal). CMS are mainly used by TV channels and more recently by VOD aggregators.

Those CMS share the following functionalities:

- Content workflow management
- Content ingest, preparation, and checking
- Content browsing and offline editing
- Work order management
- Metadata update capacity
- Content delivery

The global perimeter of the CMS platform includes a centralized content preparation system and a centralized content referential (Metadata Management System (MMS)).

- The Content Preparation System:
  - Must provide all media processing necessary to do the content deployment.
  - Will realize importation/Ingest, Encoding/Transcoding, Image Processing, Quality Control (manual or automatic), Content Checking, Content Packaging, Media Deployment.
- MMS:
  - Provides to actors of the Content Management ecosystem the metadata associated to an asset.
  - These metadata are to be provided to the MMS by the Data Supplier Actors and updated by the Content Preparation System.
  - These metadata are all the information that the Content Management may require in order to deliver content to final consumers.

- The MMS will manage the following metadata types: technical, editorial, contractual, settlements, exploitation.
- All metadata have to be published into a Content Referential from which they will be served to Systems Data Consumers Actors:
  - Service platforms
  - Service animation system
  - Infocentre enabler
  - Settlements enabler
  - Search and recommendation enablers
- Enablers (Settlement, Infocentre, Search&Reco) may receive the whole content catalogue or get the metadata they required from the MMS by calling the published services.

19.5.2 Data Mediation

A mediator is an interface between target services and existing information sources (that are distributed and possibly heterogeneous), which gives its users the illusion of a centralized and homogeneous information system. It allows them to ask domain-level queries and takes in charge in their place the access to the relevant sources in order to obtain the answers to their queries.

A mediator is particularly relevant when:

- Information stored by the sources is extremely volatile
- Information stored by the sources cannot be freely exploited (essentially for commercial reasons)
- The target services are subject to change

Most of mediator systems have in common the need to describe the domain model covered (application *domain*) and the content of the information sources (sources *description*). Different formalisms like XML or relational algebra can be used in order to cover this need.

19.5.2.1 Semantics

In recent years, the problem of information integration has received a lot of attention. In particular, a **semantic*-based approach** has been proposed. This approach differs from the others by the use of ontology[†] (using for instance semantic Web languages such as: Resource Description Framework Schema (RDFS), subsets of Web Ontology Language (OWL) like OWL DL (Description Logic), etc.) and logical mechanisms

---

* A semantic analysis retrieves a logical sense from a group of independent data.
† Ontology: explicit and formal specification of a consensual data model applying to a given domain (e.g., cinema). It is an evolved form of semantic representation of a domain.

to describe both the domain model and the contents of the information sources. In a nutshell, semantic mechanisms allow taking into account the meaning of a data within a given domain, and processing a user query in the light of this meaning.

The advantages of this approach have been pointed out:

- Such logical formalism has been specially designed for modeling and reasoning on data descriptions
- Its power of description is well suited for a natural conceptual modeling of the domain and the information sources
- It is currently considered as the most powerful and flexible approach to make databases interoperable
- Semantically relevant answers to simple or complex requests

Some studies showed moreover the interesting workload gain of the semantic mediation approach compared with a "basic" mediation (also called "vertical integration"), highlighting the fact that

$$\exists k \in N, \quad \forall n \in N, \quad Rn \geq Rk \Rightarrow W_{sm}(n) \leq W_{vi}(n)\exists$$

with W_{sm} and W_{vi} the workload functions of semantic and vertical integration, respectively, and Rn the possible values for requests number, which is a function of the number of sources and the data to be retrieved (Figure 19.6).

The area demarcated by each curve represents the total amount of workload for the corresponding mediation type. The delta area ($\Delta W = W_{vi} - W_{sm}$) between the two curves becomes positive after two requests, and will then irreversibly increase with the number or requests.

A semantic mediator basically presents the architecture shown in Figure 19.7.

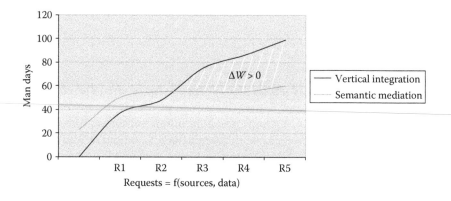

Figure 19.6   Vertical integration versus semantic mediation.

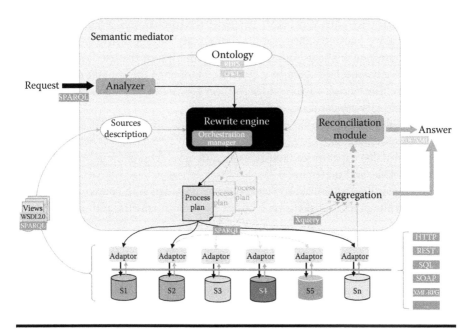

Figure 19.7   A semantic mediation architecture.

Semantic mediation, allowing the retrieval of data from distributed and hetero-geneous sources, is an academic-emerging technology which starts to be widely considered by databases community experts as the optimal solution for flexible and evolutive integration of large sets of different databases.

19.5.3 Context-Awareness*

Content distribution services deployed over telecom networks are still basically set up on a two-layer model, with connectivity networks at the lower level and applica-tion overlays at the upper level especially observed in the fixed networks where both layers are running relatively independent from each other.

Yet, current strategic development trends dictate the necessity to migrate toward integrated networks with increased real-time interactions between layers in both fixed and mobile networks.

In the literature the term "Context Awareness Principle" is used to refer in a global manner to such dynamic interactions between users, terminals, networks, and services. Information that could be seen as particular metadata (content localization, available

---

* See Towards highly-coupled Networks/Services Cooperation for Content Distribution white paper (Y. Le Louedec, H. Moustafa, B. Mathieu, Y. Couvreur, T. Labbé).

bandwidth, delay, terminal type, etc.*) is part of the mechanisms providing such an increased dynamic interaction between the connectivity and applicative layers.

Context awareness allows monitoring and gathering of the user and his environment contexts and feed them in a dynamic manner to the content distribution system. Such information could include the user context as well as the network, the terminals, and services contexts.

This enables distinguishing each user in a separate manner, and tackling not only the user satisfaction ("QoE") but also the resources level (access network capabilities, terminal(s) capabilities) and the semantic level (content and service metadata, user profile, and preferences).

Consequently, personalized services could be provided with content adaptation according to the exact user preference, to his QoE requirements, and to the different contexts.

Although the context-awareness concept is promising in enhancing the personalization, how to use the acquired context information during the content delivery process is still a challenge. Another challenge is how to integrate the different functional entities composing the context-aware system in content distribution architectures composed of several players. A useful and promising approach is to deploy a distributed context-aware system with its functional entities being integrated along the different layers of the content distribution architecture and hence being distributed among the different actors.

A network context gathering module within the provider network gathers the different context information on the network and hence evaluates the network status (in terms of bandwidth, load, etc.). This module is controlled by the network provider and inputs the network context information to the service providers (E2E as well as overlay providers), precisely to a context-aware management module for these latter ones.

The E2E and overlay providers also gather their different services context information through their service context gathering modules (that could eventually exchange the service context information) and maintain it in their context-management modules. As for the user context information, it is gathered by the concerned user context-gathering module and is maintained in the context-management-specific module for both the E2E and overlay service providers. It is the context-aware management module that reasons the different gathered context information on the network, services and users and adapts the content accordingly. Content adaptation takes place through a special module, the context-based content adaptation module, and allows content delivery in a network and service aware manner while considering the users' context and QoE.

---

* See also Towards highly-coupled Networks/Services Cooperation for Content Distribution white paper for more details.

19.6 Conclusion

Since the metadata are identified as a core element for groundbreaking services, the mastering of value-added data produced by different players (both inside and outside the operator's environment) is a major stake. As a result, a new strategical marketplace has emerged around the metadata paradigm, involving multiple players (content providers, data creator, aggregator, resaler, etc.) which have boosted the technological innovation on metadata exploitation systems. One can point out that metadata derived directly from the multimedia contents are already well exploiting, whereas metadata related to context and network are still in an experimental phase. In between, technologies like semantic analysis and mediation are in maturation with already interesting applications allowing linked media services (see the next chapter).

Finally, if it is quite difficult to presume what will be the architecture(s) involving cross usages of metadata in the next few years, it is obvious that metadata exploitation on every layer will be a major asset to leverage future media networks applications.

Acronyms

ADI	Asset Distribution Interface
API	Application Programming Interface
ARIB	Association of Radio Industries and Business
ATSC	Advanced Television Systems Committee
CMS	Content Management System
DI	DIGITAL Item (DI)
DLNA	Digital Living Network Alliance
DVB	Digital Video Broadcasting
EPG	Electronic Program Guide
IPTV	Internet Protocol TeleVision
ISAN	International Standard Audiovisual Number
LAN	Local-Area Network
MMS	Metadata Management System
MPEG	Moving Picture Experts Group
MXF	Material eXchange Format
OWL	Web Ontology Language
PVR	Private Video Recording
RDFS	Resource Description Framework Schema
SQL	Structured query Language
STB	Set-Top Box
TVOD	Television On Demand
VOD	Video on Demand
WAN	Wide-Area Network
XML	Extensible Markup Language

Chapter 20

Semantically Linked Media for Interactive User-Centric Services

Violeta Damjanovic, Thomas Kurz, Georg Güntner,
Sebastian Schaffert, and Lyndon Nixon

Contents

20.1 Introduction 446
   20.1.1 Related Projects and Events 446
   20.1.2 Technology Drivers: Hypervideo Platforms, Semantic Web
      and LOD Initiative 448
20.2 Multimedia Analysis, Integration, and Annotation Models 451
   20.2.1 Multimedia Annotation Models 453
20.3 Interlinked Media and User-Centric Approach 454
   20.3.1 Vocabularies 454
   20.3.2 Search Engines and Browsers 454
   20.3.3 User Behavior and User Web Profiles 456
20.4 The LMF Approach 457
   20.4.1 The LMF Conceptual Model 458
      20.4.1.1 RESTful Resource Management 458
      20.4.1.2 Extended Content Negotiation 458
      20.4.1.3 Semantic Indexing 459
      20.4.1.4 Semantic Enhancement via Linked Data 459
   20.4.2 The LMF Architecture and Implementation 459

20.5 Showcase Application: LMF to Support Semantically Interlinked
Media...461
20.5.1 Video Annotation via LMF...461
20.5.2 ConnectMe Synchronized and User-Centric Multimedia
Annotation via LMF...462
20.6 Further Work 464
20.7 Conclusion 465
Acronyms...466
References 466

20.1 Introduction

Nowadays media assets are getting increasingly digital and social. They open up new possibilities for value-added Web services and new user experiences through linking media content coming from elsewhere on the Web. Today's TV is still a family centerpiece of many homes, thus mediating among TV and the Web while offering novel opportunities for future commerce, games, entertainment, and social interaction [1]. In addition, today's TV is getting further into a new service with a great expectation to become available soon on a variety of devices and to connect smoothly with other online services, such as social networks, shopping websites, and many more.

At the same time, growth in a number and diversity of devices (across multiple industries) leads to the increase in cross-platform services and their interoperability challenges. Therefore, a range of difficulties in seamlessly combining and connecting video and other media content on the Web, such as ambiguity in meaning of concepts, difficulty in properly annotating video with concepts, granularity of extracted information about concepts, meaningful presentation of related multimedia content alongside video, and so on [2] will challenge the ongoing investigation in this area.

20.1.1 Related Projects and Events

At the time of writing this chapter, there exist several projects interlinking TV media (video, audio) and the Web. For example, recently launched *Mozilla's Popcorn* project shows a great interest in linking video on the Web to the rest of the Web's media content. In addition, the EU FP7 *NoTube* project brings the following promises [3]:

■ Novel deployment of existing methods for information integration in the combined TV-Web environment
■ Extensions of existing user and context modeling techniques to meet demands of the distributed TV-Web world

---

* The Mozilla's Popcorn project: http://popcornjs.org/
† The EU FP7 NoTube project: http://notube.tv/

- Implicit determination of topics of user interest via their Social Web activities
- Development of novel reasoning services for personalized content recommendation
- Integration of the second screen as interaction and information channel for TV viewers

The *NoTube* project uses Semantic Web technologies to connect Web-based TV media content via Linked Open Data (LOD)* (both technologies will be described in more detail in the following subsection). The *NoTube* project aims to provide future social TV that supports online interaction between people while watching TV. Web-based interaction requires global channel and TV program identifiers Uniform Resource Locators (URLs) that allow users to refer to the things they are talking about, as well as common application programming interfaces (APIs) for communication between TVs or Set Top Boxes and second screen devices like tablets or smartphones. The *NoTube* also investigates different ways of using Linked Data to enhance the user experience by helping people to decide what to watch, discover more information related to a TV program, and help them to have smarter, more focused conversations around Web-based TV media content. For that purpose, the *NoTube* project implements the *egtaMETA*,† a set of semantically defined attributes that describe advertising material and clusters, which are based on the following set of information:

- Descriptive information (e.g., the title of the advertising spot)
- Exploitation information (e.g., what is the period during which it shall be used)
- Credits (key persons and companies involved in the creation, postproduction, and release of the advertising spot)
- Technical information (e.g., information about the file format and its audio, video, and data components)

In addition, recently launched the *ConnectME project*‡ follows three core principles such as:

- *Hypervideo*, with the goal to make video as clickable as text is today. Until now, hypervideo has failed to make a significant impact, mainly due to need to manually prepare the hypervideo, including the selection of objects in video frames and associations to additional content. Thus, the *ConnectME* project will provide multimedia annotation and delivery of synchronized multimedia presentations to the end users.

---

* The Linked Open Data: http://linkeddata.org/
† The egtaMeta Website: http://tech.ebu.ch/egtameta
‡ The ConnectMe project: http://www.connectme.at/

- *Linked data*, with the goal to bring the concepts that can be used to annotate video, which are themselves resources on the Web with defined structured metadata about these resources.
- *Semantics*, meaning that information about media concepts and Web content can be processed by machines in a more dynamic and automated fashion, reducing the manual effort needed to generate connected media experiences.

20.1.2 Technology Drivers: Hypervideo Platforms, Semantic Web and LOD Initiative

Hypervideo (also known as interactive video) is emerging on the Web as an extension of existing video hosting platforms. The real shift is not perceived only in enhancing TV on the Web, but also in integrating the Web into new TV media-based experience [2]. One approach involves *widgets*, which are lightweight self-contained content items that make use of both open Web standards and the Internet Protocol TV (IPTV) back-channels to communicate with the Web. Another approach provides full Web experience on TV set, such as Apple's fledged browser on iPhone, or GoogleTV* that combines TV, the Web and apps, and number of ways to search across them all. In addition, the potential of augmenting the camera view in a mobile device in order to enrich a mobile device user's experience of what is around him, have developed into so-called—Augmented Reality (AR). Current AR applications have been considered as beginning of a significant future mobile applications market that is based on providing users with additional information about their immediate surrounding or locality.

The next prominent technology to support the future Web TV is *the Semantic Web*. This is an initiative of the World Wide Web Consortium (W3C) inspired by the vision of its founder, Tim Berners-Lee, of having a more flexible, integrated, automatic, and self-adapting Web, that provide a richer and more interactive experience for the end users [3a,4]. The W3C has developed a set of standards and tools to support this vision, and after several years of research and development, these are today usable and could create a real impact. Two major working groups of the W3C around these technologies are the Resource Description Framework (RDF) Working Group[†] and the Web Ontology Language (OWL) Working Group.[‡] RDF is a framework for representing information on the Web, while OWL is designed for use by applications that need to process the content of information instead of just presenting information to humans. OWL facilitates machine interpretability of Web content than is supported by XML, RDF, and RDF Schema (RDFS) by providing additional expressive power along with a formal semantics. In sum, the Semantic Web is a broader W3C-driven initiative which provides the data and

---

* GoogleTV Website: http://www.google.com/tv/features.html
† W3C RDF WG: http://www.w3.org/2011/rdf-wg/wiki/Main_Page
‡ W3C OWL WG: http://www.w3.org/2007/OWL/wiki/OWL_Working_Group

query formats that are also used by Linked Data (e.g., RDF and SPARQL Protocol and RDF Query Language (SPARQL)), and develops this further through the use of ontologies. While combined with ontology, semantic data (e.g., data in a form of RDFS and OWL) can be reasoned about and queried by using logic rules.

Today, there are a growing number of data sets published on the Web according to the *Linked Data principles*. The majority of them are part of the LOD cloud. As LOD connects data and people across different platforms in a meaningful way, one can easily assume that expert search and profiling systems would benefit from harnessing LOD. For example, professional podcasts with guest experts, video lectures, as well as online slide presentations would have been a valuable data source for expert profiling, if the data about the hosted resources and their authors were available in RDF [5]. LOD allows querying the whole Web like a huge database, thus surpassing the limits of closed data sets and closed online communities. It opens new possibilities for capturing the essence of human knowledge, experience and activities through the online and social traces that people leave on the Web.

In 2006, Tim Berners-Lee presented the Linked Data design issues [6] as a best practice for exposing, sharing, and connecting data and knowledge on the Web that can be further discovered by both humans and machines. These issues apply the following: (a) Use URIs as names for things; (b) Use HTTP URIs so that people can look up those names; (c) When someone looks up a URI, provide useful information, using the standards such as RDF and SPARQL; and (d) Include links to other URIs so that they can discover more things. The main idea behind these rules is to publish data in a structured form that is further interlinked with data on the Web.

Since the initiation of the LOD community project in 2007, the amount of data published according to these principles is steadily growing. An example of the enterprise that uses Linked Data to publish a large amount of media content online is the BBC broadcasting corporation, which publishes media content online, via eight national TV channels plus regional programming, 10 national radio stations, 40 local radio stations, and an extensive website, http://www.bbc.co.uk [7]. The BBC identified the lack of integrating at a data level and a lack of semantically meaningful predicates making it difficult to repurpose and represent data within a different context. In other words, they found it very difficult to search information the BBC has published about any given subject, nor easily navigate through across BBC domains following a particular semantic thread. Thus, the BBC launched the *BBC Programmes* to address these issues. It provides Web identifiers for every program the BBC broadcasts. Each Web identifier has multiple content-negotiated representations ensuring that data used to generate pages is reusable in different formats, for example, RDF/XML, JavaScript Object Notation* (JSON, that provides a lightweight, flexible yet strict method of storing and transmitting data), plain XML, with the aim to enable building enhanced program support applications. The BBC also launched the *Programme Ontology* [8] that exposes the data model driving the BBC website as a formal OWL

---

* JSON: http://www.json.org/

ontology, allowing anchoring data feeds within a domain model. Apart data about TV-related programs, the BBC processes music contents via linking information about an artist, releases and their reviews, to those BBC programs that have played them. The BBC is using a variety of different ontologies and vocabularies for publishing music as Linked Data.

At the same time, multimedia annotation of online information (e.g., audiovisual annotation) may be interlinked to any other kind of related information on the Web. However, this requires that the external Web content is annotated (i) in terms of the concepts, which are identified as occurring in the base multimedia content, and (ii) with sufficient information to guide the process of selection, adaptation, personalization, and packaging with other content in a meaningful multimedia presentation as a part of multimedia augmentation.

The work presented in this chapter sets out to show benefits and drawbacks of using the Linked Data cloud as the source of future multimedia content. Linked Data is sharable, extensible, and easily reusable [9]. It provides internationalization facilities (e.g., Unicode support) for data and user services. Resources can be described in collaboration with other multimedia collections and linked to data contributed by other communities or individuals. The use of identifiers ensures that the diverse descriptions are all talking about the same thing. Hence, the benefits of using Linked Data are multiple: it will enrich knowledge about media assets through linking among several domain-specific knowledge bases; it will give an opportunity to organization to improve the value proposition of describing their media assets, and so on.

Nowadays, the process of publishing data has been supported by various tools that are specifically created to support either publishing RDF as Linked Data, or generating RDF from relational sources. At the moment of writing this chapter, there are several frameworks that can be used to publish Linked Data on the Web. A detailed comparison of those frameworks is given in Ref. [10] by considering specific aspects such as data access, content negotiation, update mechanisms, caching strategies, semantic full-text search, SPARQL endpoint, SPARQL result serialization, and so on. In addition, there are some tools to extract annotations from content and thus to find candidates for linking concepts out of the LOD cloud. Other tools have the possibility to pipeline different extractors and annotators, such as GATE,* Apache Unstructured Information Management Architecture (UIMA†), Apache Stanbol,‡ and so on. So far, there are less tools offering extraction methods for different content types like image [11] or video [12]. Thus, in this chapter we present the Linked Media Framework (LMF), which is an ongoing open source development used to demonstrate the concepts of semantic lifting and interlinking of media resources. Based on the above-mentioned comparison of existing frameworks to publish Linked Data on the Web, one can conclude that the LMF follows

---

* http://gate.ac.uk/
† http://uima.apache.org/index.html
‡ http://projects.apache.org/projects/stanbol.html

approaches similar to Virtuoso* or Talis platform, but it also provides resource- and user-centerd updates, and RDF versioning.

In this chapter, after discussing the main motivation for investigating semantic Linked Media to support advanced social interaction on the top of user-centric services, we start with the analysis of various multimedia, integration, and annotation models, to identify those that could implement practical solutions for our specific use case. In the next section, we describe our proposed implementation that is called the LMF, together with the LMF conceptual model, architecture and particular software components implemented so far. Then, we describe two showcases that employ the LMF for video annotation: (i) type-dependent video annotation, and (ii) synchronized and user-centric multimedia annotation. We conclude this chapter by suggesting the directions for our future work on the development of the LMF and its showcase applications, in a manner that will strongly contribute to the domain of semantically enhanced media interlinking.

20.2 Multimedia Analysis, Integration, and Annotation Models

Various solutions to multimedia analysis have ability to link together fields such as video, image, printed document, audio, and text analysis. *Video analysis* solutions analyses digital video archives for specific data, behavior, objects, events, and attitude. They may be grouped into the following categories:

- *Structural video analysis* that decomposes video frame or frame sequences into parts that are characterized by uniformity of a certain feature or a group of features.
- *Temporal video decomposition* that can be based on simple shot boundary detection [12,13], subshot detection (e.g., detection of various events and environments within a single shot), scene detection (e.g., a subsequent grouping of shots to form scenes).
- *Spatial segmentation* that includes decomposition of single frames or frame sequences into regions with different depicted objects, and includes object detection, color segmentation, and motion segmentation [14].
- *Content-based video analysis* that identifies objects and entities on a higher level of abstraction by parsing video, audio, and text from video data. Content-based approaches comprise various object detection tasks, among other those used for text detection, segmentation, and enhancement, face detection and recognition.
- *Concept detection* that bridges the semantic gap by detecting the presence (or absence) of semantic concepts in multimedia data.

---

* http://semanticWeb.org/wiki/Virtuoso

Current solutions to *image analysis* are primarily based on the correlations between the tag and visual information space, which are established when the users suggest tags for the uploaded visual content. In Ref. [15] the authors introduce the concept of "flickr" distance which measures the semantic relation between two concepts using their visual characteristics. In Ref. [16] the authors make the assumption that semantically related images include one or several common regions (objects) with similar visual features. In Ref. [17] the problem of object recognition is viewed as a process of translating image regions to words, much as one might translate from one language to another. Similarly, Li et al. [18] propose a fully automatic learning framework that learns models from noisy data such as images and user tags from the Flickr, which is one of the best online photo management and sharing application in the world social sharing tool.

Audio analysis solutions are usually focused on speech indexing [19,20]. In the research labs, the focus is on data that (i) has well-known characteristics, (ii) form relatively homogenous collections, and (iii) is annotated in large quantities for speech recognition purposes. In real life, the exact characteristics of archival audio are often unknown and far more heterogeneous. Similarly, a number of commercial solutions covering the complete *printed document analysis* workflow appeared (e.g., ABBYY* [21], Nuance [22], I.R.I.S. [23]), but none of them are capable of higher-level analysis. The ongoing FP7 IMPACT project[†] develops novel technologies to improve the conversion of historical printed material to digital resources. It will advance the state of the art in image preprocessing, segmentation methods, and document layout analysis.

In the field of *text mining*, Information Extraction (IE) from textual documents recognizes that complete understanding of natural language text is focused on extraction of a small amount of information from text with high reliability [24]. Most IE methods assume goal-standard data that differs in structure and quality from texts extracted via Optical Character Recognition (OCR) and Automated Speech Recognition (ASR) methods. However, the construction of a model for noisy data such as audio transcripts constitutes a new challenge and requires approaches different to those based on large amounts of goal-standard text as shown in Ref. [25].

Nowadays, the dynamics and integration of social media have turned a big part of scientific community into analyzing and examining the emergent knowledge that derives from those media. Analysis of the structure and dynamics of social tagging systems is necessary for studying the patterns that arise from the interactions among individual objects. It helps to build and explore models that may develop a deeper understanding of processes in the underlying interactions and provide better analysis of the collective behavior of users.

Most of today's Web 2.0-related technologies and methods to facilitate user access to multimedia content do not include personalization and recommendations based on user activity and content similarity, content structuring, topics, and trend detection.

---

* ABBYY Group: http://www.abbyy.com/
[†] http://www.impact-project.eu/

For example, clustering digital objects by time can be applied successfully for event detection [26]. Most of the current approaches to analysis of tag patterns rely on a static basis [27,28]. There are a limited number of efforts that study the evolution and dynamics in tagging activities by using an explicit temporal dimension.

20.2.1 Multimedia Annotation Models

The description of digital media resources with metadata properties has a long history in research and industry. Over the years there came up many standards differing in complexity and completeness that leads to interoperability issues in search, retrieval, and annotation of digital media resources. To address this problem, the W3C launched the Media Annotation Working Group,* which aims to improve interoperability between multimedia metadata formats on the Web. They listed some relevant formats in the group report including basic standards like Exchangeable Image File Format (EXIF†), Extensible Metadata Platform (XMP‡) or Dublin Core (DC§) as well as higher-level description formats like MPEG-7. Altogether the group cited 18 multimedia metadata formats and six container formats and selected a greatest common subset of 28 properties building the *Core ontology* for multimedia metadata [29]. Within the recommendation they also prepared a mapping table for all standards and a client-side API [30] to access metadata information. As the Core ontology was initiated to tear down the walls between several metadata standards, its properties are extremely limited. To achieve a further more accurate description like how a media object is composed of its parts and what the parts represent, there is a need to have refined ontologies like COMM [31]. The COMM ontology is about reengineering of MPEG-7 using Descriptive Ontology for Linguistic and Cognitive Engineering (DOLCE), MultiMedia Metadata Ontology (M3O) [32], or Reusable Intelligent Content Ontology (RICO) [33] that is a conceptual model and a set of ontologies to mark up multimedia content embedded in Web pages. Even though the existing higher-level ontologies fulfill a lot of requirements for media annotation they are still not widely accepted because of its complexity, which is a big hurdle for a Web user. A model which may be mentioned here because of its diversity to apply in the domain of media annotation is the *Open Collaboration Annotation model* [34]. In combination with media-based ontologies, it is a promising solution in the Web of Data. This list of annotation and metadata models is far from being complete but gives an overview on the most important representatives that can be used for our purpose.

---

* W3C Media Annotation Working Group: http://www.w3.org/2008/WebVideo/Annotations/
† EXIF 2.2 Specification by JEITA: http://www.exif.org/Exif2-2.PDF
‡ XMP Specification: http://www.adobe.com/content/dam/Adobe/en/devnet/xmp/pdfs/XMP SpecificationPart2.pdf
§ DCMI Metadata Terms: http://dublincore.org/documents/2008/01/14/dcmi-terms/. The latest version of DCMI Metadata Terms is available at http://dublincore.org/documents/dcmi-terms

20.3 Interlinked Media and User-Centric Approach

A natural way to model the content of social media sites is to use a *hypergraph* that connects users, resources, and tags. However, the hypergraph model is rarely used in practice due to its complexity and lack of efficient techniques for analyzing its structure. Instead, the tripartite graph model can be used as an approximate representation for folksonomies [27], but further simplifications of the model, for example, bipartite graphs and one-mode networks can also be considered for tackling specific problems. However, these simplifications of the model are not sufficient in the context of Linked Data representation and sharing, what for an RDF-based representation of data sources needs to be employed. Apart from graph-based data representation and interlinking, additional *indexing structures* are also foreseen such as inverted indices, tree-based indices (e.g., R-Tree), or sophisticated indexing mechanisms to support spatio-temporal content indexing.

20.3.1 Vocabularies

With respect to the interoperability and interlinking of social media platforms, various *vocabularies* such as Friend Of A Friend (FOAF) [35], Semantically Interlinked Online Communities (SIOC) [36], CommonTag, have been provided to enable common representation of social content across applications, following the vision of Social Semantic Web that is given in Ref. [37]. However, while they are already adopted as reference vocabularies, they are still not widely deployed on the Web. Thus, there is a need for applications to enable easy translation of the existing platforms to Web 2.0 data, as well as for frameworks to enable automated, effortless publishing of Web media content based on those vocabularies.

20.3.2 Search Engines and Browsers

Within the traditional Web search scenarios, user actions are commonly interpreted as implicit feedback about the relations between queries and resources. Such relations have been widely employed for capturing query semantics [38,39], while providing extraction of taxonomies [40,41], as well as identifying other semantic relations between queries [42]. At the same time, there exists a direction of research that interprets click-through data [43], while another focuses on distinct user actions, considering them in isolation rather than as a whole, aiming to identify specific indicators such as viewing, bookmarking, printing, saving, annotating, rating, and so on [44]. With the advent of Web 2.0, rich type of user interactions enable to aggregate feedback of many users for finding the common core of all provided data.

Information retrieval (IR) and Semantic Search bring the wider benefits for exploiting documents for retrieval purposes. They both include three distinct phases: (i) Query construction, (ii) Query processing, and (iii) Result presentation;

each of them differs significantly within most of the known approaches. For example, extensions from traditional IR to Semantic Search applications is done via controlled vocabularies or semantic auto-completion during query construction. The main intention of this approach is to build a semantic layer on top of traditional IR systems and to create a searchable index which can be used as a baseline for efficient ranking and retrieval methods.

The very first systems to support semantic search and browsing are already available and evaluated: *Sindice,** as described in Ref. [45], is a scalable index of the Semantic Web. It crawls the Web for RDF Documents and Microformats and indexes resulting resource URIs, Inverse Functional Properties (IFPs) and keywords. A human user can access these documents through a simple user interface, based on indexes mentioned above.

Sigma† is rather a semantic information mash-up enabled by Sindice, than it is a self-contained semantic search service. It works as a Web of Data browser in which the user can start from any entity (found by a full text search) and then browse to the resulting page. The resources index is built out of sites which use RDF, RDFa, or Microformats.

The *Open Link Search*‡ lists entities with a user-defined text pattern occurring in any literal property value or label. It also supports Entity URI lookup. The search can be redefined by filtering type, property value, and so on. It is also possible to execute SPARQL queries by using the SPARQL endpoint. Some queries are predefined, and can easily be altered via text input fields.

Falcon§ [46] is a service for searching and browsing entities on the Semantic Web. It is a keyword-based search engine for the Semantic Web URIs that provides different query types for object, concept, and document search. It gives the facility of faceting over types by dynamically recommending ontologies. The recommendation is based on a combination of the Term Frequency-Inverse Document Frequency (TF-IDF) technique and the popularity of ontologies.

Watson¶ offers keyword-based querying to obtain a URI-list of semantic documents in which the keywords appear as identifiers or in literals of classes, properties, and individuals. Search options make it possible to restrict the search space to particular types of entities (classes, properties, or individuals) and particular elements within the entities (e.g., local name, label, comment).

Semantic Web Search Engine (*SWSE***) is a search engine for the RDF data. The IR capabilities of SWSE are more powerful because of the inherent semantics of RDF and other Semantic Web languages.

---

* Sindice Webpage: http://sindice.com/
† http://sig.ma/
‡ http://lod.openlinksw.com/
§ http://iws.seu.edu.cn/services/falcons/objectsearch/index.jsp
¶ http://watson.kmi.open.ac.uk/WatsonWUI/
** http://swse.org/

Swoogle* allows a user to search through ontologies and browse the Web of Data. It uses an archive functionality to identify and provide different versions of the Semantic Web documents.

Like it is described above, each considered semantic search service provides a certain amount of functionalities. Some of them are part of two or more services, whereas others are exclusive to one certain engine. The search engines do not consider a semantic similarity of queries and content, which definitely could increase the quality of result. There are applications in the area of Semantic Web which match some of these requirements in certain ways, for example, the interlinking frameworks.

20.3.3 User Behavior and User Web Profiles

Recently, *user behavior and Web profiles* have been extensively studied since both have potential to reveal important information space on resources the users interact with. The advent of Web 2.0 employs social networks and social media that offer new ways of communication and a very personalized user experience. At the same time, social networks can be considered as one of the biggest data sources for *user profiles on the Web*. A user profile represents a collection of the data that characterizes a user and may consist of *straightforward properties* like name and address data, a list of interests, and *collected properties* like user models derived from user activities or similar. In other words, user Web profiles are collections of data characterizing the user that can serve different purposes, for example, user personalization, recommendations, support users in sharing content with friends, or integrating content from different sources.

The current situation on user security and privacy protection is very problematic: user profile data are not owned by the users themselves but by the respective service providers. Thus, having full control over information about user Web profile, which is additionally enriched by metadata and linked together to allow the rise of novel mash-up applications, can give the users information where their data are stored, and how and by whom it is used.

In Open Source area, there are several related projects. For example, the OpenSocial† project provides a collection of APIs and gadgets for exchanging user profile information between social media applications, but does not allow to control where user data are stored. Approaches like Diaspora‡ and StatusNet§ aim to give the users full control over their profile information by letting them choose the server on which they want to store the data, but as they introduce new proprietary data silo, the users consequently lose interests to move from existing platforms to fully new set of services. Recently, there have been several new research approaches investigating data schemes like FOAF or SIOC that partially address user profile representation,

---

* http://swoogle.umbc.edu/
† http://code.google.com/apis/opensocial/
‡ https://joindiaspora.com/
§ http://status.net/

than FOAF + SSL* and WebID† that address identification and exchange of user profile information in a secure way. The problem of controlled share of user profiles among different social media websites has not yet been investigated.

In addition to having properly mastered user Web profiles, multimedia is also underrepresented in the current Web of Data [10]. This is due to the lack of integrated means to describe, publish, and interlink multimedia content. Our approach to support semantic social TV media content generation, social distribution, and consumption is called the LMF. In this chapter, we describe the LMF conceptual model, its architecture, and software components implemented so far.

20.4 The LMF Approach

The major obstacle of today's tools for publishing data according to the Linked Data principles [6] can be seen as a strong argument to differentiate information resources (e.g., a JPEG file) from noninformation resources (e.g., the description of the image content). Thus, in our work, we use the term *content* to denote the information resource and *metadata* for the noninformation resource [10]. We call the interlinked graph of resources—*Linked Media*, while framework that bridges the gap between the document Web and the Web of Data is *the LMF*.

The LMF stores and retrieves content and metadata for media resources and resource fragments in a unified way. LMF implements the Linked Data extensions proposed in Ref. [10], as well as integration of concepts from Linked Media, Media Management, and Enterprise Knowledge Management, in a way that supports semantic annotation of media content, metadata storage, indexing, and search. We call our extensions to Linked Data principles— *Linked Media principles* (for more details: http://code.google.com/p/kiwi/wiki/ LinkedMediaPrinciples):

1. Enrich content by interlinking resources to data from the Web of Data
2. Integrate additional data with existing metadata
3. Index the integrated metadata and the underlying content; refresh indexes if updates of content or metadata occur
4. Publish metadata and content as *Linked Media*
5. Provide search and query interfaces to retrieve content and data

In addition, we identified the specific requirements for the LMF platform that cover each of the above-identified principles in a sufficient way:

- Data source candidates for linking must be determined within the Linked Data

---

* http://www.w3.org/wiki/Foaf%2Bssl
† http://www.w3.org/2005/Incubator/Webid/wiki/Main_Page

- Both local and remote resources must be seamlessly integrated
- The data from the LOD cloud must be held up-to-date
- There must be an index that can handle resource content as well as metadata
- The framework should provide the functionality of a Linked Data Server
- Interfaces (Web services and graphical user interfaces (GUIs)) must allow access to indexed data

20.4.1 The LMF Conceptual Model

The LMF conceptual model has been defined based on the above set of requirements and a set of extensions to the Linked Data. Primarily, it is concerned with two dimensions of extending the Linked Data principles, such as (i) Linked Data updates, and (ii) managing media content and metadata. It implements concepts such as RESTful Resource Management, Extended Content Negotiation, Semantic Enhancement via Linked Data, and Semantic Indexing.

20.4.1.1 RESTful Resource Management

The LMF implements several extensions to Linked Data by applying REpresentational State Transfer (REST) principles [47]. The central principle of REST is "the existence of resources (sources of specific information), each of which is referenced with a global identifier (e.g., a URI in HTTP)." Therefore, combining REST and Linked Data is a natural choice. For example, in addition to the HTTP GET command, which is used to retrieve a resource, we can use POST, PUT, and DELETE in the following way: (i) POST creates a resource represented by the URI used in the request; (ii) PUT replaces the content or metadata of the resource represented by the URI used in the request with the data contained in the request body; and (iii) DELETE removes the resource represented by the URI used in the request and all associated content and metadata. In the case of GET, the extension implements for the *Access header* to determine what to retrieve and redirects to the appropriate URL, while for PUT, the *Content-Type header* determines what to update and redirect to the appropriate URL.

20.4.1.2 Extended Content Negotiation

Linked Data is currently concerned about data and does not take into account media content that is often associated with the resource. Hence, it is not possible to distinguish between a document of type RDF/XML *as content* and the metadata about this content *as data*. The Content Negotiation concept distinguishes between an RDF representation and a human readable (HTML-based) representation in the GET request. LMF extends this principle in a way to support arbitrary formats based on the MIME type in Accept/Content-Type headers, and

distinguish between content and metadata based on the *rel* extension for Accept/
Content-Type headers:

- A media type of `type/subtype; rel=content` indicates that the client requests or sends *the content* associated with the resource in the requested format.
- A media type of `type/subtype; rel=meta` indicates that the client requests or sends *the metadata* associated with the resource in the requested format.

20.4.1.3 Semantic Indexing

The LMF holds RDF data in a triple store that implements a Sesame-Sails repository. In parallel, it manages a full text index for metadata and textual content based on Apache SOLR.* Apache SOLR is an open source enterprise search platform with major features include powerful full-text search, hit highlighting, faceted search, dynamic clustering, database integration, rich document handling, and geospatial search. This allows the complex graph-based queries to be issued via SPARQL, and allows for textual queries in SOLR search syntax, in such a way that provide client libraries for both indexes (e.g., SOLR Client Libraries and SPARQL Clients) to be used natively. In addition, we developed *SOLR Shortcuts* language that makes the SOLR search index to be dynamically adaptable and scalable.

20.4.1.4 Semantic Enhancement via Linked Data

As described in Ref. [48], there are four models that can be used to find relevant linking partners, such as automatic interlinking, emerging interlinking, user contributed and game-based interlinking. Currently, the LMF uses automatic methods to get recommendations for possible links or linkable fragments and special GUIs for user-based confirmation.

20.4.2 The LMF Architecture and Implementation

The LMF is grounded on Service-Oriented Architecture (SOA) that is implemented via CDI Weld† that is the reference implementation of the Java standard for dependency injection and contextual life cycle management. Communication between the LMF and the outside world happens via RESTful Web services. The LMF consists of the following modules:

LMF core: It implements the Linked Data Server with the proposed extensions that enable handling resources via the *ResourceWebService*. In addition to triple management, the LMF Core provides transaction management and versioning.

The LMF's Linked Data Server follows a 3-tier architecture: (i) data layer is a relational database that implements the LMF triplestore including resources,

---

* http://lucene.apache.org/solr/
† http://seamframework.org/Weld

content, and triples, as well as versioning and user management information, (ii) second layer includes system services for managing transactions and versioning. It also contains services to manage resources (content and metadata); and (iii) the presentation layer consists of several RESTful Web services to manage transactions, resources, users, and configurations from the outside world.

LMF Search: It offers semantic search over resources that is based on Apache SOLR. It includes extendable rule-sets for RDF, SKOS (Simple Knowledge Organization System), DC, and GeoNames. The LMF search module can be accessed via REST API that conforms to the OpenSearch* standard and is compatible with the Apache SOLR search. For example, text-based semantic index is built upon an SOLR multicore index for multilingual indexing. The reindexing of resources is pushed by several events in the core module such as *resource_created*, *resource_updated*, and so on. In addition, the LMF REST API is implemented as frontend that can serve to a number of existing commercial systems in a way that they can expose their content and metadata following Linked Data principles.

LMF SPARQL: It provides an SPARQL end point for querying the data that are contained within the LMF. As mentioned before, the LMF implements a Sesame-Sails repository, which allows us to issue SPARQL queries on LMF data and retrieve results in several serialization formats. Because of the Sesame binding ability, it currently offers only querying but as soon as Sesame allows SPARQL update, the LMF will provide the same.

When working with Linked Data, sometimes it becomes necessary to offer local caching of remote resources. Thus, the LMF implements *Linked Data Caching mechanism* which is based on the existing Web Caching approaches, such as `Expires:` and `ETag:` HTTP headers.

Furthermore, the LMF's components interact by sending events. For example, upon completion of transaction, all transaction data are handed over to the LMF search and SPARQL components for indexing.

In addition, the LMF includes an (optional) rule-based reasoner[†] called sKWRL (simple rule-based reasoning language for RDF) that is highly customizable and allows for user-defined rules to be evaluated over triples in the LMF Triple Store. The rules can be uploaded and stored in the LMF via an easy-to-use Web service. The evaluation strategy is an incremental forward-chaining reasoning with truth maintenance and is reasonably efficient even for big data sets. The truth maintenance can be used to provide explanations for inferred triples as well as for efficient updating of the reasoning information.

The sKWRL reasoner [11] has originally been developed for the EU FP7 KiWi project, but has been reimplemented in context of the LMF by using more efficient evaluation strategies and a simplified language with no negation.

---

* http://www.opensearch.org/Home
[†] LMF Reasoner: http://code.google.com/p/kiwi/wiki/Reasoning

To create semantic annotations via linking to existing resources, the LMF enhancer uses interlinking pipelines [49]. In order to provide semantic annotations via linking to existing resources, the Apache UIMA Framework that provides a platform for the integration of NLP (Natural Language Processing) components and the deployment of complex NLP pipelines has been used. UIMA allows usage of a series of extractors, annotators, and interlinkers based on specific content issues such as MIME type, language, and so on. The enhancement pipeline can either be triggered by system events or by using a RESTful API.

As one of the most valuable data sources on the Web are user Web profiles, the LMF will also implement several adapters to existing social networking websites and services by using their public and proprietary APIs to access and integrate these profiles. In addition, user Web profiles will be enriched by metadata and linked to allow the rise of novel mash-up applications. In sum, user Web profiles supported by the LMF will give the users full control over the data pools in which these data are stored, answering question on how and by whom they can be used.

The latest LMF source code is available as Open Source at the project website http://code.google.com/p/kiwi.

20.5 Showcase Application: LMF to Support Semantically Interlinked Media

In the following, we discuss two showcase applications that support semantic social TV media content generation, social distribution, and consumption via LMF. The first demo integrates Linked Data about various Social Media content types via PopCorn.js* framework. The other interesting approach to demoing the LMF functionality might be implemented via *hypervideo* that is a displayed video stream containing embedded user-clickable anchors [47]. Finally, the second demo plays with the synchronized and user-centric multimedia annotation via LMF.

20.5.1 Video Annotation via LMF

Here, we annotate video resources with additional content coming from the external Social Media platforms (e.g., Twitter). The annotation itself follows the Open Annotation Collaboration model [48], which allows linking both temporal and special video fragment to external and internal content like related videos or websites. To benefit from added content, we firstly cache and index external content, which offer an advanced search functionality that includes search for similar or related information integrated from different Social Media websites, Linked Data sources, and even plain text websites. To display video content and annotation in parallel, we use PopCorn.js that is an event framework for HTML5 <video>

---

* http://popcornjs.org/

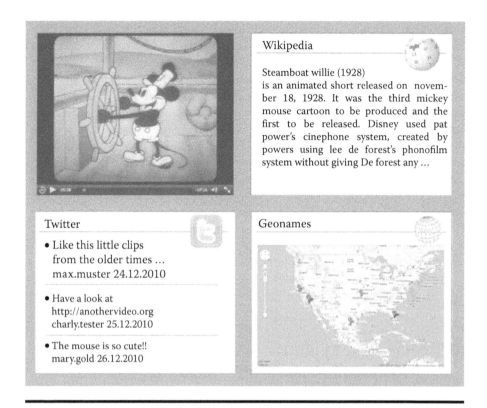

Figure 20.1   Type-dependent video annotation via LMF.

elements. It contains an easy-to-use and simple extendable widget mechanism. It utilizes the native `HTMLMediaElement` DOM (Document Object Model) Interface, its properties, methods, and events. As shown in Figure 20.1, LMF-based video annotation support a lot of extensions for type-dependent annotation representation (e.g., widget for Wikipedia articles, Twitter feeds, etc.).

20.5.2 ConnectMe Synchronized and User-Centric Multimedia Annotation via LMF

The *ConnectMe* project will provide multimedia annotation and delivery of synchronized multimedia presentations to the end users via a Web-based tool that is available to the content owner. It will implement intuitive user interfaces at the device side, so that the selection of the on-screen objects and browsing of content associated to them in the resulting presentation can be done in a nondisruptive and intuitive fashion [2]. In that sense, three separated showcase applications will be developed such as (i) an enhanced WebTV experience in the Web browser, (ii) personalized IPTV (Internet Protocol) TV, and (iii) mobile TV. This reflects the reality that "television" today is no longer tied to the TV set, but that Web and TV converge also in the personal computer and mobile device, as well as become part of standard IPTV offerings. Each showcase will run through the semantic media services platform that is called the ConnectMe Framework (shown in Figure 20.2), that has to provide a set of annotated services with a role to facilitate a new interactive user-centric media experience on top of the convergence of TV/video and the Web, for example, content extraction, reasoning about concepts, retrieval of Web content, ranking and selection of content, packaging in a multimedia presentations, adaptation to context and devices, delivery. These services will be supported by the LMF [2] in order to connect media via the following three steps:

- *Channel annotation:* It groups videos into channels, and then annotates video segments with links to semantic concepts. The annotation in the ConnectMe Framework is not just temporal but also spatial in reference to the location of the object in the video frame. The channel annotation can be driven by automatic classification over various media features (e.g., visual features, audio, subtitling, crowdsourced tagging) or tool tips that are choosing concepts from preselected sources.
- *Channel enrichment:* It consists of (i) preselection of content sources, (ii) preview mode for annotated video, and (iii) on-the-fly video enrichment with selected content.
- *Channel provision:* It provides services around video analysis, annotation, and distributed multimedia metadata querying to be delivered to the ConnectMe Framework for further annotation.

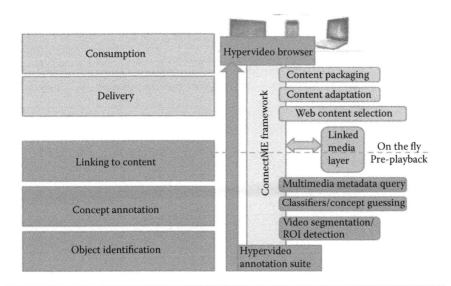

Figure 20.2   The ConnectMe Framework.

Figure 20.3   The ConnectMe Hypervideo Browser.

In reference to the ConnectMe Framework, which is supported by back-end services including those from the LMF, it will act as a back-end to the *Hypervideo Annotation Suite*, a Web-based tool in which the content owner can semiautomatically annotate their video, select the content sources they wish to use, and see a preview of the resulting enrichment. Each step is eased through the use of Linked Data and semantic technology to automate the process as much as possible, while still ensuring a final manual check so that content owners always have final control over the expected result. A dedicated *Hypervideo Browser* will be able to interpret the video stream with the synchronized enrichment information (clickable regions, content to display) in order to provide an user interface to the enriched video, in which users can select regions and access additional information, as shown in a mock-up (see Figure 20.3).

20.6 Further Work

The LMF team is working on a number of optional modules to extend the functionality of the Linked Media Server (for more details see: http://code.google.com/p/kiwi/), before perform the first benchmarking and evaluation procedures:

LMF Permissions implements and extends the WebID and WebACL (Access Control List) specifications for standards-conforming authentication and access control in the LMF. These extensions follow Linked Data principles and still provide required security aspects.

LMF Enhancer offers semantic enhancement of content by analyzing textual and media content. It is built upon Apache UIMA, Apache Tika, and our own Apache Stanbol* framework.

LMF Media Interlinking implements support for multimedia interlinking based on the work in the W3C Multimedia Fragments WG[†] and the W3C Multimedia Annotations Working Group (WG).[‡]

LMF Versioning implements versioning of metadata updates. It is itself already carried out by LMF Core, but the management of versions will be carried out by this module.

20.7 Conclusion

Currently, due to the lack of integrated means to describe, publish, and interlink multimedia objects and data, these contents remain underrepresented in the Web of Data. At the same time, media-related services are still limited with regard to their personalization, user profiling, social, and community aspects. These require more advanced data analysis, reasoning and understanding, as well as common data schemas, and APIs to support data exchange and integration across different sources. Semantic technologies, related standards, initiatives and recommendations hold promise for interoperability in various domains of knowledge and content technologies. These technologies bring new types of services that can simultaneously force the extension of existing and emergence of new TV-related infrastructure, specification and standards.

In the context of diversification and proliferation of location- and interest-specific Social Media communities, the interconnection between them and their platforms becomes especially important [50]. In that sense, Social Media service extensions lead to creating *advanced open interfaces and frameworks* to make the content of Social Media websites available for embedding in other contexts and Really Simple Syndication (RSS) feeds, and to track new information that was made available by the user. For many Social Media websites with specific aims (such as TV), it is of substantial interest to tap into *the local knowledge held by members of hyperlocal social communities*, which may be unavailable to professional content creators. Here, there is also an important role for the use of mobile devices to capture such information on the spot and virtually in real time, to report high-profile events. Social Semantic Web technologies bring the necessary data understanding and integration that can enhance current TV services to be more personal and social, and better adapted to the end-user individual interests and needs.

---

* http://incubator.apache.org/stanbol/

[†] W3C Multimedia Fragments WG: http://www.w3.org/2008/WebVideo/Fragments/

[‡] W3C Multimedia Annotations WG: http://www.w3.org/2008/01/media-annotations-wg.html

Acronyms

ACL	Access Control List
ASR	Automated Speech Recognition
AR	Augmented Reality
DC	Dublin Core
DOLCE	Descriptive Ontology for Linguistic and Cognitive Engineering
DOM	Document Object Model
EXIF	Exchangeable Image File Format
FOAF	Friend Of A Friend
IE	Information Extraction
IFPs	Inverse Functional Properties
IPTV	Internet Protocol TV
IR	Information retrieval
JSON	JavaScript Object Notation
LOD	Linked Open Data
LMF	Linked Media Framework
M3O	MultiMedia Metadata Ontology
NLP	Natural Language Processing
OCR	Optical Character Recognition
OWL	Web Ontology Language
RDF	Resource Description Framework
RDFS	RDF Schema
REST	REpresentational State Transfer
RICO	Reusable Intelligent Content Ontology
SIOC	Semantically Interlinked Online Communities
SOA	Service-Oriented Architecture
SPARQL	SPARQL Protocol and RDF Query Language
SWSE	Semantic Web Search Engine
TF-IDF	Term Frequency—Inverse Document Frequency
UIMA	Unstructured Information Management Architecture
W3C	World Wide Web Consortium
XMP	Extensible Metadata Platform

References

1. *W3C Issues Report on Web and Television Convergence*. 2011. Online available: http://www.w3.org/2011/03/Webtv-pr.
2. The ConnectMe project proposal (accepted): Connected Media Experiences. The COIN Programmline "Kooperation und Netzwerke," 2010. http://www.connectme.at.
3. Nixon, L., Aroyo, L., Miller, L.: NoTube—The television experience enhanced by online social and semantic data. *At the 1st IEEE International Conference on Consumer Electronics (ICCE 2011)*, Berlin, Germany, 2011, pp. 269–273.

3a. Berners-Lee, T., Hendler, J., Lassila, O.: The semantic Web. *Scientific American.* Available at: http://www.scientificamerican.com/article.cfm?id=the-semantic-Web.

4. Mathews, B.: Semantic Web technologies. *JICS Technology and Standards Watch*, 2010.

5. Stankovic, M., Wagner, C., Jovanovic, J., Laublet, P.: Looking for experts? What can linked data do for you? In *Proceedings of the LDOW2010*, April, Raleight, USA.

6. Berners-Lee, T.: *Linked Data—Design Issues.* Online available at: http://www.w3.org/DesignIssues/LinkedData.html, 2006.

7. Raimond, Y., Scott, T., Oliver, S., Sinclair, P., Smethurst, M.: Use of semantic Web technologies on the BBC Web sites. In D. Wood (Ed.): *Linking Enterprise Data.* Springer 2010, pp. 263–283.

8. Raimond, Y., Sinclair, P., Humfrey, N. J., Smethurst, M.: *Programmes Ontology.* Online ontology, September 2009. Available at: http://purl.org/ontology/po/. Last accessed April 2010.

9. Library Linked Data Incubator Group wiki: Draft Report, 2011. Online available: http://www.w3.org/2005/Incubator/lld/wiki/DraftReportWithTransclusion.

10. Kurz, T., Schaffert, S., Bürger, T.: LMF—A framework for linked media. *Workshop on Multimedia on the Web (MMWeb 2011) in conjunction with i-Know and i-Semantics 2011 Conference*, 8th September 2011, Graz, Austria. (to appear).

11. Kotowski, J., Bry, F.: A perfect match for reasoning, explanation, and reason maintenance: OWL 2 RL and semantic Wikis. *Proceedings of the 5th Semantic Wiki Workshop (SemWiki2010)*, collocated with the 7th ESWC2010. 2010.

12. Saez, E., Benavides, J. I., Guil, N.: Reliable real time scene change detection in MPEG compressed video. *Proceedings of International Conference on Multimedia and Expo (ICME 2004)*, Taipei, Taiwan, 2004, Vol. 1, pp. 567–570.

13. Qian, X., Liu, G., Su, R.: Effective fades and flash-light detection based on accumulating histogram difference, *IEEE Transactions on Circuits and Systems for Video Technology*, **16**(10): 1245, 2006.

14. Bailer, W., Hoeller, F., Messina, A., Airola, D., Schallauer, P., Hausenblas, M.: State of the Art of Content Analysis Tools for Video, Audio, and Speech. PrestoSpace, Del. 15.3, 2005.

15. Wu, L., Hua, X. S., Yu, N., Ma, W. Y., Xi, W., Fan, W.: Optimizing Web search using Web click-through data. *Proceedings of the ACM International Conference on Information and Knowledge Management (CIKM 2004)*, Washington DC, USA, 2004, pp. 118–126.

16. Sun, Y., Shimada, S., Taniguchi, Y., Kojima, A.: A novel region-based approach to visual concept modeling using Web images. In *ACM Multimedia*, Vancouver, Canada, 2008, pp. 635–638.

17. Barnard, K., Duygulu, P., Forsyth, D. A., de Freitas, N., Blei, D. M., Jordan, M. I.: Matching words and pictures. *Journal of Machine Learning Research*, **3**: 1107–1135, 2003.

18. Li, L. J., Socher, R., Fei-Fei, L.: Towards total scene understanding: classification, annotation, and segmentation in an automatic framework. *In IEEE Conference on Computer Vision and Pattern Recognition*, Miami, FL, USA, 2009, pp. 2036–2043.

19. Goldman, J., Renals, S., Bird, S., deJong, F. M. G., Federico, M., Fleischhauer, C., Kornbluh, M. et al: Accessing the spoken word. *International Journal on Digital Libraries*, **5**(4): 287–298. ISSN 1432-5012. 2005.

20. Ordelman, R. J. F., deJong, F. M. G., van Leeuwen, D. A.: Speech indexing. In Blanken, H. M., de Vries, A. P., Blok, H. E., Feng, L. (Eds.): *Multimedia Retrieval.* Springer, pp. 199–224, 2007.

21. ABBYY FineReader OCR Product Line. Available: http://finereader.abbyy.com/.
22. OmniPage Professional, Available online: http://www.nuance.de/ Last visited: 04 July, 2011.
23. I.R.I.S. Company Homepage: http://www.irislink.com/Last visited. 04 July, 2011.
24. Witten, J. H., Frank, E.: *Data Mining: Practical Machine Learning Tools and Techniques*. Morgan Kaufmann, San Francisco, 2nd ed., 2005.
25. Passant, A., Laublet, P., Breslin, J., Decker, S.: A URI is worth a thousand tags: From tagging to linked data with MOAT. *International Journal on Semantic Web and Information Systems*. Special Issue on Linked Data, Sept. 2009 – IGI Global.
26. Cooper, M., Foote J., Girgensohn A., Wilcox L.: Temporal event clustering for digital photo collections. *ACM Transactions on Multimedia Computing, Communications and Applications*. 2005.
27. Mika P.: Ontologies are us: A unified model of social networks and semantics. *In Proceedings of the 4th International Semantic Web Conference (ISWC 2005)*, Galway, Ireland, 2005, pp. 522–536.
28. Specia, I., Motta, E.: Integrating folksonomies with the semantic Web. *In Proceedings of the 4th ESWC*, (ESWC 2007), Innsbruck, Austria, 2007, pp. 624–639.
29. Stegmaier, F., Lee, W., Poppe, C., Bailer, W: API for media resources 1.0. W3C Working Draft. 12 July 2011. http://www.w3.org/TR/2011/WD-mediaont-api-1.0-20110712/.
30. Arndt, R., Troncy R., Staab S., Hardman L.: COMM: A core ontology for multimedia annotation. *Handbook on Ontologies*, 2nd Edition (pp. 403–422). Springer. 2009.
31. Saathoff, C., Scherp, A.: Unlocking the semantics of multimedia presentations in the Web with the multimedia metadata ontology. *World Wide Web Conference (WWW 2010)*, Raleigh, NC, USA, April 2010, pp. 831–840.
32. Bürger, T., Simperl, E.: A conceptual model for publishing multimedia content on the semantic Web. *4th International Conference on Semantic and Digital Media Technology (SAMT)*, Graz, 2009, pp. 101–113.
33. The RICO Model. Online available: http://www.tobiasbuerger.com/icontent/ricoon-tologies.html.
34. The Open Collaboration Annotation Model. Online available: http://www.openan-notation.org/.
35. Brickley, D., Miller, L.: Friend of a Friend. 2000. http://foaf-project.org.
36. Breslin, J., Harth, A., Bojars, U., Decker, S.: Towards semantically-interlinked online communities. *In Proceedings of the 3rd ESWC. (ESWC 2006)*, Budva, Montenegro, 2006, pp. 71–83.
37. Breslin, J., Passant, A., Decker, S.: *The Social Semantic Web*. Springer, 2009, ISBN 364201173X.
38. Dai, H., Zhao, L., Nie, Z., Wen, J. R., Wang, L., Li, Y.: Detecting online commercial intention (OCI). *Proceedings of WWW2006, (www 2006)*, Edinburg, Scotland, 2006, pp. 829–837.
39. Gravano, L., Hatzivassiloglou, V., Lichtenstein, R.: Categorizing Web queries according to geographical locality. *Proceedings of International Conference on Information and Knowledge Management (CIKM 2003)*, New Orleans, USA, 2003, pp. 325–333.
40. Chuang, S. L., Chien, L. F.: Towards automatic generation of query taxonomy: A hierarchical query clustering approach. *Proceedings of ICDM 2002 (ICDM 2002)*, Maebashi City, Japan, 2002, p. 75.
41. Baeza_Yates, R., Calderon-Benavides, L., Gonzales, C.: The intention behind Web queries. *SPIRE*, pp. 98–109. 2006.

42. Baeza_Yates, R., Tiberi, A.: Extracting Semantic Relations from Query Logs. *In Proceedings of ACM SIGKDD Conference on Knowledge Discovery and Data Mining (KDD 2007)*, San Jose, CA, USA, 2007, pp. 76–85.

43. Fox, S., Karnawat, K., Mydland, M., Dumais, S., White, T.: Evaluating implicit measures to improve Web search. *ACM Transactions on Information Systems (TOIS)*, **23**(2):147–168, 2005.

44. Claypool, M., Le, P., Waseda, M., Brown, D.: Implicit Interest Indicators. *In Proceedings of the 6th International Conference on Intelligent User Interfaces (JUI'01)*, Santa Fe, New Mexico, USA, 2001, pp. 33–40.

45. Oren, E., Delbru, R., Catasta, M., Cyganiak, R., Stenzhorn, H., Tummarello, G.: Sindice.com: A document-oriented lookup index for open linked data. *International Journal of Metadata, Semantics and Ontologies*, **3**(1):37–52, 2008.

46. Cheng, G., Qu, Y.: Searching linked objects with falcons: Approach, implementation and evaluation. *International Journal on Semantic Web and Information Systems*, **5**(3):49–70, 2009.

47. Fielding, R. T.: *Architectural Styles and the Design of Network-based Software Architectures*. PhD Thesis, University of California, Irvine, 2000.

48. Bürger, T., Hausenblas, M.: Interlinking multimedia—Principles and requirements. *Proceedings of the First International Workshop on Interacting with Multimedia Content on the Social Semantic Web, co-located with SAMT08 (SAMT 2008)*, Koblenz, Germany, 2008.

49. Bürger, T.: Insemtives Deliverable 2.2.2/2.2.3. Insemtives Project. FP7-ICT-2007-3, 2009.

50. Bruns, A., Bahnisch, M.: Social media: Tools for user-generated content. *Social Drivers behind Growing Consumer Participation in User-Led Content Generation. Volume 1 – State of the Art*. Smart Services CRC Pty Ltd. Australian Technology Park, Locomotive Workshop. March 2009. Online available: http://bit.ly/j0uPzI.

Chapter 21

Telepresence: Immersive Experience and Interoperability

Bruno Chatras

Contents

21.1 Introduction 472
21.2 Technical Overview and Requirements 472
    21.2.1 Overview 472
    21.2.2 Signaling Requirements 473
    21.2.3 Transport Requirements 476
    21.2.4 Other Aspects 477
21.3 Session Initiation Protocol-Based Systems 477
    21.3.1 Overview 477
    21.3.2 Streams Properties and SDP 479
    21.3.3 Media Transport 480
21.4 Interoperability and Standardization 481
21.5 IMS and TP 483
21.6 Future Directions 485
Acronyms 485
References 486

21.1 Introduction

Telepresence (TP) represents a pivotal evolution of the videoconferencing market in terms of user experience and growing opportunity. The attractiveness of TP is currently boosted by expected financial and environmental savings brought by remote collaboration solutions in general. TP refers to an advanced videoconferencing technology where participants feel as if they were sitting in the same room. It is much more than high-quality audio and video conferencing and provides a so-called "immersive" or "being there" experience.

The business sector is currently the main share of the TP market, especially in-company usage. Multisite companies see TP as one of the solutions they can apply for saving time and money by having their employees meeting without traveling. The growing opportunity is reinforced by the "green" trend as TP and remote collaboration in general help enterprises lower their carbon footprint. Other potential opportunities exist in the distance education market and telemedicine, ranging from remote diagnosis to remote surgery. Bringing TP to the residential market is currently a challenge due to the cost of systems that can provide a truly immersive experience. Still, things may rapidly develop with the rapid consumer adoption of connected televisions (TVs) with large-sized high definition (HD) screens. The availability of standard solutions enabling any HD video device to participate to a TP conference will also be instrumental in widening the TP market to residential customers and small offices.

This chapter starts with an overview of TP systems and of the features that distinguish them from conventional videoconferencing. Technical constraints and requirements associated with these features are then described with emphasis on Session Initiation Protocol (SIP)-based systems, signaling issues, and barriers to interoperability. A review of ongoing standardization efforts is also provided. The chapter ends with a discussion on the potential benefits of using the 3rd Generation Partnership Project (3GPP) IP Multimedia core network Subsystem (IMS) as a platform for supporting TP systems involving a wide range of fixed and mobile devices.

21.2 Technical Overview and Requirements

21.2.1 Overview

While videoconferencing has been around for over 40 years with low user adoption rate, TP emerged on the market in the early 2000s as a technology breakthrough in the evolution of videoconferencing. The term TP is associated to high-fidelity wideband audio sound and high-definition video, but TP is much more than that: TP refers to a technology where participants feel as if they were in the same meeting room. This includes the ability to display life-size images of the conference rooms and participants, with spatial audio sound and eye-to-eye contact feeling.

The trade secret of TP solutions for providing the "immersive experience" relies on a combination of technology elements and design tricks that give the users this unique feeling of being in the same room as the participants in remote locations.

A TP conference is a special videoconference that connects two or more TP endpoints through multimedia communication, including at least the exchange of high-quality audio and video streams with associated control streams. A TP endpoint can be a specifically equipped and designed meeting room (a.k.a. a TP room) gathering several participants or a simpler personal device with video capabilities, although not all device types can provide a truly immersive experience to end users. TP rooms are typically composed of multiple video cameras, microphones, loud speakers, and large size display monitors, each of which serves a different purpose. Some devices capture and render images and sound from conference participants. Others may be used for presenting a meeting agenda or any other document or for playing back a video for display to the remote participants.

When TP emulates a genuine face-to-face business meeting, the cameras and screens are arranged to provide a panoramic view of the room. Each camera captures images from one region of the room. Slightly different arrangements are used when the TP technology is applied to distance education or other use cases.

21.2.2 Signaling Requirements

TP, as any videoconferencing system, has signaling requirements beyond those required to set up a basic call between two endpoints. This includes, for example, the need to exchange sufficient information about the media streams to be able to make intelligent decisions about which monitor should be used to display a particular stream. Indeed, in a dual-monitor system, a videoconferencing system will typically display the "main" video stream representing remote participants on a large-sized HD monitor and the "presentation" stream on smaller monitors located near the participants. Means to exchange information about the stream roles depend on the signaling protocol used between the endpoints. Videoconferencing systems based on the H.323 [1] series of ITU Telecommunication Sector (ITU-T) Recommendations use a mechanism defined in ITU-T Recommendation H.239 [2]. Videoconferencing systems based on the Internet Engineering Task Force (IETF)-defined SIP use a mechanism defined in RFC 4796 [3].

Additional characteristics beyond the stream role such as aspect ratios of cameras and display monitors or the camera field of view may need to be signaled between TP endpoints, as well. For example, if the aspect ratios in different sites are not the same, some technique needs to be applied to make up for the difference.

TP adds further signaling requirements on the top of videoconferencing. In the simplest case, a TP conference is established between two rooms equipped with the same number of screens and cameras. A typical and widespread configuration consists of three screens and associated camera with stereo sound in each room (see Figure 21.1). The cameras and screens are arranged to provide a panoramic view of

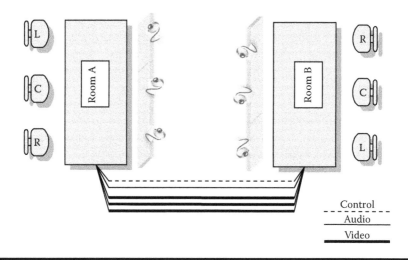

Figure 21.1   Typical TP rooms configuration.

the remote room. Each camera captures a different portion of the scene (left, center, right) and generates video streams that have to be displayed by the remote site on the appropriate screens (i.e., the output of the left camera has to be displayed on the screen at the right side of the remote site). Audio streams generated at one site need of course to be mapped at the remote site in the same way as the video so as to produce a consistent and life-like experience.

This relatively simple configuration already gives rise to specific signaling requirements compared to traditional videoconferencing. Each site will receive multiple video streams representing remote participants and will have to decide which one is to be displayed on which screen. Appropriate labels should therefore be exchanged between TP rooms to map video streams on spatial positions.

Moreover, needless to say, TP requires synchronization of audio and video streams, played out by different devices but representing the same speaker. So, in some configurations, to enable pairing audio streams and video streams, additional information need to be signaled between rooms.

The above example is relatively simple. However, there are more complex configurations, where each site has a different number of screens and cameras than the other site [4]. Whether these configurations still provide immersive experience can be questioned. Nevertheless, these configurations exist in practice and have to be managed. This creates additional technical challenges.

Let us take an example where one of the TP rooms has a single camera and screen and the other is a typical room with three cameras and screens. The main challenge is for the one-screen room to handle video streams received from the three-camera room. One strategy is to select only one of the three video streams at a given time (e.g., the stream with the image of the current speaker) and display it on the single screen. Another strategy would be to compose the three streams into a single image for display on this single screen. Other approaches are possible as well. For example, the three-screen site may have an extra camera that can automatically focus on the current speaker, in which case the one-screen room can request that only the video stream produced by this camera be sent to it. To choose a strategy, the receiving TP system needs to have information about the capabilities of the remote site and the content of the streams it receives. In the first case, means for dynamically detecting which video stream is associated with the current speaker are required so that the receiving room can automatically switch to that stream and forward it on its single screen. If the second strategy is selected, video streams have to be tagged with spatial indications so as to enable the receiving system to compose a meaningful image. If the third strategy is used, a tag has to be associated to the video stream representing the current speaker so that the receiving room can select it during session establishment as the only video stream it wants to receive.

Another variant is the multiroom conference (see Figure 21.2), where several conference rooms are connected to a bridge known as a Multipoint Conferencing Unit (MCU), which may be hosted in the same premises than one of the conference rooms or in a third-party location (e.g., network operator's premises).

In such configurations, it is usually not possible to provide a "being there" experience simultaneously to all participants in all rooms. Several policies can be used at each site for selecting what to display on the local screens. Well-known policies already in use for traditional videoconferencing apply to TP as well, in particular those known as "site switching" and "segment switching." With site switching, all the camera images from the site where the current speaker is located are forwarded to all other sites. Therefore, at each receiving remote site, all the screens display camera images from the current speaker's site and the other sites are not visible. With segment switching, only the image representing the speaker is sent so that the

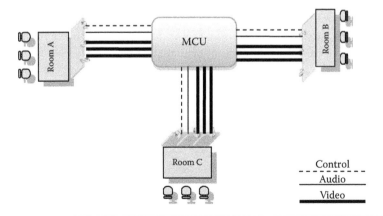

Figure 21.2   Multipoint TP conference.

remaining screens can be used to display images from other remote sites. Other policies are also possible, for example, one could compose a single-screen image from the three images received from all three-screen sites and dedicate a screen to each remote location. This approach does not preserve of course full-size image display of the participants and therefore is harmful to the immersive experience but is the solution enabling more sites to be seen simultaneously. Whatever approach is followed, both TP endpoints need to have information about the capabilities of the remote site (e.g., whether segment switching is supported) and the content of the streams they exchange to select and apply a particular strategy.

A slightly different configuration is used when the TP technology is applied to distance education. There is typically a main site (e.g., a specifically equipped auditorium) with multiple cameras, one of which is focused on the professor. Other cameras may be used to capture views of the audience (e.g., overall view, zoom on a student asking a question, etc.). The configuration of the remote sites can be identical to that used for a business meeting but the role of the screens and cameras is different. All remote sites will typically display a full-size image of the professor on one of the screens while the two others may show different views of the entire auditorium or a close-up view of a student asking a question. TP endpoints in this case need to know whether a received video stream corresponds to the professor, a student asking a question, a classroom or auditorium view, a slideware, and so on.

21.2.3 Transport Requirements

The quality of experience is of paramount importance for TP solutions to become a widely used substitute to genuine face-to-face meetings. This quality does not only depend on the design of the meeting rooms and the quality of audio and video codecs. The Quality of Service (QoS) provided by the transport network connecting TP rooms also plays a major role.

TP rooms are typically composed of multiple devices most of which generate and receive high-quality media streams. This obviously increases bandwidth demand on the networks compared to conventional single-screen video conferencing systems. It is generally admitted that TP systems usually consume about 5 Mbps per screen/camera for today's technology. So a three-screen system requires 15 Mbps of network bandwidth to operate. Moreover, as any real-time application, TP is sensitive to latency (delay), jitter (delay variation), and packet loss. Lack of bandwidth or inappropriate QoS parameters is likely to result in speech distortion and image degradation, thereby ruining the "being there" experience.

The solutions to guarantee a suitable QoS are highly dependent on the network technology used to transport TP traffic and whether this traffic is carried over dedicated resources or has to share bandwidth with other applications. Some networks will handle QoS at Layer 2 of the OSI model and some others at Layer 3, using the DiffServ technology [5] or a combination of both. It should be noted that there is no standard DiffServ marking for TP traffic. However, a survey of the literature available on the subject shows that the recommended marking for this type of traffic is either Assured Forwarding (AF) or Class Selector 4 (CS4) per hop behavior (PHB) which are the recommended markings for videoconferencing applications and real-time interactive services as per RFC4594 [5], respectively.

21.2.4 Other Aspects

The requirements previously discussed can be fulfilled by including appropriate information in signaling messages and by ensuring that the network is properly engineered to provide suitable QoS to TP traffic. However, achieving a true immersive experience requires more than advanced signaling capabilities and bandwidth.

The position of cameras is critical to improve eye-to-eye contact between users. Conference users need to get the feeling that remote participants are looking at them in the eyes. Ideally, the camera should be aligned at the eye level of the participants, which requires sophisticated systems. In most solutions available today on the market, a nearly eye-to-eye contact is obtained by placing cameras closely above the screens in order to obtain a small angle between camera, eyes, and screen [6].

Room acoustics and lighting have to be carefully considered as well. Ideally, rooms that are expected to participate to the same TP conference should have identical wall paint colours, specially designed tables, and symmetrical furniture to create a reasonable illusion that the remote participants are in the same room. In other words, in a point-to-point conference, TP rooms should appear as if they were two halves of the same room. Furthermore, many TP systems use various tricks to hide the cameras, microphones, speakers, and so on from the users, as these items would not be present in a real face-to-face meeting and should ideally not be visible in a TP meeting.

21.3 Session Initiation Protocol-Based Systems

21.3.1 Overview

TP as any videoconferencing service requires signaling protocols to setup sessions between endpoints and negotiate the characteristics of the media and control streams they are willing to exchange. Most videoconferencing systems available today on the enterprise market are still using the protocols defined in the ITU-T H.323 series of recommendations. However, there is a clear market trend toward using the IETF-defined SIP and its sister protocol the Session Description Protocol (SDP). This chapter focuses on SIP-based systems. A comparison of H.323 and SIP videoconferencing support can be found in ref. [7]. SIP [8] is a signaling protocol enabling to establish multimedia sessions between two or more network endpoints.

Over the last decade, SIP has become the predominant protocol for Voice over IP (VoIP). SIP borrows its basic design principles from Hyper Text Transfer Protocol (HTTP): text-based (human readable) and request–response paradigm. SIP messages are either requests (e.g., REGISTER, INVITE, etc.) or responses (e.g., 180 Ringing, 200OK, 486 Busy Here, etc.), indeed following the HTTP model. SIP separates session control (who connects to whom) from session descriptions (how they communicate: voice, video, other media, or control channel). SIP delegates session descriptions management to the SDP and carries SDP messages in the payload of its own messages.

TP makes a relatively basic usage of SIP and does not require any extension to it. A point-to-point TP conference is set up as a SIP session between two SIP user agents, each of which represents a TP room or another type of TP endpoint. One of the endpoints initiates the session by sending an SIP INVITE request toward the remote site. Before reaching their destination, SIP messages may go through a number of intermediate SIP servers (e.g., for routing purposes) but these servers are not required to implement TP-specific logic (Figure 21.3).

Multipoints TP conferences rely on the conferencing framework specified in RFC4353 [9]. A TP conference involving an MCU is seen as a tightly coupled conference with the MCU playing the role of the conference focus for each conference. All streams to and from all endpoints in the conference are flowing through the MCU, which also maintains a SIP signaling relationship with each of these endpoints.

When used for controlling videoconferencing and TP conferences, SIP works in concert with a number of other protocols (see Figure 21.4), including but not limited to:

- SDP: To signal and negotiate media stream properties
- RTP: To convey the audio and video streams and possibly other data streams
- RTCP: To convey QoS feedback and codec control commands
- BFCP: To enable floor control

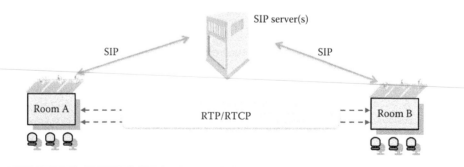

Figure 21.3 SIP-based architecture for TP.

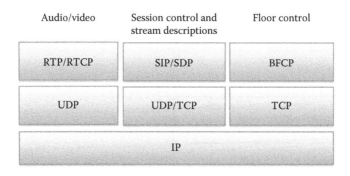

Audio/video   Session control and   Floor control
stream descriptions

Figure 21.4   Basic protocol stack for videoconferencing.

As the name implies the Binary Floor Control Protocol (BFCP) specified in RFC4582 [10] enables floor control—not to be confused with flow control—a well-known feature of video conferencing that enables controlling access to shared conference resources (e.g., requesting or granting permission to speak or present).

21.3.2 Streams Properties and SDP

Despite its name, the SDP [11] is more a format than a protocol. It is a format for describing and negotiating media properties. When used in combination with SIP, SDP messages are embedded in the body of SIP messages during the establishment of a session, or at a later stage if media properties have to be modified during the course of the session.

A SIP device opening a SIP session with a remote endpoint will typically include an SDP Offer in the INVITE request it sends. The SDP Offer will indicate how many media streams it can send and receive and describe their properties (media format, bandwidth, etc.). The remote endpoint will send back an SDP Answer indicating the media streams that it can send and is willing to receive, taking the contents of the SDP Offer into account. With advanced HD videoconferences, SDP is not only used to announce and negotiate codecs and bandwidth applicable to media streams but also additional properties that endpoints need to be aware of when making decisions about rendering them on output devices.

Below is an example of the content of an SDP Offer that is not specific to TP and could apply to any advanced HD videoconference involving two rooms equipped with a single camera and screen for capturing and rendering images from the participants, one of them generating an additional media stream representing presentation slides. In this example, the full-band audio codec G.719 is used for the audio stream, the video stream representing the speaker is encoded using H.264 and the video stream representing the presentation is encoded using H.263.

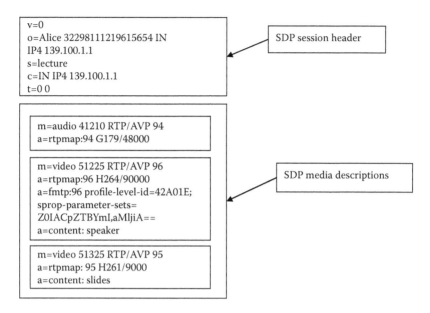

The "content" attribute defined in RFC4796 [3] is used to assign roles to streams so as to enable the remote site to decide whether to display a video stream on the main screen or an auxiliary screen. In a multiscreen, multicamera TP configuration, more media description blocks (starting with an m=line) would be required, each of which representing the media streams sent by the multiple cameras and microphones capturing video and audio from the conference participants. Although no standard solution exists yet, it is expected that each of these media lines will have to be associated with a new attribute representing its spatial position, in addition to its role.

TP could benefit from an advanced usage of SDP to cope with heterogeneous systems and asymmetric room configurations. For example, use of RFC5939 [12] enables structuring an SDP Offer in such a way that the same TP room would be able to successfully establish a basic videoconference with a basic video endpoint and a true "immersive" TP session with another TP room. A TP room initiating a session to an unknown system would typically include information that can be understood by any device as part of the RFC5939 "actual configuration" and information that can only be understood by "true" TP systems as part of the RFC5939 "potential configurations." The basic video device would just ignore the "potential configurations" portion of the SDP Offer and reply as if only the "actual configuration" portion had been received.

21.3.3 Media Transport

SIP-based systems use the IETF-defined Real-Time Transport Protocol (RTP) to carry audio and video streams between communication endpoints. They also use the Real-Time Transport Control Protocol (RTCP), a sister protocol of RTP aimed at providing feedback on the QoS, by periodically sending statistics information to senders of RTP traffic. Beside basic feedback, video communications, including TP, use RTCP for reporting lost pictures and for sending codec control messages such as the Full Intra Request (FIR) command—also known as "video fast update request" defined in RFC5104 [13]—which requests the media source to send back a decoder refresh point to reset the decoder to a known state. In the context of TP, RTP is used with either the Audio/Visual Profile (AVP) defined in RFC3551 [14] or preferably the early feedback profile (AVPF) defined in RFC4585 [15]. Using one profile or the other will determine the type of feedback that a receiver can send using RTCP. Other profiles may be used as well if media flow encryption is required.

Beyond voice and video, a TP system may handle other media types than can be transported over RTP as well or use a different transport. One could think of presentations as an obvious example of non-RTP media. However, presentations in today's systems are usually handled as a particular type of video (a PC is typically used as a virtual content camera) and therefore do not require TP systems to handle any specific media type. On the other hand, other media types may be required for integrating data collaboration applications (e.g., collaborative white-board or shared document editing) in a TP conference. In ref. [16], the Message Session Relay Protocol (MSRP) [17] is used as a media transport protocol for that purpose since this protocol can carry any arbitrary (binary) Multipurpose Internet Mail Extensions (MIME) compliant application data. Haptic device data are another example where a media type different from audio and video would have to be transported should TP be used in the context of Telemedicine or Telerobotics [18].

21.4 Interoperability and Standardization

Interoperability between videoconferencing systems supplied by different manufacturers is already a well-known challenge. TP adds even more complexity due to the number of media streams involved and the need to interconnect heterogeneous configurations. Several competing products based on open standards (SIP or H.323) exist today on the market place, but lack interoperability, due to different interpretations of the based standards, or the use of proprietary extensions made to these base protocols, or because they rely on implicit assumptions about the conference rooms configurations.

There are actually two issues. The first one is due to different implementations using different options among those permitted by the standards to perform the same thing. The second one is due to the lack of standard protocol support for some TP-specific features.

The first case is best illustrated by the problem created if one TP room uses an SIP session per device and tries to connect to another room that uses a single SIP session for handling all media streams sent and received by all devices. Multiplexing of media streams is another example: Some systems will map all media streams onto a single RTP session while others will implement a one-to-one mapping between media streams and RTP sessions. To get rid of this type of issue, strictly speaking there is no need for additional standards but the industry need to specify protocol profiles specifically tailored for TP and to document best current practices on how all protocols involved in TP should work together.

The second case is best illustrated by the lack of standard solution to associate a spatial position to a stream so that the receiving side can decide where to display the corresponding images. Some TP systems have created proprietary SDP extensions, others make implicit assumptions (e.g., the order of media descriptions in SDP Offers and Answers is interpreted as representing the spatial ordering of streams), and others handle this by adding proprietary extensions at the RTCP/RTP level, etc. Obviously, none of these systems can interwork with the others, unless additional equipment, which translates from one vendor to another, is added to the architecture.

In this context, the IETF and the International Telecommunication Union (ITU) have both launched a standardization initiative in 2011 to deliver specifications aimed at ensuring interoperability between TP systems.

Within the IETF, TP specific requirements are addressed in the *clue* working group (http://datatracker.ietf.org/wg/clue/charter/). This working group has been charted to create specifications for SIP-based conferencing systems to *"enable communication of information about media streams so that a sending system, receiving system, or intermediate system can make reasonable decisions about transmitting, selecting, and rendering media streams."* Most of the work that this group will perform is likely to consist in providing an information model for representing endpoints and stream properties, and deliver protocol specifications enabling such information to be exchanged between endpoints. At the time of writing this chapter, the IETF was assessing a number of protocol solutions, ranging from extending SDP with new attributes to defining a new protocol for use in a control channel established on the media plane using SIP and SDP.

Within the Telecommunication Sector of the ITU, Study Group 16, the historical home for H.323 standards, has created a Question (i.e., a Working Group in ITU-T parlance) in charge of TP (http://www.itu.int/ITU-T/studygroups/com16/sg16-q5.html). This Question will focus on H.323-based TP systems that are facing similar interoperability issues than those based on SIP.

The International Multimedia Telecommunications Consortium (IMTC) is also playing a role in videoconferencing and TP standardization (http://www.imtc.org/). From its creation, this organization has been focusing on facilitating interoperable multimedia conferencing solutions, starting with those based on ITU standards and now addressing IETF-based solutions as well. In 2011, the IMTC has published two Best Practices documents intended to help manufacturers of videoconferencing equipment implement SIP-based standard features in an interoperable manner [19,20]. It is likely that the IMTC will expand this work toward TP.

Furthermore, at the end of 2011, 3GPP created a work item as part of the Release 12 specification effort, aimed at incorporating support for TP in the IP Multimedia Subsystem (IMS) specifications.

21.5 IMS and TP

IMS stands for IP Multimedia core network Subsystem, a concept developed and specified by the 3GPP. The original aim of 3GPP was to specify a SIP-based multimedia architecture, enabling mobile operators to control usage of IP transport resources in their packet switched domain based on the general packet-radio service (GPRS) and offer value-added services. Since then, the IMS has gained tremendous momentum as THE carrier-grade architecture based on SIP. The rapid spread of fixed-broadband access and the obsolescence of TDM switching equipment, combined with the sustained interest in fixed and mobile convergence have made IMS increasingly relevant to fixed operators as well. The latest versions of the IMS specifications include all extensions required to make them truly suitable for use in both mobile and fixed (wireline) environments.

There are numerous good reasons for using IMS as means to connect TP endpoints. Each of these reasons considered in isolation from the others might not provide a sufficiently strong driver to justify this approach, but all together they become quite appealing.

- *Multipurpose platform*: A large number of fixed-line operators have already deployed an IMS infrastructure for supporting VoIP in both the residential and enterprise market and a growing number of mobile operators have plans to use it as well as part of their migration strategy to long-term evolution (LTE). Using the IMS in support of TP will leverage this investment by enabling a new class of services on the same SIP-based infrastructure.
- *Multimedia conferencing [21]*: The IMS already provides full specifications of signaling procedures and media processing for multimedia conferencing. TP support can be easily added on top of this existing basis.
- *Mobility*: Mobility is a given of the IMS. Mobility of TP rooms might appear to the reader as a gadget. However, one should consider that building a TP solution over the IMS also means that TP rooms will no longer be bound to a specific network access point, which can greatly improve resilience of the overall solution in case of the failure of an access point (e.g., failure of an SIP edge proxy server).
- *Resource control*: IMS, when used in combination with a policy control platform, enables easy and dynamic adaptation of transport resources to application requirements, thereby ensuring appropriate QoS according to the Service Level Agreement (SLA) between the enterprise owning TP rooms and the network operator.

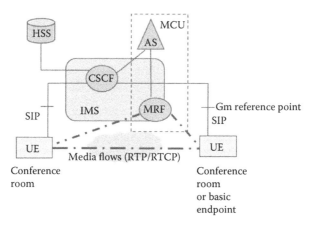

Figure 21.5 IMS-based architecture for TP.

Furthermore, IMS will facilitate extension of TP beyond the enterprise market by enabling any device that complies with the IMS user-network interface specification to connect a TP conference. This potentially includes a wide variety of devices: mobile devices with video facilities (Smartphone, tablets, etc.), PC with soft-video client, TV set connected to a Home Gateway, and so on.

The IMS architecture [22] is centered on a number of SIP-based session servers known as Call Session Control Functions (CSCF), complemented by other functional entities required for interworking with circuit-switched networks, for playing tones and announcements and for bridging calls into conferences. The Proxy CSCF (P-CSCF) is the first IMS point of contact for the user equipment (UE). The Serving CSCF (S-CSCF) is the entity that maintains registration and session state for each user. It also provides access to services hosted in Application Servers (ASs) that are selected based on users' specific criteria contained in the user profile stored in a logically centralized database known as the Home Subscriber Server (HSS). The Interrogating CSCF (I-CSCF) enables retrieving the S-CSCF's identity from the HSS when an incoming session is set up.

Figure 21.5 illustrates an IMS-based architecture for TP. Each conference room is modeled as a single "complex" UE instance, very much like a home network or a corporate network. MCU is modeled as a combination of an AS and a Media Resource Function (MRF). The communication between the AS and the MRF uses SIP and an additional control channel as specified in RFC6230 [24] for controlling the actual MRF resources (e.g., the bridge functionality). The MRF is split into Multimedia Resource Function Controller (MRFC) and Multimedia Resource Function Processor (MRFP). When the MRFC and the MRFP are implemented in separate physical units, a profile of the ITU-T H.248 protocol [23] is used to enable communication between them. It enables the MRFC to have fine-grained control on the media stream resources provided by the MRFP.

It is assumed that there is one controller device within the conference room that registers with the IMS on behalf of the room. Sessions to/from other conference rooms or to/from an MCU are setup by this controller device. The controller device is connected to the IMS at the Gm reference point, as any other type of UE. It may communicate with other devices inside the room, in which case the communication may be based on SIP or on other technologies. Individual devices do not register with the IMS. Media flows sent and received by other devices (cameras, screens, computers, etc.) may go through the controller device or be exchanged directly between devices and the remote sites or MCUs.

21.6 Future Directions

TP represents an important evolution of the videoconferencing market in terms of user experience and growing opportunity but is currently facing a number of challenges due to lack of interoperability and the overall cost of TP solutions. Ongoing standardization activities will hopefully improve the situation enabling TP systems from different manufacturers to connect to each others. These standardization efforts currently focus on replicating the level of functionality available with today's proprietary solutions, with business meetings, and remote education as the main use cases. There are, however, more advanced flavors of TP that are likely to emerge in the standardization arena in the coming years. This includes providing standards for integrating three-dimensional (3D) [25] and haptic technology in TP systems and adding new signaling capabilities required to support not only more advanced forms of business meetings (e.g., those including tools for remote collaboration with application sharing) but also other types of applications such as telesurgery, telerobotics, or gaming.

Acronyms

3D	Three dimensional
3GPP	3rd Generation Partnership Project
AF	Assured Forwarding
AS	Application Server
BFCP	Binary Floor Control Protocol
CS	Class Selector
CSCF	Call Session Control Function
GPRS	General Packet-Radio Service
HD	High Definition
HSS	Home Subscriber Server
HTTP	Hyper Text Transfer Protocol
IETF	Internet Engineering Task Force
IMS	IP Multimedia core network Subsystem
IP	Internet Protocol

ITU   International Telecommunication Union
ITU-T   ITU Telecommunication Sector
LTE   Long-Term Evolution
MCU   Multipoint Control Unit
MIME   Multipurpose Internet Mail Extensions
MRF   Multimedia Resource Function
MRFC   MRF Controller
MRFP   MRF Processor
MSRP   Message Session Relay Protocol
NAT   Network Address Translation
PC   Personal Computer
PHB   Per Hop Behavior
QoS   Quality of Service
RTCP   RTP Control Protocol
RTP   Real-time Transport Protocol
SDP   Session Description Protocol
SIP   Session Initiation Protocol
SLA   Service Level Agreement
SSRC   Synchronization Source
TCP   Transmission Control Protocol
TP   Telepresence
TV   Television
UE   User Equipment

References

1. ITU-T H.323, Packet-based multimedia communications systems, 2009.
2. ITU-T H.239, Role management and additional media channels for H.300-series terminals, 2005.
3. IETF RFC 4796, The Session Description Protocol (SDP) Content Attribute, 2007.
4. A. Romanow, S. Botzko, M. Duckworth, R. Even, and T. Eubanks, IETF, Use Cases for Telepresence Multi-streams, draft-ietf-clue-telepresence-use-cases-01.txt, July 2011 (work in progress).
5. IETF RFC 4594, Configuration Guidelines for DiffServ Service Classes, 2006.
6. ITU-T Technology Watch Briefing Reports No2; Telepresence: High-Performance Video Conferencing, November 2007.
7. N. Oertel, Comparison of H.323 and SIP Video-Conferencing support, presented at the *First SIP Forum Interoperability Workshop*, Vancouver, 2007.
8. IETF RFC 3261, SIP: Session Initiation Protocol, June 2002.
9. IETF RFC 4353, A Framework for Conferencing with the Session Initiation Protocol (SIP), 2006.
10. IETF RFC 4582, The Binary Floor Control Protocol (BFCP), November 2006.
11. IETF RFC 4566, SDP: Session Description Protocol, July 2006.

12. IETF RFC 5939, Session Description Protocol (SDP) Capability Negotiation, September 2010.
13. IETF RFC 5104, Codec Control Messages in the RTP AudioVisual Profile with Feedback (AVPF), February 2008.
14. IETF RFC 3551, RTP Profile for Audio and Video Conferences with Minimal Control, July 2003.
15. IETF RFC 4585, Extended RTP Profile for Real-time Transport Control Protocol (RTCP)-Based Feedback (RTP/AVPF), July 2006.
16. Wei Zhang et al., An extended data collaboration mechanism for XCON Multimedia Conference, presented at *Ninth International Conference on Hybrid Intelligent Systems*, 2009.
17. IETF RFC4975, The Message Session Relay Protocol (MSRP), September 2007.
18. H. Hawkeye King, B. Hannaford, J. Kammerl, and E. Steinbach, Establishing multi-modal telepresence sessions using the session initiation protocol (SIP) and advanced haptic Codecs, presented at *IEEE Haptic Symposium*, 2010.
19. SIP Video Profile Bandwidth, Flow Control and Intra-frame Request Use Cases & Proposed Best Practices, IMTC, May 2010.
20. Best Practices for Role-Based Video Streams (RBVS) in SIP, IMTC, May 2011.
21. 3GPP TS 24.147, Conferencing using the IP Multimedia (IM) Core Network (CN) subsystem; Stage 3, 2011.
22. 3GPP TS 23.228, IP Multimedia Subsystem (IMS); Stage 2, 2011.
23. 3GPP TS 29.333, Multimedia Resource Function Controller (MRFC)—Multimedia Resource Function Processor (MRFP) Mp interface; Stage 3, 2011.
24. IETF RFC 6230, Media Control Channel Framework, 2011.
25. Z. Zhou, X. Chen, L. Zhang, and X. Chang, Internet-wide multi-party tele-immersion framework for remote 3D collaboration, *Proceedings of the IEEE International Symposium on Virtual Reality Innovation*, Singapore, 2011.

Chapter 22

E-Health: User Interaction with Domestic Rehabilitation Tools

Lynne Baillie

Contents

22.1 Introduction 490
22.2 Background 490
    22.2.1 Access and Use of Technology among Older Adults 490
22.3 Current Care and Monitoring in the Home 492
22.4 Future Trends for Rehabilitation in the Home 494
    22.4.1 Game Consoles 494
    22.4.2 Tracking Technology 494
    22.4.3 Visualizations 495
    22.4.4 Digital Television 496
22.5 Conclusion 497
References 498

In this chapter, we report upon the current utilization and trends regarding the use of technology to encourage and engage users in their own rehabilitation in community and home settings. Current evidence suggests that technology can be used to assist rehabilitation in various settings. The findings from these studies, including our own, demonstrate that the use of technology in the home for e-health, including rehabilitation is in the process of taking off. The chapter starts by outlining the need for such technology in the home and reflects upon the types of technology currently available and how successful the deployment of this technology in the home has been. We then look at some new trends in the use of technology in the home such as video conferencing, wii fit, Kinect, wireless sensors, and so on.

22.1 Introduction

There is a growing need to develop new technology for home rehabilitation and this is because of Europe's aging population, for example, by 2020, more than 25% of the Europeans will be aged 60+ (Lutz et al., 2003). However, people within this age group are highly diverse along a range of dimensions, including socioeconomic status, familial status, health, and extent of social engagement. In political terms, this age group has had impact on policies and practices at both national and local levels. This, coupled with their potential as consumers and economic players, means that they offer both opportunities and challenges for the governance of Europe and EU economies. According to writers such as Castells (1996), the rapid expansion of new technology, allowing capital and information to be passed round the world at amazing speed, has transformed existing social and economic relations.

As a result, European and national governments are committed to using digital technology to support a move away from hospital-based care to rehabilitation in the home or community (Scottish Government, 2007). It is envisaged that enhancing older people's access to new technologies in the home will improve their quality of life in a number of ways, including those connected with health, engaging with lifelong learning, and connecting with local communities to enhance social capital. While some quantitative data are available on the type of technologies to which older people have access and which they use, little information is available on the cultural meanings and understandings attached to the use of new technologies.

22.2 Background

22.2.1 Access and Use of Technology among Older Adults

Historic access to Information and Communications Technology (ICT) has been quite limited; with regard to this, a survey was conducted by Selwyn et al. in 2003, which included interesting data on access and use. The survey used a lengthy structured survey instrument to gather data on older adults' use of ICT. The survey was administered in four local authorities in the United Kingdom. The final sample comprised 1001 adults. It was found that only two out of five older people had access to a computer in the home, and less than a quarter of respondents aged 60 or over had used a computer in the last 12 months. The level of access to the Internet was low, and mainly through computers (15%) rather than the newer Internet-enabling technologies, such as digital televisions (1%) or mobile phones (0.3%). The most frequently-cited location for access to a computer was the home of a relative, followed by the respondent's own home. Libraries and friends' homes were also cited as possible places of access. The authors suggest that "access by association" is a significant feature of older people's access to ICT.

In terms of the social factors influencing access, Wilhelm (2000) and Murdock (2002) suggest a distinction between those who have "core access," implying easy access to computers and support in the home, allowing effective use and development of skills. A second group has "peripheral home access," their use of ICT limited by aging equipment and lack of support. An additional group suggested by Selwyn et al. (2003) are those who have "peripheral family access," characterized by dependency on family members to gain access. A further group is described as having "peripheral public access" through terminals in libraries and other external locations such as Internet cafes. Finally, the group who is denied access entirely is described as excluded. Selwyn et al. (2003) found that in their sample, 17% were excluded from ICT access and 24% had ready access at home. The majority of older adults were reliant on some form of outside home peripheral access, generally supplied by family members. Those with ready access were more likely to be male, aged 61–70, married and with post-16 education. Interestingly, there was no association between illness/disability and access to ICT. On the basis of their own work and a review of the wider literature, Selwyn et al. (2003) suggest that a crude distinction between technology users and nonusers may be inappropriate, since the majority appear to gain access through the support of family members. Older people appear to be enthusiastic users of some forms of technology, particularly radio and television, but are less enthusiastic about newer forms of technology. They also use home access, rather than public access, so expanding access in libraries is less helpful to this group than to younger people. Finally, an important point to note is that as those who are currently in the 50+ age group move into the 60+ age group, use of new technologies will be greater because this group will have been accustomed to using ICT in the workplace and for leisure purposes. It is therefore important to compare "younger old people" with "older old people."

It is anticipated that unlike in the past, the new 60+ age group will know about and potentially use a wide range of technologies to assist them to age in place. This conclusion is based on the findings of McCreadie and Tinker (2005). They investigated the use and experience of diverse forms of Assistive Technology (AT) among people over the age of 70 years in the United Kingdom. The forms of AT examined included both specialist items (e.g., grab-rails, wheelchairs, social alarms) and standard items that have specific assistive applications (remote controls, telephones, smoke-detectors, etc.). What McCreadie and Tinker found was that the acceptability of AT depends on the interactions between the felt need for assistance and the recognition of product quality. Felt need depended on user characteristics such as housing type and the interaction of these variables. Four user attributes were found to be particularly relevant:

1. Disability
2. Living arrangements

3. Care needs
4. Personal motivations and preferences

Interestingly, none of the older people in this study were described as techno-phobic. Those who chose not to use AT essentially made pragmatic judgments. If the AT was straightforward, reliable, and met a need, respondents were positive that they would use the technology. Another very interesting finding was that the acceptability of AT to an individual depends on the extent to which it alters the character of their home. The purpose of a house is to provide an environment that will help to meet basic needs such as eating, sleeping, washing, dressing, relaxing, as well as meeting needs for social activities. Even this short list suggests that houses have to be more flexible than most other buildings. Because houses have to meet more needs, they also, if inadequate, have the potential to be more psychologically damaging.

Houses, like our clothes, are extensions of our personality and possible and future selves. It is perhaps, therefore, unsurprising that what older people often dread is having to leave their homes and move into a care home. Although many may be prepared to alter their homes substantially in order to remain at home, if the provision of AT requires unacceptably radical alteration, the changes and sense of intrusion of AT may alter the individual's sense of their home environment. Finally, McCreadie and Tinker note that there is a potential tension relating to AT and that in order to better understand what is meant by an "enabling environment," we need to listen more to what older people themselves have to say about the natural and the built environment, and design technologies from this informed perspective.

In conclusion, many people who are moving into retirement age, for example, people aged 55–64, will have a relatively high level of experience of information and communications technologies (ICTs). People in the older age group (e.g., 65–74) may have lesser knowledge of and engagement with new technologies, although as reported earlier, there will also be a high degree of diversity within this group. From our review we can conclude that older people are willing to embrace technology if it meets a need and that there is a desire from older people to live in their own homes. Home rehabilitation based on mobile and fixed ICTs has the potential to satisfy this demand; however, what type and level of technology may be best for rehabilitation in the home is unknown and therefore a review of past, current, and possible future technologies for such a use are reviewed in the next section.

22.3 Current Care and Monitoring in the Home

Traditional rehabilitation is hospital-based and performed on a one-to-one basis with patient and their therapist. This form of one-to-one care is becoming more scarce due to the expense of one-to-one care (Rand et al., 2004; Flynn et al., 2007), and the availability of therapists (Burdea, 2002) resulting in residents in rural areas having limited access to therapists. Patients are therefore being increasingly urged to continue the rehabilitation process at home. However, a study by Shaughnessy et al. (2006), found that <1/3 of the patients actually performed the rehabilitation exercises in the home. It is therefore interesting to review what technology could be utilized in the home to facilitate home rehabilitation.

Telemonitoring is one area in which technology has been introduced into the home in order to support the wellness of people. However, the main focus of most telemonitoring projects has been to support patients and care workers in monitoring a life-threatening (Benatar et al., 2003) or chronic illness (Giraldo et al., 2002). One of the main issues regarding telemonitoring systems, therefore, is the requirement for the home to be equipped with multiple sensors, devices, and complex network setups (Lukowicz et al., 2002; Sachpazidis, 2002). A benefit can clearly be seen for employing technology with this level of complexity and pervasiveness when the purpose is to monitor patients who have chronic or life-threatening diseases, but the rationale is less persuasive when the person has only low monitoring or only temporary rehabilitation needs.

Another issue is that most research initiatives in the area of e-health for the elderly in the home have been restricted to solutions for discrete narrow-age segments with specific needs, for example, people with severe disabilities and heavy monitoring needs (Lukowicz et al., 2002; Sachpazidis, 2002). However, gerontologists see aging less as a staged sequence and more as a continuous, incremental process (Dharmarajan and Ugalino, 2000), which is characterized by:

1. Gradual loss of skills—motor, sensorial, cognitive skills (e.g., memory)
2. Increasing frailty and (chronic) illnesses
3. Increasing social exclusion and problems of isolation

Therefore, equipment used in telemonitoring may not always be suitable for rehabilitation or aging in place, but rather a selection of supportive technologies would better support people seamlessly through their old age.

Currently the tools that are provided to help patients undertake their rehabilitation are: websites, videos, and paper-based handouts and manuals. Although helpful, they cannot confirm that the patient will follow the exercises, or whether the exercises will be performed correctly. Researchers have employed different technologies in order to address this issue. For example, Egglestone et al. (2009) provide a detailed review on the use of robotic devices, virtual rehabilitation, haptic user interfaces, and sensor-based motion capturing systems. Another recent study by Liolios et al. (2010) provides us with a review on the development of (wireless) body area networks and their applications in healthcare and in assistive environments. Therefore, in recent years, researchers have explored the use of low-cost options that could offer the patients an economically feasible solution for home-based rehabilitation that could potentially limit the requirement for a heavy telemonitoring presence. The next section of the chapter reviews some of these technologies.

22.4 Future Trends for Rehabilitation in the Home

22.4.1 Game Consoles

Sony Playstation was the first game console to provide users with a gesture recognition interface using the EyeToy camera (Sony, 2011a,b). Studies on the suitability of the console for home-based rehabilitation (Rand et al., 2004; Flynn et al., 2007; Yavuzer et al., 2008) have confirmed the potential effectiveness of the EyeToy games (EyeToy Play, Sony, 2011a,b) as part of the treatment of upper extremity-related motor function. They have also mentioned however, limitations of the technology with regards to inability to grade the level of the environment, the exclusion of patients with severe weaknesses, and the often necessary presence of a therapist.

The Wii, released by Nintendo in 2011, includes the Wiimote (remote controller) with built-in sensors that track position, orientation, and motion. This console has been employed by many researchers (Deutsch et al., 2008; Medical College Georgia, 2009; Murgia et al., 2008; Attygalle et al., 2008; Leder et al., 2008; Galego and Simone, 2007; Alankus et al., 2010) for rehabilitation who have reported positive outcomes at motor impairment and functional levels, however, as with the Playstation study, many of the studies excluded some patients either due to the need of a full-range motion or due to the need to grasp the Wiimote. Some studies used games provided by the console (Medical College Georgia, 2009; Alankus et al., 2010) and others integrated the Wiimotes to other existing games (Galego and Simone, 2007) or systems (Deutsch et al., 2008; Murgia et al., 2008; Attygalle et al., 2008; Leder et al., 2008). Alankus et al. (2010) reported the successful translation of the basic motions regained into functional improvements that impacted the patient's daily life.

The release of Microsoft's Kinect in late 2011 opened up new areas for research since it captures video data in 3D allowing for an even more natural interaction using gestures (and spoken commands). No studies have yet been published on the suitability of this console in stroke (or other motor-related) rehabilitation, however, based on the innovative interaction interface it provides and the already promising results acquired from the studies presented earlier, it is considered a suitable technology. An issue that should be addressed is its accuracy, something that Schönauer et al. (2011) have already mentioned. The recent study by Wilson (2010) on alternative uses of Kinect sensors is promising as that might provide researchers alternative options for home-based rehabilitation.

22.4.2 Tracking Technology

Several motion capture tracking systems have been developed over the last years with potential applications in home rehabilitation. A wireless magnetic motion capture system consists of magnetic markers and a pickup coil array (Hashi et al., 2006). However, this system is limited by the size of the pickup coil array and unsuitable for home rehabilitation as a large surface area is required to mount the pickup coils for full body motion capture. Also a variety of video motion capture systems have also been developed. For instance, Zhuang et al. (1999) used image sequence tracking and human body modeling for human motion capture but factors such as privacy and occlusion limit their application in home rehabilitation. Frey (1996) showed that an inertial tracking system would adequately satisfy the requirements of human body motion capture without the drawbacks of the aforementioned systems.

Subsequently, research into using wireless magnetic, angular rate, and gravity sensor modules for human motion tracking were launched (Aylward and Paradiso, 2007).

Wireless inertia measurement units have been used to capture 3D knee joint angle motion (Kobashi et al., 2009) and Zhoua et al. (2008) used similar sensor units to capture upper limb motion. Moreover, Smit (2007) used low-cost orientation sensors for near real-time full body motion capture in the laboratory. A real-time human arm motion detector to aid the home-based rehabilitation of stroke patients was developed by Zhou and Huosheng (2007) using two tri-axial inertial sensors. However, the application of their proposed system in real rehabilitation applications is yet to be verified. In another work, the concept of visualization of motion was shown to help in medical diagnostics and rehabilitation of patients (Zsolt and Istvan, 2006).

In summary, wireless orientation sensors have the potential to be used in home rehabilitation applications because of the availability of low power radio technologies (Pantelopoulos and Bourbakis, 2010) allowing users unrestricted movements. This can be achieved by using low-cost microelectromechanical sensors and efficient orientation algorithms (Lympourides et al., 2009; Young, 2010; Madgwick et al., 2010). The effectiveness of these systems can be improved by combining them with other tracking systems such as radio frequency (Marins et al., 2001), capacitive/proximity sensors (Aylward et al., 2006), and vibrotactile actuators (Alahakone et al., 2009).

Other studies (Burke et al., 2008, 2009a,b) have also used simple Web cameras as tracking devices in order to facilitate stroke rehabilitation. The simplicity of such a system (a PC with a Web camera, and a marker), coupled with the promising results of the studies, display a considerable alternative that is both cheap and easy to set up. A preliminary study conducted by the makers of the SilverFit softkinetic rehabilitation system revealed that the participants often performed more exercise than was required of them as they found most of the games very addictive (Doyle et al., 2010).

The system utilized a large screen and a depth sensing camera capable of translating natural human movement into on-screen actions. Even though this type of depth sensing camera technology is very accurate, it is bulky, expensive, and often requires a professional to initialize and operate it.

22.4.3 Visualizations

The importance of visual feedback to inform patients of the actions they performed and the outcomes of these actions (Shaughnessy et al., 2006) has not been sufficiently addressed in traditional rehabilitation. Most therapists and patients use mechanical devices* that only sometimes include biometric data,† leading to the lack of monitoring the patient's progress that is necessary for a positive outcome of the treatment (Burdea, 2002). Chua et al. (2007), have denoted the positive effect of the visual feedback for a patient and their family in understanding the effectiveness of the rehabilitation process. Jung et al. (2006) suggested that visualizing the changes of the patient in real time may affect their motivation in participating in the exercises and increase their confidence and self-efficacy in performing activities of daily living in the real world. Motivation plays a key role in the home rehabilitation process and may affect the use of a technology (Xu et al., 2006; Burke et al., 2009a,b; Egglestone et al., 2009).

22.4.4 Digital Television

The vision of digital interactive TV foresees users engaging with interactive services across a variety of contexts and user interfaces. Following this idea, this chapter looks at how digital interactive TV could support rehabilitation in the home. In the traditional living-room context, interacting with your TV is a lean back and highly immersive activity. However, there is a move to "Interactive" TV, with new program concepts moving to higher levels of interaction, program input, personalization, and sharing. This agenda setting gives rise to the concept of using the TV as a portal to access rehabilitation advice and content.

This section elaborates on how interactive TV could be extended from traditional living-room TV to a rehabilitation aid.

From its very beginnings, broadcast TV had elements of audience involvement. Prime examples are quiz shows during which viewers mentally compete with each other or live sports coverage, where people congregate in front of the TV to jointly watch a game. Consequently, television has become an established provider of common ground necessary for socialization and bonding among viewers (Lull, 1990). In recent times, the proliferation of networked set-top boxes and triple-play concepts has enabled the widespread diffusion of convergent interactive TV services that blurs the boundaries between content consumption and communication.

The main trend that supports the idea of rehabilitation TV is the advances in networked communications. For example, many people are now used to being offered a proliferation of TV channels, all with on-demand content and the ability to time-shift media consumption (as enabled by PVRs), and the increasing share of Internet usage for media access (Goldenberg, 2007).

Another driver is the advanced diffusion of networked communication technologies such as mobile voice telephony, SMS, Instant-messaging, and voice-over-Internet protocol. This trend not only has led to an increased usage of networked communication

---

* For example, products from http://www.activeforever.com/c-11-stroke-and-neuro-rehab.aspx
† For example, http://www.activeforever.com/p-661-omron-hem-711ac-automatic-blood-pressure-monitor-with-intellisense.aspx

but also to the frequent utilization simultaneous with TV consumption (Pilotta et al., 2004). For example, over 50% of U.S. youngsters regularly chat on the phone or browse the Internet, effectively multitasking while watching TV (USC, 2007). Furthermore, the emergence of the Web 2.0 zeitgeist has caused growth of activity in online communities and networks, which people use to connect with each other by sharing content and experiences. Consequently, television itself is no longer marketed as a distinct, isolated channel, but as one center of gravity in your home that is an interconnected ecosystem that provides access to a large variety of forms of content and interaction.

O'Brien et al.'s (1999) ethnographic study of set-top boxes in networked homes has shown that, even when being indoors, inhabitants can be very mobile and pursue various activities in parallel. Their findings suggest that home users require more flexible interfaces in order to naturally distribute activities across different people and spaces. Accordingly, some recent research projects address the opportunities of mixed device setups at home. Cesar et al. (2008) developed a prototype system featuring handheld devices as a secondary screen for iTV in order to control, enrich, and share content. In similar ways, the "PresenceRemote" proposed by Sokoler and Svensson (2008) features a Personal Digital Assistant (PDA) serving as an enhanced remote control for senior citizens. Another study by Taylor et al. (2011) revealed how video-conferencing via the user TV over an Internet connection enabled users to stay at home to attend a virtual exercise class with other users during pulmonary rehabilitation.

The premise behind this study was that, because many patients lived far away from rehabilitation centers, patients (and in some cases, physiotherapists) were required to make frequent long journeys to the classes (or patients' homes). The technology used in the study enabled active involvement in rehabilitation by the participants who remained motivated.

22.5 Conclusion

The ability to initiate and maintain movement to perform the physical tasks of normal daily activity are crucial factors for a healthy and fulfilling life. Across the lifespan injury, illnesses, and aging factors affecting the musculoskeletal or neurological systems can reduce one's capacity to live an independent life. We are currently suffering an epidemic of long-term health conditions, a significant number impacting on functional ability and ability to work. Projections of population profiles for the EU to 2018/2020 reveal the scale of the impending problems. As the population ages, this situation will worsen, leading to an increasing burden on community health services and social service support. An EU-wide "silver" review has recommended that Europe-wide actions are required to tackle such problems. Appropriate AT of the sort outlined in this chapter has the potential to enable older adults to age in place. We, as designers and developers of such technology should take on board the barriers to acceptance, outlined in the background section, of such new enabling technology and design it in such a way as to make sure that it is: useful, usable, and used.

References

Alahakone, A.U., and Senanayake, S.M. A Real Time vibrotactile biofeedback system for improving lower extremity kinematic motion during sports training. In the *Proceedings of the International Conference on Soft Computing and Pattern Recognition, 2009. SOCPAR '09.* Malacca, Dec. 2009.

Alankus, G., Proffitt, R., Kelleher, C., and Engsberg, J. 2010. Stroke therapy through motion-based games: A case study. *Proceedings of the 12th International ACM SIGACCESS Conference on Computers and Accessibility (ASSETS '10).* ACM, New York, NY, USA, pp. 219–226.

Attygalle, S., Duff, M., Rikakis, T., and He, J. 2008. Low-cost, at-home assessment system with Wii remote based motion capture. *Conf. Proc. Virtual Rehabilitation*, pp. 168–174.

Aylward, R., Lovell, S.D., and Paradiso, J.A. 2006. A compact, wireless, wearable sensor network for interactive dance ensembles. *Proceedings of the International Workshop on Wearable and Implantable Body Sensor Networks BSN*, Cambridge, MA, USA, pp. 65–70.

Aylward, R., and Paradiso, J.A. 2007. A compact, high-speed, wearable sensor network for biomotion capture and interactive media. *6th International Symposium on Information Processing in Sensor Networks IPSN*, Cambridge, MA, USA, pp. 380–389.

Benatar, D., Bondmass, J., Ghitelman, J., and Avitall, B. 2003. Outcomes of chronic heart failure. *International Medicine*, 16, 347–352.

Burdea, G. 2002. Key note address: Virtual rehabilitation—Benefits and challenges. In: *1st Intl. Workshop on Virtual Reality Rehabilitation (Mental Health, Neurological, Physical, Vocational)*, pp. 1–11.

Burke, J.W., McNeill, M.D.J., Charles, D.K., Morrow, P.J., Crosbie, J.H., and McDonough, S.M. 2009a. Optimising engagement for stroke rehabilitation using serious games. *The Visual Computer*, 25(12), 1085–1099.

Burke, J.W., McNeill, M.D.J., Charles, D.K., Morrow, P.J., McDonough, S.M., and Crosbie, J.H. 2009b. Serious games for upper limb rehabilitation following stroke. *IEEE International Conference in Games and Virtual Worlds for Serious Applications (VS Games '09)*, March 23–24, 2009, pp. 103–110.

Burke, J.W., Morrow, P.J., McNeill, M.D.J., McDonough, S.M., and Charles, D.K. 2008. Vision based games for upper-limb stroke rehabilitation. *International Machine Vision and Image Processing Conference (IMVIP)*, September 3–5, 2008, pp. 159–164.

Castells, M. 1996. *The Information Age: Economy, Society and Culture.* John Wiley & Sons, New York.

Cesar, P., Bulterman, D., and Jansen, A. 2008. Usages of the secondary screen in an interactive television environment: Control, enrich, share, and transfer television content. *Proceedings of 6th European Conference EuroITV 2008*, Salzburg, Austria, July 3–4, Springer-Verlag, Berlin, Heidelberg, pp. 168–177.

Chua, S., Lim, G.H., and Ghista, D.N. 2007. A quantifiable assessment device for stroke patients. *Proceedings of the 1st International Convention on Rehabilitation Engineering &*

Assistive Technology: In Conjunction with 1st Tan Tock Seng Hospital Neurorehabilitation Meeting (i-CREATe /07). ACM, New York, NY, USA, pp. 123–128.

Deutsch, J.E., Borbely, M., Filler, J., Huhn, K., and Guarrera-Bowlby, P. 2008. Use of a low-cost, commercially available gaming console (Wii) for rehabilitation of an adolescent with cerebral palsy. *Physical Therapy*, 88(10), 1196–1207.

Dharmarajan, T.S., and Ugalino, J.T. 2000. The aging process. In D. Dreger, and B. Krumm (Eds.). *Hospital Physician Geriatric Medicine Board Review Manual* (Vol. 1, Part 1, pp. 1–12). Wayne, PA: Turner White Communications Inc.

Doyle, J., Bailey, C., Dromey, B., and Scanaill, C.N. 2010. BASE—An interactive technology solution to deliver balance and strength exercises to older adults. *Proceedings of Pervasive Computing Technologies for Healthcare (PervasiveHealth), 2010, 4th International Conference*.

Egglestone, R. S., Axelrod, L., Nind, T., Turk, R., Wilkinson, A., Burridge, J., Fitzpatrick, G. et al. 2009. A design framework for a home-based stroke rehabilitation system: Identifying the key components. *PervasiveHealth 2009, 3rd International Conference on Pervasive Computing Technologies for Healthcare 2009*, April 1–3, 2009, London, UK.

Flynn, S., Palma, P., and Bender, A. 2007. Feasibility of using the Sony PlayStation 2 gaming platform for an individual poststroke: A case report. *Journal of Neuro. Physical Therapy: JNPT*, 31(4), 180–189.

Frey, W. 1996. Application of inertial sensors and flux-gate magnetometer to real-time human body motion capture. *Masters Thesis*, Naval Postgraduate School, Monterey, CA.

Galego, B., and Simone, L. 2007. Leveraging online virtual worlds for extremity rehabilitation. *Bioengineering Conference, 2007. NEBC '07*. IEEE 33rd Annual Northeast, Long Ilsand, New York.

Giraldo, C., Helal, S., and Mann, W. 2002. mPCA–A mobile patient care-giving assistant for Alzheimer patients. *Proceedings of the First International Workshop on Ubiquitous Computing for Cognitive Aids*, Gothenburg, Sweden, September, 2002.

Goldenberg, S. 2007. Digital video recorders and micro-social networking: Recreating the shared watching experience of television. Paper presented at the *5th European Conference, EuroITV 2007*, Amsterdam, The Netherlands, May 24–25, Springer Berlin/Heidelberg.

Hashi, S., Toyoda, M., Yabukami, S., Ishiyama, K., Okazaki, Y., and Arai, K.I. 2006. Wireless magnetic motion capture system for multi-marker detection. *IEEE Transactions on Magnetics*, 42(0), 3279–3281.

Jung, Y., Yeh, S.C., and Stewart, J. 2006. Tailoring virtual reality technology for stroke rehabilitation: A human factors design. In *CHI '06 Extended Abstracts on Human factors in Computing Systems (CHI EA '06)*. ACM, New York, NY, USA, pp. 929–934.

Kobashi, S., Tsumori, Y., Imawaki, S., Yoshiya, S., and Hata, Y. 2009. Wearable knee kinematics monitoring system of MARG sensor and pressure sensor systems. *IEEE International Conference on System of Systems Engineering SoSE*, pp. 1–6.

Leder, R.S., Azcarate, G., Savage, R., Savage, S., Sucar, L.E., Reinkensmeyer, D., Toxtli, C., Roth, E., and Molina, A. 2008. Nintendo wii remote for computer simulated arm and wrist therapy in stroke survivors with upper extremity hemipariesis. In *Proceedings of the Virtual Rehabilitation Conference, 2008*. Vancouver, BC.

Liolios, C., Doukas, C., Fourlas, G., and Maglogiannis, I. 2010. An overview of body sensor networks in enabling pervasive healthcare and assistive environments. *Proceedings of the 3rd International Conference on PErvasive Technologies Related to Assistive Environments (PETRA '10)*. F. Makedon, I. Maglogiannis, and S. Kapidakis (Eds.). ACM, New York, NY, USA, Article 43, 10pp.

Lukowicz, P., Anliker, U., Ward, J., Troster, G., Hirt, E., and Neufeld, C., 2002. AMON: A wearable medical computer for high risk patients. Paper presented at the Sixth International Symposium on Wearable Computers (ISWC 2002), Seattle, WA, USA, October 2002.

Lull, J. 1990. *Inside Family Viewing: Ethnographic Research on Television Audiences.* London, Routledge.

Lutz, W., O'Neill, B.C., and Scherbov, S. 2003. Europe's population at a turning point. *Science,* 299(5615), 1991–1992.

Lympourides, V., Arvind, D.K., and Parker, M. 2009. Fully wireless, full body 3-d motion capture for improvisational performances. *Proceedings of the Workshop on Whole Body Interaction SIGCHI,* 2009.

McCreadie, C., and Tinker, J. 2005. *The Acceptability of Assistive Technology to Older People Ageing and Society.* Cambridge, Cambridge University Press.

Madgwick, S.O.H., Vaidyanathan, R., and Harrison, A.J.L. 2010. An efficient orientation filter for IMU and MARG sensor arrays, Department of Mechanical Engineering, University of Bristol. [Available online].

Marins, J., Yun, X., Bachmann, E., McGhee, R.B., and Zyda, M.J. 2001. An extended Kalman filter for quaternion-based orientation using MARG sensors. *Proc. Int. Conf. Intell. Robots Syst.,* November 2001, Vol. 4, pp. 2003–2011.

Medical College of Georgia. 2009. Wii-hab may enhance Parkinson's treatment. http://www.physorg.com/11 June 2009.

Microsoft. 2011. Kinect full body interaction. http://www.xbox.com/kinect, last visited April 2011.

Murdock, G. 2002. Tackling the digital divide: Evidence and intervention. Paper presented given to British Educational Communications and Technology Agency seminar, The Digital Divide, 19 February, Coventry, Warwickshire.

Murgia, A., Wolff, R., Sharkey, P.M., and Clark, B. 2008. Low-cost optical tracking for immersive collaboration in the CAVE using the Wii Remote. In *Proceedings of the International Conference on Disability, Virtual Reality and Associated Technologies (ICDVRAT),* Porto, Portugal, 2008.

Nintendo. 2011. Wii. http://uk.wii.com/, last visited April 2011.

Norman, D.A. 1991. *The Design of Everyday Things.* Addison-Wesley Publishing Company, Inc., Reading, MA.

O'Brien, J., Rodden, T., Rouncefield, M., and Hughes, J. 1999. At home with the technology: An ethnographic study of a set-top-box trial. *ACM Transactions on Computer-Human Interaction (TOCHI),* 6, 282–308.

Pantelopoulos, A., and Bourbakis, N.G. 2010. A survey on wearable sensor-based systems for health monitoring and prognosis. *IEEE Transactions on Systems, Man, and Cybernetics, Part C: Applications and Reviews,* 40(1), 1–12.

Pilotta, J.J., Schultz, D.E., Drenik, G., and Philip, R. 2004. Simultaneous media usage: A critical consumer orientation to media planning. *Journal of Consumer Behaviour,* 3(3), 285–292.

Rand, D., Kizony, R., and Weiss, P.L. 2004. Virtual reality rehabilitation for all: Vivid GX versus Sony PlayStation II EyeToy. *Virtual Reality,* 131(2), 87–94.

Sachpazidis, I. 2002. @Home: A modular telemedicine system. Paper presented at the conference on Mobile Computing in Medicine. Heidelberg, Germany, April 2002.

Schönauer, C., Pintaric, T., and Kaufmann, H. 2011. Full body interaction for serious games in motor rehabilitation. *Proceedings of the 2nd Augmented Human International Conference (AH '11).* ACM, New York, NY, USA, Article 4, 8pp.

Scottish Government. 2007. Co-ordinated, integrated and fit for purpose: A delivery framework for adult rehabilitation in Scotland. Available from: http://www.scotland.gov.uk/Publications/2007/02/20154247/0.

Selwyn, N., Gorard, S., and Furlong, J. 2003. *Older Adults' use of Information and Communications Technology in Everyday Life*. Cambridge, Cambridge University Press.

Shaughnessy, M., Resnick, B.M., and Macko, R.F. 2006. Testing a model of post-stroke exercise behavior. *Rehabil Nurs.*, 31(1), 15–21.

Smit, P.C. 2007. *The Capture of Skeletal Motion from Sourceless Orientation Sensors*, MSc dissertation, Glasgow, Glasgow Caledonian University.

Sokoler, T., and Svensson, M.S. 2008. PresenceRemote: Embracing ambiguity in the design of social TV for senior citizens. *Proceedings of 6th European Conference EuroITV 2008*, Salzburg, Austria, July 3–4, Springer-Verlag, Berlin, Heidelberg, pp. 158–162.

Sony. 2011a. Peripherals—EyeToy USB camera. http://uk.playstation.com/ps2/peripherals/detail/item51697/EyeToy-USB-Camera/, last visited April 2011.

Sony. 2011b. Playstation Games—EyeToy Play. http://uk.playstation.com/ps2/games/detail/item35282/EyeToy-Play/, last visited April 2011.

Taylor, A., Aitken, A., Godden, D., and Colligan, J. 2011. Group pulmonary rehabilitation delivered to the home via the Internet: Feasibility and patient perception. *Proc. CHI*.

USC Center for the Digital Future. 2007. Sixth Study of the Internet by the Digital Future Project. http://www.digitalcenter.org.

Wilhelm, A. 2000. *Democracy in the Digital Age: Challenges to Political Life in Cyberspace*. Routledge, London.

Wilson, A.D. 2007. Using a depth camera as a touch sensor. In *Proceedings of ITS '10 ACM International Conference on Interactive Tabletops and Surfaces*, ACM, New York, NY, USA, 2010.

Xu, W., Chen, Y., Hari Sundaram, H., and Rikakis, T. 2006. Multimodal archiving, real-time annotation and information visualization in a biofeedback system for stroke patient rehabilitation. *Proceedings of the 3rd ACM Workshop on Continuous Archival and Retrival of Personal Experences (CARPE '06)*. ACM, New York, NY, USA, pp. 3–12.

Yavuzer, G., Senel, A., Atay, M.B., and Stam, H.J. 2008. "Playstation EyeToy games" improve upper extremity-related motor functioning in subacute stroke: A randomized controlled clinical trial. *European Journal of Physical and Rehabilitation Medicine*, 44(3), 237–244.

Young, A.D. 2010. Wireless realtime motion tracking system using localised orientation estimation. PhD thesis, submitted to University of Edinburgh.

Zhou, H., and Huosheng, Hu. 2007. Upper limb motion estimation from inertial measurements. *International Journal of Information Technology*,13(1), 1–14.

Zhoua, H., Stoneb, T., Huc, H., and Harrisb, N. 2008. Use of multiple wearable inertial sensors in upper limb motion tracking. *Technical Note Medical Engineering & Physics*, 30, 123–133.

Zhuang, Y., Liu, X., and Pan, Y. 1999. Video motion capture using feature tracking and skeleton reconstruction. *Proceedings of International Conference on Image Processing ICIP*, ISBN 0-7803-5467-2, 4, Kobe, Japan, pp. 232–236.

Zsolt, K. and Istvan, L. 2006. 3D motion capture methods for pathological and non-pathological human motion analysis, *IEEE Proceedings of 2nd Information and Communication Technologies ICTTA*, Budapest, Hungary, pp. 1062–1067.

Chapter 23

Societal Challenges for Networked Media

Pierre-Yves Danet

Contents

23.1 What Are the Grand Societal Challenges?.. 504
23.2 What Is the NEM Sector? 507
23.3 How NEM Sector Could Help in Some of the Challenges? 507
23.4 What NEM Technologies Could Bring in Addition? 508
    23.4.1 Energy Efficiency 508
    23.4.2 Intelligent Transport System 509
23.5 Conclusion 509

The Networked Electro Media sector, actually structured by the Networked Electro Media (NEM) European Technology Platform (ETP), is one of the most important contributors to the Digital Agenda as far as multimedia technologies should answer to most of the Grands Societal Challenges that Europe will face in the coming years.

NEM delivers sustainable European leadership in the convergence of media, information, and communication technologies, by leveraging the innovation chain to deliver rich user/citizen experiences and services using NEM technologies to solve societal challenges. The European Knowledge Society must tackle these challenges through the application of the best analysis, most powerful actions, and increased resources than can be brought to bear. The way that we address these challenges must create innovative and sustainable solutions in areas such as:

■ Global warming
■ Tightening supplies of energy, water, and food
■ Aging societies
■ Public health, pandemics
■ Security

It is obvious and well known that Networked Media should help in these fields but it is also obvious that NEM technologies could offer more comfort in complementary new services.

23.1 What Are the Grand Societal Challenges?

The report of the European Research Area Expert Group (EG) on "*The Role of Community Research Policy in the Knowledge-Based Economy*" was prepared for the European Commission (EC) Research DG over the first nine months of 2009. The group was asked to review, assess, and interpret the existing evidence on the state of the knowledge-based economy in Europe as well as on the effectiveness, in terms of roles, objectives, and rationales, of the main existing research policy instruments and to come up with recommendations on how to frame and articulate the Community research policy in the post-2010 period. The Terms of Reference (ToR) of the EG explicitly referred to the need for an economic assessment that would bring forth new ideas, analyses, and the so-called "evidence-based recommendations for actions," hence, the dominance in this EG of experts from the academic, business, and policy-making community with a strong economic background. The EG started its reflective work with an internal discussion brainstorming on the major challenges the European Union would be likely to face over the next 10 to 15 years. These major challenges, grouped under the notion of "drivers," were pulled together under five headings which formed the main sections of the EG report and led, in a final section, to a number of conclusions and policy recommendations.

The *first driver* believed by the EG, had affected European research over the last decade, and was likely to affect it even more over the next 10 years; it is the trend toward globalization and concentration of research in Europe and the rest of the world.

The *second driver* considered by the EG, in more detail, was the notion of *Societal Challenges* (often referred to as "*Grand*" Challenges). The notion of *Societal Challenges*, which the EG preferred to use, applies to major social problems that cannot be solved in a reasonable time and/or with acceptable social conditions, without a strong and, in the European case, coordinated input requiring both technological and nontechnological innovation, and at times, though not necessarily always, advances in scientific understanding.

A Societal Challenge dimension would, in other words, add a new objective to public policy, whereby research and innovation are seen not as ends in themselves but as a means to a wider goal defined as a societal benefit. The aim is to foster those activities that have greatest impact on achieving the societal challenge, and not necessarily to increase research and improve innovation across the board. The relevant actors not only include, of course, private companies in various sectors, but also institutions involved in innovation in the public sector as well as public services, and in setting demand side and regulatory and market frameworks that support innovation.

The second set of recommendations of the EG deals therefore with the question on how to achieve compatibility between such "grand" societal top-down initiatives and a more market-driven resource allocation logic that would allow for "multiple decentralized experiments." In practice, the EG follows the line here as set out in the Lund declaration. Meeting the Societal Challenges will require among others: strengthening *frontier* research initiated by the research community itself and taking a lead in the development of enabling technologies in particular along the lines of the so-called *lead-market initiative* such as in the case of "green technologies." Attention should be given here to measures that can enhance the effectiveness of both public and private research and development investment in the wide and diverse array of "green" technologies facilitating knowledge sharing, adaptation, and diffusion of innovations.

The *third driver* considered by the EG is the need for Community research policies more based on so-called *merit-based competition* than collaboration across the EU.

A series of recommendations are also made with respect to the trend toward open innovation, considered by the EG as a *fourth key driver.*

The final *fifth driver* considered by the EG is the one of regional specialization and cohesion policies.

Considering these high-level societal challenges, the European Research Agency has identified six topics that the European Research should study:

Global warming: Due to air pollution (industry, cars, home heating, …) CO_2 is beating the earth protection layer against the sun. This implies an increase in the overall temperature which will have considerable impact in our future life (storms, under sea area extension, dry area extension, etc.) which have.

Tightening supplies of energy: Fossil energy will be less and less available; there is not only a need to find some new resources but also a need to save energy.

Water and food: Due to the enlargement of the world population, it is and it will be more and more difficult to have sufficient food and water for everybody.

Aging societies: Due to medical advances, people are living longer and there will be a need to help people to stay at home.

Public health, pandemics: It is in our basic instinct to live longer, medicine is making great progress but there are always new virus arising, that need great effort in research but also in public infrastructure which cost more and more and difficult to fund.

Security: Due to unemployment, burglary and antisocialism are on the rise and this implies a higher crime rate.

From these Grand Challenges the Information Society and media clusters have identified six main areas where Information Communication Technology (ICT) could contribute:

Smart energy grid: Energy grids will increasingly face risks of overdemand and blackout. Internet connectivity, computing power, digital sensors, and remote control of the transmission and distribution system will help to make grids smarter, greener, and more efficient; this should be complemented by smart metering which also makes energy consumption more efficient.

Smart environmental information systems: The use of sensor networks for collecting real or near real-time environmental data is a growing field of application. It requires Internet connectivity for data management, dissemination, and integration in complex information systems; it should be of benefit to services such as disaster management.

Smart systems for transport and mobility: Putting "intelligence" into roads and cars with Intelligent Transport Systems (ITS) using sensor networks, radio frequency tags, and positioning systems for optimization. The new cooperative system linked to the Internet provides a solution to interconnect these diverse technologies and improve travel and mobility through real-time management of public and private transport resources, traveller information, and decision-making tools.

Smart health care systems and ambient assistant living: Current research experiments aim to develop technologies for "ambient" environments capable of assisting patients and satisfying their information and communication needs. These technologies combine devices (sensors, actuators, special hardware, and equipment), networks, and service platforms to harness information about medical conditions, patient records, allergies, and illnesses.

Smart culture and knowledge: European culture is very rich and European people are so creative that we will be soon overwhelmed by information and archives. Despite advances in search engine technology, there will be a need to help people manage their content, including for "deleting" exhaustively their information wherever it is stored.

Smart content for entertainment industry: Entertainment area is an essential driving force for innovation and user acceptance of smart technology allowing economic growth in the coming years. Smart content enhanced with metadata combined with intelligent access mechanisms are example of key technologies allowing the end user to enjoy with immersive experience.

23.2 What Is the NEM Sector?

NEM sector focuses on an innovative mix of various media forms, delivered seamlessly over technologically transparent networks, to improve the quality, enjoyment, and value of life. NEM sector represents the convergence of existing and new technologies, including broadband, mobile, and new media across all ICT sectors, to create a new and exciting era of advanced personalized services. These services are delivered over a wide variety of complementary access networks, including satellite, terrestrial, cable, twisted-pairs, optical fiber, community installations, and microwaves infrastructures. The services are delivered in a seamless and interactive way to a variety of end-user terminals and devices, including fixed and handheld terminals. A main goal is to empower end users in creating their media and communication environments, including user-generated content, in which the quality of access to value-added content and services is the key enabling factor.

This sector is organized in Europe through the NEM ETP (http://www.nem-initiative.org).

23.3 How NEM Sector Could Help in Some of the Challenges?

The European Knowledge Society must tackle these challenges through the application of the best analysis, most powerful actions, and increased resources than can be brought to bear.

A Societal Challenge dimension would add a new objective to the public policy, whereby research and innovation are seen not as ends by themselves, but as contributors to a wider goal, defined as a societal benefit. The aim should be to foster those activities that have greatest impact on achieving the societal challenge, and not necessarily to increase research and improve innovation across the board.

In several of those areas, the *Networked and Electronic Media community* would be able to contribute to the common effort and bring in its expertise and technologies.

However, *Smart culture and knowledge* is the one with the greatest synergy with NEM's core research focus. In the last version of our Digital Agenda we have identified 30 research topics which have been ranked by the NEM community through a recent survey. The six most important topics are as follows:

1. User satisfaction and quality of experience
2. Home and extended home networks
3. Network architecture
4. Intelligent delivery
5. Representation of content
6. Tools for content creation and manipulation

23.4 What NEM Technologies Could Bring in Addition?

ICT is one of the most important contributors to support these societal challenges. For this reason, we are now working on the ICT vision regarding these societal challenges though across ETP work group. Networks!, ISI, EPOSS, Photonics21, and NEM are actively working on a common white paper that should be available at the beginning of 2011.

As a first step, the group will work on the already-identified subjects described above and when the results of the ERA Experts Group think tank are available we will update our paper accordingly.

The methodology used by the working group will follow the above steps:

1. Establish a formal liaison with the ERA EG in order to get an up-to-date list and definition of the Grand Challenges.
2. Collect the existing material in each ETP.
3. Analyze the Grand Societal Challenges in order to identify fields related to X-ETP activities.
4. Brainstorm ideas where ICT ETPs could contribute.
5. Select from the X-ETP FI SRA the research topic addressing those fields.
6. Propose a white/position paper on potential view.
7. Identify possible gaps in our SRA for the next version.
8. Propose cross-ETP projects that could be submitted.

23.4.1 Energy Efficiency

NEM sector is also active in Energy efficiency since NEM is in tackling the Home Network services which could help people to decrease their electricity consumption. In addition, the NEM agenda is also addressing immersive communication which could help people to communicate in more realistic modes, hence avoiding travels. The conclusion of the position paper points out the climate change is one of the most important challenges the world will face in the near future. Effective actions are therefore required for preventing mankind to confront a myriad of disasters and natural catastrophes. A general consensus was reached: ICTs, and hence NEM technologies, can significantly contribute to solve this problem. To achieve that objective NEM members are supporting the following activities:

■ Enable advanced ICT for devices and make their functionality available in an open and service-oriented way
■ Enable cooperation among devices and enable correlation to the user's tasks
■ Enhance energy efficiency by enhancing user awareness and dynamically enable them to adjust their lifestyle requirements to optimize energy consumption

With these targets in mind, the following positions should be encouraged:

- Promotion of the launching of services whose mission will be the provision of information to users about consumption in conjunction with other services developed under the framework of the digital home
- Encouraging the deployment of advanced metering and home network management services

23.4.2 Intelligent Transport System

NEM has recently published a position paper pointing out that multimedia services should be part of this challenge since location services combined with multimedia information could help people in their transportation situation. On the one hand, the range of possible applications from the NEM sector should improve the information available to transport users and operators, to make them more aware of the implications of their use and operation of the transport system, and thus to support transport policy objectives. This information will help travelers make more informed decisions about how, when, where, and whether to travel. In this case, the ubiquitous and immersive use of audiovisual information is key for the satisfaction of the expectation of travelers regarding reliable ITS information.

Road traffic congestion and road fatalities have been identified as major challenges that Europe's transport system needs to overcome. Conventional approaches such as the development of new infrastructure have not provided the necessary results required by the magnitude of these challenges. Innovative solutions are therefore clearly needed.

NEM technologies can definitively contribute to foster implementation of ITS-based solutions, as they will gradually provide a range of new services to citizens and also enable improved real-time management of traffic movements. Additionally, there will be obvious benefits for transport operators and clients, since the new systems will provide public administrations with rapid and detailed information on infrastructure and maintenance needs. Furthermore, NEM technologies will provide new, more easily used, and comfortable services to passengers, and increase safety and security.

23.5 Conclusion

Grand societal challenges are key elements which will structure the European research in the next decade. It is true in Europe but also in the World and Network Media sectors should give their contribution in several domains such as culture, energy savings, and transport.

It is obvious that the future research program will tackle these challenges and the Networked media projects should also bring their contribution for the well-being of the European community as well as in other worldwide regions.

Index

Note: n = Footnote

A

Access Control List (ACL), 464
Access network (AN), 15, 303, 320. *See also*
    Core network (CN)
  capacity, 20
  heterogeneity management, 21
  in mobile TV architecture, 58
  for user-centric IPTV architecture, 303
Access Network Adaptation Module
    (ANAM), 315–316
Access Network Discovery and Selection
    Function (ANDSF), 70
Access Network Monitoring Module
    (ANMM), 312, 313
Access Point (AP), 38, 406, 483
Access-centric CDN models, 175
Accommodation, 277, 284. *See also*
    Convergence
Action Engine Module (AEM), 305, 310
  for cross-layer adaptation, 309
  logic, 310, 326
  responsibility, 308
Adaptation engines, 313. *See also* Multimedia
    Content Management System
    (MCMS); Monitoring engines
  ANAM, 315–316
  MSAM, 313–314
  TNAM, 314–315
Adaptive IPTV components, 297
  advances in cross-layer adaptation,
    300–301
  cross-layer management/monitoring,
    299–300

  QoE/PQoS mechanism, 298–299
  QoE-enhanced IMS, 298
Adaptive streaming, 64, 219
  HTTP, 202
  in server–client architectures, 354
Advanced Audio Coding (AAC), 61
Advanced Domain Name System (Advanced
    DNS), 213
Advanced Television Systems Committee
    (ATSC), 431
Advanced Three-dimensional Television
    System Technologies (ATTEST),
    349
Advanced Video Coding (AVC), 61, 352
  multiview amendment, 353
  stereo high profile, 352
Aging, 493
Aging societies, 505. *See also* Rehabilitation in
    home
  in Europe, 490
Akamai technologies, 189, 311,
    218–219
Alliance for Telecommunications Industry
    Solutions (ATIS), 13, 109
Ambient lighting effects, 285–287
Android 2.2, 38
Anycast
  in ICN, 171, 172
  requests to local iBox, 261, 262
Anywhere, anytime TV services
  place-shifting TV, 228
  remote control of IPTV services, 229
  session continuity services, 228–229
  time-shifted TV, 228

Application
    acceleration, 214–215
    developers, 176
    layer multicast, 191
    providers, 65, 66
Application Layer Traffic Optimization
    (ALTO), 114, 126. *See also* Future
    internet; ENVISION system
    architecture, 127, 128
    CDNs, 127
    elemental methods, 135–136
    protocol, 128, 129–130
    WG, 114
Application programming interface (API), 14,
    66, 421, 447
    client-side, 115
    LMF REST, 460
    OpenSocial project, 456
Application server (AS), 103, 104, 485
    communication with MRF, 484
    for IPTV application, 303
    Sh interface, 306
Arte, 89
Asset Distribution Interface (ADI), 431
Asset Location Function (ALF), 106
Assistive Technology (AT), 491
Association of Radio Industries and Business
    (ARIB), 96, 431
Assured Forwarding (AF), 477, 485
Asymmetric Digital Subscriber Line (ADSL), 4
Audio analysis solutions, 452
Audio and Video channels (A/V channels), 16
Audio Visual (AV), 58, 430
    content chain, 426–427
    metadata, 427
    perception measure, 226
Audio/Visual Profile (AVP), 481
Audiovisual content. *See* Multimedia Content
    Management System (MCMS)
Audiovisual metadata, 427
    chain, 430
    families, 428–429
    types, 427–428
Augmented Reality (AR), 448
Authentic content providers. *See* Secure
    transmission medium
Authentication, authorization, and accounting
    (AAA), 132, 392, 404
Automated Speech Recognition (ASR), 452
Autonomous Systems (AS), 137
Auto-stereoscopic displays, 92
Average Revenue Per User (ARPU), 9

B

Back to Back User Agent (B2BUA), 306
Backend services, 435–436. *See also* Frontend
    services
Back-office node, 193
Basic channel service, 16. *See also* Enhanced
    selective service; Interactive data
    service
Belgacom, 9, 10
Best effort (BE), 46
Billing issues, 23
Binary Floor Control Protocol (BFCP), 478, 479
Binocular disparity, 278, 279
    HVS sensitivity for, 281–282
    visual fatigue variation, 284–285
Binocular stereopsis, 278–279
Bit Error Rate (BER), 343
Bit Sliced Arithmetic Coding (BSAC), 60
Bit-level Objects (BO), 156
Bit-Torrent, 366
Black-box objective methods, 226
Block Error Rate (BLER), 324
Bloom filter, 158
Blu-ray Disc Association (BDA), 97
Body Area Network (BAN), 411
Bottleneck connection rate, 367, 368
Broadband
    access technologies, 10
    network, 4
Broadband Contents Guide (BCG), 17
Broadcast (BC), 4, 104
Broadcast Program Guide (BPG), 384
Broadcast transmission (BCAST
    transmission), 37
Broadcast/Multicast Service Centre
    (BM-SC), 111
Browsing interaction, 425
Business models, 64, 66, 395. *See also*
    Content—personalization
    advertising, 67
    based on content and context fusion, 396
    based on content storage and behavior,
    396–397
    based on scalable content compression and
    transmission, 397
    ICN, 174
    MMS transmission and reception, 67
    publishing to app store, 67
    SMS reception, 67
    SMS transmission, 66
    user context, 67

Business to Business (B2B), 16, 18, 395, 396
Business-to-Business-to-Client (B2B2C), 395,
      396, 397

C

Cache behavior
    cache hierarchy, 200
    dynamic URL handling, 200–201
    high-level principle, 199
    memory management, 199, 200
    stale content detection, 201–202
Cache nodes. *See* Surrogates
Cache poisoning, 202–203
Caching
    content, 125
    flash crowd effect, 168
    in ICN, 166, 167
    in-network caching, 167–168
    mean rate of request, 169
    outside network, 167
    transparent, 190
Call Session Control Function (CSCF), 304,
      484, 485
CallerID
    localization, 408
    on TV, 230–231
Capability and Preference Information
      (CPI), 394
Capital Expenditure (CAPEX), 30,
      41, 51
Care and monitoring in home, 492–493
Carriers, 192
Carrier-to-Noise ratio (CNR), 311
Cascading Style Sheets (CSS), 86, 90
Cellular networks, 63–64
Channel annotation, 463
Channel enrichment, 463
Channel provision, 463
Chat interaction, 426
Chunk generation process, 367
Class Selector 4 (CS4), 477
Click interaction, 425
Client–server approach-based TV
      systems, 416
Cloud
    computing, 41, 48, 115
    design motivations, 50–51
Cloud computing infrastructure (CCI), 43
Cluster Controller Function (CCF),
      106, 107
Code Division Multiple Access (CDMA), 61

Collaboration Interface between Network and
      Application interface (CINA
      interface), 131, 134
    collaboration scenario, 137, 138
    communication protocol, 135–137
    discovery mechanism, 135
    functional entity implementation, 134
    multicast triggering call flow, 138, 139
    security mechanisms, 135
Color Moving Picture Quality Metric
      (CMPQM), 261
Color-plus-depth format, 362
    LDV format, 351
    3D video format, 350
Comments and votes interaction, 426
CommonTag, 454
Communication. *See also* Information and
      Communications Technology
      (ICT)
    bidirectional, 229
    CINA interface, 134
    CINA protocol, 135–137
    devices, 4, 5, 376
    LMF, 459
    models, 171–172
    multiparty, 231
    NFC, 24
    T-communication, 18
Community WiFi, 64
Composite Capability/Preference Profiles
      (CC/PP), 394
CONNECT [Content-Oriented Networking:
      A New Experience for Content
      Transfer], 182
Connected TV, 78, 89–90. *See also*
      Standardization
ConnectME
    framework, 463
    Hypervideo Browser, 447, 464
    linked data, 448
    project, 447, 462
    semantics, 448
Consortium for Telecommunications
      (CNIT), 181
Constraint map, 136–137
Consumer domain, 11
Consumer Electronic Show (CES), 92
Consumer electronics (CE), 90, 92
    companies, 349
    PVR, 18
Consumer Electronics Association (CEA),
      86, 87

Content, 182
  Bridge Alliance, 174
  consumers, 176–177, 180
  context information, 392, 415
  delivery call flow, 197, 198
  delivery processes, 211, 212, 257–258
  deployment/distribution, 194
  identifier, 412–413
  information, 413
  ingestion/acquisition, 106, 194, 195, 199
  management service, 214
  owners, 173–174, 464
  personalization, 72
  preparation, 194, 438
  prepositioning, 194
  protection, 20–21, 85
  storage to dynamic object, 205–206
  types, 190
Content Adaptations Interactions (CAI), 43
Content Delivery Function (CDF), 12, 106
  in delivering content, 108
  functional architecture, 109
Content delivery network (CDN), 105, 127,
        189, 210, 211, 396. *See also*
        Technology innovations
  advanced DNS, 213
  Akamai technologies, 218–219
  cache poisoning, 202–203
  carriers, 192
  CDNI, 205
  comparison, 192
  content management service, 214
  content storage, 205–206
  content types, 190
  CSPs, 192
  denial of service attacks, 203–204
  end-to-end internet to client-to-content
        paradigm, 206
  industrialization, 204–205
  ISPs, 192
  Limelight content delivery platform, 220
  Limelight networks, 219
  market overview, 211, 212
  media delivery services, 212, 213
  multicast, 191
  multi-CDN, 214
  network operators, 217–218
  online video platform, 214
  OTT, 191
  P2P, 190
  pure-play service providers, 216–217
  reporting and analytics, 214

  security mechanism, 204
  services evolution, 215–216
  taxonomy, 191
  technology vendors, 218
  transparent caching, 190
  unauthorized access to content, 203
  value-added services, 214–215
  Yankee group CDN scorecard, 213
Content delivery network architecture (CDN
        architecture), 105–106, 193, 257
  functional entities, 106
  future activities, 107
  and IPTV subsystem relationship, 106–107
  QoE-aware, 262
Content Delivery Network Controller
        Function (CDNCF), 106, 107
Content delivery network functioning (CDN
        functioning). *See also* Content
        delivery network (CDN)
  cache behavior, 195, 199–202
  CDN operations, 195
  content ingestion/acquisition, 194, 195
  content preparation, 194
  content prepositioning, 194
  content request redirection, 195–199
  deployment/distribution, 194
  high-level architecture, 193
  logging, 202
Content Delivery Network Interconnection
        (CDNI), 183, 205
Content delivery network services evolution
        (CDN services evolution), 215.
        See also Content delivery network
        (CDN)
  Limelight's traffic growth, 215
  Limelight's value-added services, 216
Content Delivery System Manager (CDSM), 218
Content Delivery System–Internet Streamer
        (CDS–IS), 218
Content Distribution Architecture (CDA), 261
Content Distribution Architecture using
        QQAR protocol (CDA-QQAR),
        261. *See also* Content Distribution
        Network (CDN); Quality of
        experience (QoE)
  architecture, 263
  control overhead, 269–270
  iBox component, 261, 262
  QoE-aware CDN architecture, 262
  QQAR Algorithm, 262–263
  server selection method, 263–264
  simulation approach comparison, 268

simulation network topology, 267
testbeds for PSQA, 266–267
user perception, 268–269
Content Distribution Network (CDN), 256.
    See also Quality of experience (QoE)
    architecture, 256, 257
    content delivery processes, 257–258
    surrogate servers, 257
Content distribution system (CDS), 113
Content management system (CMS), 194, 433,
    438–439
COntent Mediator architecture for content-
    aware nETworks (COMET), 181
Content on Demand (CoD), 104, 384
Content Origin Function (COF), 106
Content personalization, 72
    during nomadism, 379
    uses, 402
Content provider (CP), 4, 30, 65, 174, 189, 192
    benefit from ICN, 180
    content distribution, 59
    cookie support, 84
    cope with technical fragmentation, 86
    domain, 11
    in PURSUIT, 182
    service development, 83
    user interface, 81
Content recommendation systems, 390
    accuracy, 244, 390
    capabilities, 243
Content request redirection, 195
    to cache node, 196, 197
    to CDN, 196
    CDN paradox, 197, 198
    clustering, 197
    comparison with DNS, 198–199
    content ingestion, 199
    request routing, 199
    two-level process, 195, 197
Content resolution approaches, 165
    hop-by-hop discovery approach, 166, 167
    lookup-based resolution approach, 165–166
Content service provider (CSP). *See* Content
    provider (CP)
Content transmission forms, 36. *See also*
    Open-IPTV services
    Live-TV scenario, 39, 40–41
    multicast transmission, 37
    unicast transmission, 36–37
    validations-based testbed, 38
    VOD scenarios, 39
    WEB-TV, 39

Content Type (CT), 322, 324
    header, 458
    IPLR at BLER value, 323
Content-based video analysis, 451
Content-centric communications, 142
Content-Centric Networking (CCN), 143,
    151, 206
    approaches to routing scalability,
        152–153
    communications, 151
    FIB and PARC, 151
    forwarding, 151
    name design, 160
    NDN, 152
    node, 152
Context awareness (CA), 383–384, 441
    IPTV architecture, 417–418
    requirements, 391–392
    TV architecture, 415
Context in IPTV services, 403. *See also*
        Content—personalization
    information types, 403–404
    modeling, 413–414
    obtaining contextual information, 404
    personalization approaches, 415–417
Context modeling, 413–414
Context-Aware Middleware for Ubiquitous
        System (CAMUS), 416
Contextual information, 224, 403
Contribution network, 95–96
Conventional stereoscopic video. *See*
        Stereoscopy—3D video
Convergence, 181–182, 277. *See also*
        Accommodation
    fixed/mobile, 301
    future network-service, 48
    IMS, 303
    levels, 424–425
    of multimedia services, 296
    for natural viewing, 284
    by telecom, 47
Cookie support, 84
Core access, 491
Core IMS, 103, 104
Core network (CN), 14–15, 320. *See also*
        Access network (AN)
    DiffServ/MPLS-enabled, 303
    PS, 110
Cost map, 114, 129, 136. *See also*
        Network—map
    client requests, 138
    real metrics for, 137

Cross Domain Authentication (CDA), 45
Cross-layer adaptation, 300, 301, 308, 321, 331
    characteristics, 300
    multiuser, 324–326
    PQoS-aware dynamic, 301
    for user experience optimization, 297, 307
Customer Facing IPTV Application (CFIA),
    105
Customer to Customer (C2C), 16, 18
Cyclic Redundancy Check (CRC), 62

D

Data management, 133
Data mediation, 439
    mediator, 439
    semantic mediation architecture, 441
    semantics, 439–440
    vertical integration *vs.* semantic mediation,
        440
Data routing, 256, 258, 265
Data-centric networking paradigm, 144
Data–money flow, 177
    in CDN-based model, 178
    in ICN model, 178–179
    in legacy model, 177
Data-Oriented Network Architecture
    (DONA), 153
    caching implementation, 154
    content delivery in, 153
    content forwarding in, 153
Dead reckoning module, 407
Decoding capabilities, 84
Decoupled Application Data Enroute
    (DECADE), 168
Delivery servers. *See* Surrogates
Depth
    map videos, 352
    sensation metrics, 289
Depth Distortion Model (DDM), 289
Depth Enhanced Stereo (DES), 351
Depth of Field (dof), 284
Depth perception, 279, 280. *See also* 3D video
    QoE
    binocular disparity, 281–282
    JNDD experiment, 280
    oculomotor cues, 277
    relative size cue, 283–284
    retinal blur, 282–283
    visual cues, 278
Description Definition Language, 395
Description Logic (DL), 439

Descriptive Ontology for Linguistic and
    Cognitive Engineering (DOLCE),
    453
Device/terminal context, 403
    information, 391–392
Devices Vendors, 65, 66
Diaspora, 456
Differential Quadrature Phase Shift Keying
    (DQPSK), 61
Differentiated Services (DiffServ), 260
Digital Agenda, 507
Digital Audio Broadcasting (DAB), 60
Digital Cinema Initiatives (DCI), 431
Digital entertainment devices, 10
Digital Item (DI), 431
Digital Living Network Alliance (DLNA),
    78n, 87, 411, 431
Digital Multimedia Broadcasting (DMB), 59,
    60
Digital rights management (DRM), 13, 45,
    71–72, 413
Digital television (DTV), 237, 496–497. *See
    also* Internet Protocol Television
    (IPTV)
    future of, 23–24
    harmonization, 114–115
    IPTV, 4–5
    market, 4, 5–7
    recommendation architecture, 416
Digital Terrestrial TV (DTT), 98
Digital TV. *See* Digital television (DTV)
Digital Versatile Disk (DVD), 16
Digital Video Broadcasting (DVB), 13, 96,
    394
    ATTEST, 349
    uses, 31
Digital Video Broadcasting Service discovery
    and Selection Transport Protocol
    (DVBSTP), 116
Digital Video Broadcasting—Handheld
    (DVB-H), 59, 61, 70
Digital Video Broadcasting-Multimedia Home
    Platform (DVB-MHP), 394
Digital Video Broadcasting—Terrestrial
    (DVB-T), 58, 369
Digital video recorder (DVR), 395
Distance-Vector (DV), 268
Distributed context-aware ITV solution, 415
Distributed hash table (DHT), 150, 157, 165
Distributed services networking (DSN), 112,
    113
Distribution network, 96–97

DNS Extensions (dnsext), 183
Document Object Model (DOM), 80, 90
Domain name system (DNS), 165, 198, 213
Domestic cloud, 43, 51, 52
Domestic general content manager
    (DGCM), 43
Drivers. *See* Societal Challenges, 491
Dublin Core (DC), 453
DVB IP DataCasting (DVB-IPDC), 61
DVB-IPTV service, 31, 394
Dynamic Adaptive Streaming over HTTP
    (DASH), 110, 112
Dynamic caching, 200
Dynamic network characteristics, 22
Dynamic site acceleration (DSA), 190
Dynamic subscriber behaviors, 22

E

EC-funded projects, 300
Edutainment, 66
egtaMETA, 447
E-Health, 490
Electroencephalography (EEG), 227
    analysis, 245, 246
Electromyography (EMG), 244
Electronic Learning (E-Learning), 66
Electronic Program Guide (EPG), 4, 17
    EIT use, 96
    information, 413
    personalized, 239, 378, 416
Elementary Control Function (ECF), 116
Elementary Forwarding Function (EFF), 116
Emotiv EPOC device, 245, 246
End users, 21, 179, 180
    accessing mobile TV services, 64
    application management, 133
    caching, 125
    ConnectME project use, 447
    devices, 20
    expectations, 59
    mobile device role, 65
    service quality by, 256
Endpoint service, 136
Endpoint-centric communication model, 143
End-to-end (e2e), 319, 259
    internet connectivity, 175
    QoS optimization, 270
    quality chain in, 260
Energy
    efficiency, 508–509
    grids, 506

incentives on savings, 179
supplies, 505
Enhanced selective service, 16. *See also* Basic
    channel service; Interactive data
    service
    B2B hosting service, 18
    C2C hosting service, 18
    EPG, 17
    MoD service, 17
    multiangle service, 18
    PVR, 18
    VoD services, 16, 17
Entertainment industry, 506
Environmental information systems, 506
ENVISION system, 130. *See also* Application
    Layer Traffic Optimization
    (ALTO)
    CINA interface, 131, 134–138
    collaboration, 130–131
    functional entities, 132, 133
    high-level architecture, 131–132
    interface types, 132
    ISP, 131
e-sell Through. *See* Purchased VoD
EU ICT ETICS project, 175–176
European Commission (EC), 504
European Research. *See also* ENVISION
    system
    study topics, 505–506
European Technology Platform (ETP), 503
European Telecommunications Standards
    Institute (ETSI), 59, 227
European Telecommunications Standards
    Institute/Telecommunication and
    Internet converged Services and
    Protocols for Advanced Network
    (ETSI/TISPAN), 13, 102
    CDN architecture, 105–107
    IMS-based IPTV, 102–104
    NGN-integrated IPTV, 104–105
Event information table (EIT), 96
Evolved MBMS (e-MBMS), 63, 64
Evolved Packet Core (EPC), 111
Exchangeable Image File Format (EXIF), 453
Expert Group (EG), 504
    recommendations, 505
    ToR, 504
Explicit caching technique, 125
Extended content negotiation, 458–459
Extensible Markup Language (XML), 431
Extensible Metadata Platform (XMP), 453
External Marking Modules (EMM), 312

F

Falcon, 455
Federation Identity (FI), 54
FIB, 151
Fiber Optic Service (FiOS), 6
Fiber technology, 15
Fiber to the home technology (FTTH
        technology), 19
Fiber-To-The-Node (FTTN), 15
Field of view (FOV), 355
File Delivery over Unidirectional Transport
        (FLUTE), 103, 116
File Transfer Protocol (FTP), 213
Filtering map, 136
Fixed/mobile convergence, 301
Flash crowd effect, 168
Focus Group (FG), 12, 13
Focus Group on IPTV (FG IPTV), 12,
        13, 108
Fonts, 83
Footprint map, 136
Forward Error Correction (FEC), 300
Fourth Generation (4G), 69
Frame packing, 95, 97
        SEI, 352
Frame Rate (FR), 322, 324
Frame-compatible compatible (FCC), 98
Frame-compatible formats, 93, 94
Free Viewpoint TV (FTV), 98
Free-viewpoint vision, 348
Frequency Division Duplexing (FDD), 63
Friend Of A Friend (FOAF), 454
From hourglass to lovehandles problem,
        142, 143
Frontend services, 436
Full Intra Request command (FIR command),
        481
Fully qualified domain name (FQDN),
        165, 196
Functional architecture
        CDN, 106
        IMS-based PSS, 111–112
        IPTV, 11–12
        ITU-T IPTV, 108
        NGN-based IPTV, 108–109
        protocols mapped to NGN-integrated
                IPTV, 105
        for user profiling, 384–385
Future internet. *See also* Application Layer
        Traffic Optimization (ALTO);
        ENVISION system

ALTO specification, 124
caching content, 125
collaborating applications, 124–126
ISP, 125
media applications, 122–123
P4P/ALTO initiative, 124–125
Future Internet Architecture (FIA), 152

G

Galvanic skin response (GSR), 243, 244
Game consoles, 494
Gateway GPRS Support Node (GGSN),
        63, 315
Gaussian blur, 282, 283
General Content Management (GCM), 42
General packet-radio service (GPRS), 483, 485
Geo-blocking, 213
Global Positioning System (GPS), 65–66, 407
Global Standards Initiative (GSI), 108
Global warming, 505
Globally Executable MHP (GEM), 13
Gnu's Not Unix (GNU), 344
Google TV, 81–82
Grand Challenges. *See* Societal Challenges
Graphic resolutions, 83
Graphical models, 414
Graphical user interface (GUI), 229, 458
Group-Of-Pictures (GOP), 366

H

HDMI LLC, 97
Head-Related Transfer Functions (HRTFs), 357
Health care systems and ambient assistant
        living, 506
Hierarchical B-frame prediction, 353
High definition (HD), 6, 122
High-Definition Television (HDTV), 6, 237
High-level contextual information. *See also*
        Context in IPTV services
        network states, 412
        service/content, 412–413
        used device, 410–412
        user identification, 404–406
        user mood/activity detection, 408
        user's localization, 406–408
        user's preferences, 408–409
        user's profile, 404
        user's social network credentials, 408
        user's watching habits and consumption
                history, 409–410

High-quality video, 20
High-speed Digital Subscriber Line (xDSL), 4
Home Cinema, 10–11
Home network (HN), 53, 97
　　all based, 45–46
　　service-based, 46
Home Network End Device (HNED), 394
Home Network Interfaces (HNI), 108
Home Subscriber Server (HSS), 304, 484, 485
Hop-by-hop discovery approach, 166, 167
Host Identity Protocol (HIP), 171
Host-centric Internet, 171
Host-to-host communications, 142
Hybrid broadband–broadcast approach, 368
　　adaptation decision engine module use,
　　　　369, 371
　　multiview delivery system, 370
　　terrestrial DVB, 368
　　3D content server uses, 369
Hybrid Broadcast Broadband TV (HBBTV),
　　88, 377
Hybrid caching, 200
Hybrid content recommendation solution, 390
Hybrid Fiber Cable (HFC), 15
Hypertext Markup Language (HTML), 116
　　HTML 5, 89
Hypertext Transfer Protocol (HTTP), 36, 64,
　　478
　　HTTP 1.1 protocol standard, 189
　　HTTP-based streaming, 219
Hypervideo, 447, 461
　　annotation suite, 464
　　browser, 464
　　platforms, 448

I

Image analysis solutions, 452
Image Quality Assessment (IQA), 235
IMS Core. *See* Core IMS
Information and Communications Technology
　　(ICT), 115, 490, 508
　　contributing areas, 506
　　social factors influencing access, 491
Information Extraction (IE), 452
Information identifier (ID), 170
Information integration, 439–440
Information Objects (IO), 156, 176. *See also*
　　Naming information objects
　　addressing, 149
　　content-based security, 147
　　naming, 146

　　security, 170
　　transport layer for, 150
Information retrieval (IR), 454
Information Technology Infrastructure Library
　　(ITIL), 232
Information-Centric Networking (ICN), 142,
　　206
　　anycast and multicast in ICN, 171–172
　　application developers, 176
　　business incentives, 172–173
　　caching in ICN, 166, 167–169
　　CDN providers, 174
　　communication use, 144–145
　　content addressing, 149
　　content consumers, 176–177
　　content dissemination, 145
　　content forwarding, 149–150
　　content owners, 173–174
　　content providers, 174
　　content resolution approaches, 165–166
　　content retrieval, 145, 146
　　data–money flow, 177–179
　　energy-saving incentives, 179
　　information object security, 170
　　manageability, 160
　　migration challenges, 179–181
　　mobility support in ICN, 171
　　motivation for information network,
　　　　143–144
　　naming information objects, 146–149
　　network management, 160
　　network operators, 174–176
　　open Internet marketplace, 164
　　principles, 145, 146
　　QoS considerations, 159
　　reliability, 160
　　Research Activities, 181–182
　　scalability, 160
　　security, 158–159
　　Standardization Activities, 183–184
　　transport layer for information objects, 150
Information-Centric Networking Research
　　Group (ICNRG), 183
In-network caching, 167–168
Instant Messaging (IM), 59
Instant messaging on TV, 231
Institute of Electrical and Electronics
　　Engineers (IEEE), 233
Integral imaging, 348
Integrated Mobile Broadcast (IMB), 63
Integrated Services Digital Broadcasting-
　　Terrestrial (ISDB-T), 59, 62–63

Intelligent Transport Systems (ITS), 506, 509
Interactive data service, 17. *See also* Basic
      channel service; Enhanced selective
      service
  T-commerce, 18
  T-communication, 18–19
  T-entertainment, 19
  T-information, 18
  T-learning, 19
Interactive Program Guide (IPG), 17, 237
Interactive synchronized services, 229
Interactive TV (iTV), 7, 229, 415, 496
  advertising and personalized advertisement
      insertion, 230
  interactive synchronized services, 229
  parental control services, 230
  search, discovery, and content
      recommendation services, 229
  streaming games on IPTV, 230
  user-generated content services, 230
Interactivity, 425–426
Interlinked media and user-centric approach,
      454
Internal Marking Module (IMM), 314
International Affective Picture System (IAPS),
      245
International cloud (IC), 51, 52
International Mobile Telecommunication
      2000 (IMT-2000), 60
International Multimedia Telecommunications
      Consortium (IMTC), 482
International Standard Audiovisual Number
      (ISAN), 433
International Telecommunication Union
      (ITU), 108, 233, 393–394, 482
International Telecommunication Union—
      Telecommunication Standardization
      Sector (ITU-T), 13, 226
  definition of IPTV, 31
  DSN, 112–113
  forming FG IPTV, 13
  IPTV functional architecture, 108
  IPTV-GSI, 108–109
  recommendations, 228, 233, 240
Internet, 6, 188
  cellular networks, 63, 80
  collaboration concept, 123–126
  content growth, 171
  design, 206
  framework, 207
  marketplace, 173
  media applications, 122–123

  pervasiveness, 10
  usage, 189n
Internet Engineering Task Force (IETF), 183,
      482, 485. *See also* Internet Research
      Task Force (IRTF)
  ALTO working group, 127
  CDNi, 205
  protocol design activities, 113–114
Internet Group Management Protocol
      (IGMP), 37, 45
Internet Media Guide (IMG), 17
Internet Protocol (IP), 4, 485
  address use, 171
  datagram fragmentation, 150
  flow, 61
  lookup techniques, 160
  lovehandles problem in, 143
  networks, 15
  traffic information, 160
Internet Protocol Multimedia core network
      Subsystem (IMS). *See* Internet
      Protocol Multimedia Subsystem
      (IMS)
Internet Protocol Multimedia Subsystem
      (IMS), 13, 102, 297, 472, 483
  architecture, 306, 484
  CSCF, 304
  enhancement, 303–305
  HSS, 304
  IMS architecture and elements, 306
  infrastructure for management and control
      of services, 24
  ISC interface, 305, 306
  MCMS and NGN IMS interaction,
      304–305
  MCMS specifications, 305
  mobile communication platform, 299
  multimodal management system use, 297
  PCC Function, 306–307
  QoE-enhanced, 298
  and TP, 483–485
Internet Protocol Multimedia Subsystem
      Service Control (ISC), 305
  interface, 305, 306
Internet Protocol Multimedia Subsystem-based
      architecture, 111. *See also* Non-IMS-
      based architecture
  functions, 111, 112
  PSS architecture, 111
  for TP, 484
Internet Protocol Multimedia Subsystem-based
      IPTV, 102–103

IPTV Media Functions, 107
new components, 103–104
NGN component reuse, 103
protocol specifications, 104
Internet Protocol packet Delay Variation
(IPDV), 316
Internet Protocol packet Error Ratio (IPER), 316
Internet Protocol Packet Loss Ratio (IPLR),
316, 322
Internet Protocol packet Transfer Delay
(IPTD), 316
Internet Protocol Television (IPTV), 4, 10, 31,
224, 431, 448. *See also* Nomadic
IPTV services; Open IPTV; Quality
of experience (QoE)
advantages, 224, 225
anywhere, anytime TV services, 228–229
changing QoE perception, 228
current status, 5–6
deployment challenges, 19–20
digitization, 10
domains, 11
functional architecture, 11–12
history, 4–5
Home Cinema, 10–11
initiatives to standardization, 12–14
interactive TV services, 229–230
Internet TV, 5, 11
IPTV, 14–16, 102
management models, 31
NGN integrated IPTV, 104–105, 108, 109
open, 34
personal, 34
presence, 230
revenue forecast, 9
via satellite, 24
service adaptation steps, 318–319
service architecture, 381–383
social TV services, 230–232
subscriber forecast, 8
technical challenges, 20–23
terminologies, 33–34
TS 181 016 v.3.3.1 document, 227
US market, 8
user-Centric delivery model, 381
web services use, 11
Internet Protocol Television Control
(IPTV-C), 105
Internet Protocol Television Interoperability
Forum (IIF), 13, 109
Internet Protocol Television market, 7. *See also*
Digital television (DTV)—market

Belgacom, 9, 10
evolution, 376
revenue forecast, 9
service providers investments, 9
subscriber forecast, 8
US IPTV revenues, 8
Internet Protocol Television operation, 14
access networks, 15
core networks, 14–15
STB, 15–16
video head-end, 14
Internet Protocol Television services, 16, 17,
377. *See also* Context in IPTV
services; Nomadic IPTV services
basic channel service, 16
context awareness, 383–384
enhanced selective services, 16, 17–18
existing personalization approaches,
390–391
existing standardization efforts, 393–395
interactive data services, 18–19
interest rate in, 380
IPTV service architecture, 381–383
limitations, 377
multiscreen approach, 380–381
privacy and identity management, 389–390
requirements, 391–393
services for use cases, 379
services for users, 378–379
users' expectations, 379
Internet Protocol Television standardization,
102
ETSI TISPAN, 102–105
SDO activities on IPTV, 107–109
Internet Protocol Television User Data
Function (IUDF), 105
Internet Research Task Force (IRTF), 183. *See
also* Internet Engineering Task Force
(IETF)
Internet Service Providers (ISPs), 30, 167, 258
CDNs, 192
CINA server, 138
cloud network, 30
P4P/ALTO initiative, 123
uses for P2P applications, 126
Internet TV, 5, 11
Internet Video. *See* User Generated Content
(UGC)
Internet@TV. *See* Smart TV
Interoperability, 20, 52, 481
DLNA forum, 87
requirements, 391

Interrogating CSCF (I-CSCF), 484
Inverse Functional Properties (IFPs), 455
IxLoad, 236
IxN2X ITPV QoE solution, 236

J

JavaScript Object Notation (JSON), 449
Jitter. *See* Internet Protocol packet Delay
        Variation (IPDV)

K

Key Performance Indicators (KPI), 235,
        288, 386
Key players and value chains positioning, 434
Key-Value model, 414

L

Layered depth video (LDV), 351
Least Frequently Used (LFU), 200
Least recently used (LRU), 200
Least recently used rule (LRU rule), 169
Lenticular lens sheet, 93
Light-field representation, 351
Limelight
    content delivery platform, 220
    networks, 211, 219
Linear TV, 4, 224
Link Layer Discovery Protocol (LLDP), 411,
        419
Link Layer Topology Discovery (LLTD), 411,
        419
Link State (LS), 270
Linked data, 448, 449–451, 458
    caching mechanism, 460
    SERVER, 459
Linked Media Framework (LMF), 450,
        451, 457
    architecture and implementation, 459
    Channel annotation, 463
    Channel enrichment, 463
    Channel provision, 463
    conceptual model, 458
    ConnectMe framework, 463
    ConnectMe hypervideo browser, 464
    ConnectMe project, 462
    core, 459
    extended content negotiation, 458–459
    Linked Data, 458, 459
    linked media principles, 457

LMF enhancer, 465
LMF Search, 460
LMF SPARQL, 460
LMF versioning, 465
    media interlinking, 465
    RESTful resource management, 458
    semantic enhancement via linked data, 459
    semantic indexing, 459
    sKWRL reasoner, 460
    video annotation via LMF, 461–462
Linked media principles, 457
Linked Open Data (LOD), 447
    initiative, 449–450
Live IPTV, 316
Live video broadcast server, 14
Live-TV scenario, 39
    flow control messages, 40
    RTSP commands, 40–41
    using validations-based testbed, 39
Local area networks (LANs), 190, 436
Local Content Management (LCM), 42
Logging, 202
Logic-based models, 414
Long-term evolution (LTE), 19, 63, 483
Lookup-based resolution approach, 165–166

M

Managed IPTV architectures, 102
Managed network, 31. *See also* Unmanaged
        network
    IPTV Services, 107
    nomadic access under, 46
    QoS, 87
Managed service, 35
Markup scheme models, 414
Material eXchange Format (MXF), 431
Maximum transmission unit (MTU), 150
Mean Opinion Score (MOS), 22, 305
    for adaptation scheme, 342
    CDA-QQAR, 268–269
    for CN loads, 341
    for depth in 3D video *vs.* channel
        bandwidth, 286
    for image quality *vs.* channel bandwidth,
        286
    for IPTV sessions and adaptation
        approaches, 333
    QP levels, 365
    for SBR, 323
    score evolution, 337
Measuring Normalizing Blocks (MNB), 261

Media
   assets, 446
   delivery, 213
   player, 84, 432
   transport, 480–481
Media Access Control (MAC), 405
Media Control Functions (MCF), 104, 105
Media Delivery Functions (MDF), 104
Media delivery services, 212, 213
   CDN, 214
   CDN services evolution, 215–216
Media Forward Link Only (Media FLO),
      61–62
Media Grid Engine System (MGE), 177
Media Resource action Controller (MRFC),
      303, 484
Media Resource Function (MRF), 303, 484
Media Resource Function Processor (MRFP),
      303, 484
Media Server Resource Function (MSRF),
      302, 303
Mediator, 439
Message Session Relay Protocol (MSRP), 481
Metadata, 148–149, 426. *See also* Moving
      audiovisual experience
   ADI, 431
   aggregation players, 433
   audiovisual content chain and, 426–427
   audiovisual metadata type, 427–430
   chain, 430
   creation players, 432–433
   DCI, 431
   DI, 431
   DLNA, 431
   exploitation players, 433
   families, 428–429
   IPTV, 431
   key players and value chains positioning,
      434
   MPEG-7, 430
   MXF, 431
   pure media players, 432
   segmentation, 395
   standards, 430–432
   TV Anytime metadata, 430–431
Metadata Management System (MMS), 438
Metadata processing, 434
   backend services, 435–436
   frontend services, 436
   technologies involved, 435
   TVOD, 435
   used in different technologies, 437

Metadata systems, 438
   content management systems, 438–439
   context-awareness, 441
   data mediation, 439
   semantic mediation approach, 439–441
   vertical integration, 440
Microsoft Media Services (MMS), 213
Middleware, 70
   CAMUS, 416
   infrastructure approach, 383–384
Migration challenges, 179–181
Mobile Broadcast Services Enabler Suite
      (BCAST), 61
Mobile network nodes, 63
Mobile operators, 65
   localization information, 66
   market share, 65
   QoE, 59
   roaming issue in, 54
Mobile Station International Subscriber
      Directory Number (MSISDN), 382
Mobile TV, 58
   bandwidth requirements, 69
   broadcast transmission in a heterogeneous
      environment, 68
   business issues, 72
   business models, 64–65, 66–67
   consumer adoption, 72
   device characteristics, 68
   ecosystem, 59
   evolution, 58, 59
   IPTV, 19
   mobile device support, 65–66
   open issues and challenges, 67–68
   QoS and QoE, 70–72
   service coverage, 69–70
   STB, 65
   SVC, 70
   trends, 59–60
   wireless mobility, 69
Mobile TV distribution, 63
   cellular networks, 63–64
   MBMS, 63
   WiFi networks, 64
Mobile TV standards, 60
   DMB, 60
   DVB-H, 61
   ISDB-T, 62–63
   MediaFLO, 61–62
   S-DMB, 60–61
   T-DMB, 61
Mobile user location, 65

Monitoring engines. *See also* Multimedia
Content Management System
(MCMS); Adaptation engines
  ANMM, 312, 313
  MSMM, 311–312
  TNMM, 312
Motion parallax, 278
Moving audiovisual experience, 424. *See also*
Metadata
  convergence, 424–425
  interactivity, 425–426
  recommendation, 425
  social networks, 424
Moving Pictures Experts Group (MPEG), 14,
98
  MPEG-21, 14, 395
  MPEG-7, 14, 395, 430
  MPEG-E standard, 14
Mozilla's Popcorn project, 446n
MPEG Frame Compatible (MFC), 98
Multi Protocol Label Switching (MPLS), 14
Multiaccess devices, 21–22
Multiangle service, 18
Multicast (MCAST], 191. *See also* Content
delivery network (CDN); Video-on-
demand (VOD)
  in ICN, 171
  instantiation, 137
  IP multicast traffic, 111
  routing protocols, 47
  via state-based approach, 172
  testbed, 137
  transmission, 37, 46
  tree structure, 45
Multicast Listener Discovery (MLD)
Multicast Nodes (MCN), 36
Multi-CDN platforms, 221
Multichannels view, 23
Multi-content delivery networks (Multi-
CDN), 214, 221
Multi-Identity (MI), 54
Multimedia
  annotation models, 453
  framework, 14
  services, 5, 31, 236, 296, 298
Multimedia Application Formats (MAF), 352
Multimedia Broadcast Multicast Service
(MBMS), 63, 110
  e-MBMS, 64
  user service architecture, 110, 111
Multimedia Content Management System
(MCMS), 297, 307

  adaptation, 332
  AEM, 308, 310
  architecture, 305
  module, 302
  multimodal management system, 308
  and NGN IMS interaction, 303, 304–305
  NQoS-centric adaptation/management
scheme, 308
  for optimizing user experience, 309
  PQoS-aware MCMS, 308
  specifications, 305
  video service adaptation, 322
Multimedia Home Platform (MHP), 13, 394
Multimedia Messaging Service (MMS), 66
  protocol implementation, 84
  transmission and reception, 67
MultiMedia Metadata Ontology (M3O), 453
Multimedia Research Group (MRG), 8, 9
Multimedia Resource Function (MRF), 303,
484
Multimedia Resource Function Controller
(MRFC), 303, 484
Multimedia Resource Function Processor
(MRFP), 303, 484
Multimedia Server and Resource Function
(MSRF), 302, 303
Multimedia Service Adaptation Module
(MSAM), 313–314
Multimedia Service Monitoring Module
(MSMM), 311–312
Multimedia Streaming Servers (MSSs), 257
Multiple Auxiliary Components (MAC), 352,
405
Multiple DHT (MDHT), 156
Multipoint Conferencing Unit (MCU), 475
Multipoint Control Unit (MCU), 475, 484,
486
MultiProtocol Label Switching (MPLS), 14
Multipurpose Internet Mail Extensions
(MIME), 481, 486
Multiscreen approach, 380–381
Multiscreens delivery, 219
Multiuser cross-layer adaptations, 324. *See also*
Video service adaptation
  AEM GA-based logic, 326
  allowed values for input parameters, 325
  AN degradations, 329–333, 334
  AN/CN degradations, 337, 338–343
  CN degradations, 333, 335–337
  creation function, 326, 327
  fitness function, 328
  initial population, 326

MOS evolution, 326, 327
service characteristics, 324–325
Multiview depth processing framework, 358
Multi-View plus Depth map (MVD), 351, 353
Multiview profile (MVP), 352
Multiview video, 353
    capturing, 357
    formats, 98
    streaming, 363
    transmission scheme, 366
Multiview video coding (MVC), 97, 353, 363
Music on Demand (MoD), 17

N

Name Resolution System (NRS), 149, 156, 182
Named Data Networking (NDN), 152, 182
Naming information objects, 146
    metadata, 148–149
    properties, 147–148
Natural Language Processing (NLP), 461
Near Field Communication (NFC), 24,
        382, 405
Netcast, 80
NetEm, 266
Network
    context, 392, 404
    management, 133, 160
    map, 129, 135–136, 138, 317, 318
    provider domain, 11, 418
    service instantiation, 136, 137
    service map, 136
    traffic class mapping, 320
    unmanaged, 31, 46–47
Network Abstraction Layer (NAL), 367
Network Address Translation (NAT), 45
Network as a service (NaaS), 395
Network attachment control functions
        (NACF), 108–109
Network Attachment Sub-System (NASS),
        103, 104
Network of Information (NetInf), 156
    asymmetric key-based cryptography
        algorithms, 170
    DHT lookup, 156–157
    name persistency, 156
    NRS, 156
    SAIL project, 157, 182
Network of Information (NetInf), 156, 182
Network operators, 174
    access-centric CDN models, 175
    AT&T, 217

challenges to, 123
end-to-end Internet connectivity, 175
EU ICT ETICS project, 175–176
in ICN deployment, 179
Level3, 217–218
and third parties, 396, 397
transit network operators, 175
Verizon, 217
Network Personal Video Recorder (N-PVR),
        104, 402
Network Provider Interfaces (NPI), 108
Network Quality of Service (NQoS), 297
Network service providers (NSPs), 173
Network services, 122
    challenge in, 123
    control, 133, 134
    convergence, 48
    home, 508
Network Time Shifting (nTS), 402
Network-driven service adaptation, 321, 331
Networked Electro Media sector (NEM
        sector), 507
    contributions in Societal Challenges, 507,
        508
    Digital Agenda, 507
    energy efficiency, 508–509
    ITS, 509
Next Generation Network (NGN), 12–13. *See
        also* Internet Protocol Television
        (IPTV)
    components reuse, 103, 104
    IPTV architecture, 108, 109
    Release 1 and 2, 103
Next Generation Network-integrated IPTV,
        104
    functional architecture, 105
    NGN component reuse, 104, 105
Nomadic access, 44, 45
    all based home network, 45–46
    under managed network, 46
    service-based home network, 46
    under unmanaged network, 46–47
Nomadic Access to Nomadic Service
        (NA-to-NS), 44
Nomadic device, 44
Nomadic IPTV services, 44. *See also* Open
        IPTV business model; Open-IPTV
        services
    nomadic access, 45–47
    nomadism, 44, 45
    roaming, 44, 45
Nomadic network, 44

Nomadic service, 44
　accessing, 45–46
Nomadism, 44, 45
　all based home network, 45–46
　content personalization during, 379
　nomadic access issues, 45
　service-based home network, 46
Non-IMS-based architecture, 110, 111. *See also*
　　　Internet Protocol Multimedia
　　　Subsystem-based architecture
Non-Line-Of-Sight (NLOS), 69
Normalization Video Fidelity Metric
　　　(NVFM), 261
NoTube project, 446–447

O

Object-oriented models, 414
Occlusion, 278
Ocular depth of focus, 284
Oculomotor cues, 277
Open collaboration annotation model, 453
Open Internet, 87, 107
Open IPTV business model, 41. *See also*
　　　Nomadic IPTV services; Open-
　　　IPTV services
　collaborated open architecture, 43–44
　collaborative architecture, 41–42
　current architecture status, 42–43
　design factors, 41
Open IPTV Forum (OIPF), 13, 34, 87–88,
　　　107, 394
OPEN Link Search, 455
Open Mobile Alliance (OMA), 61
Open Set-Top-Box (O-STB), 33
Open-IPTV, 31, 32. *See also* Open IPTV
　　　business model; Nomadic IPTV
　　　services
　components, 32, 34–36
　flow control for, 35
　IPTV Follow-me, 34
　IPTV terminologies, 33–34
　models, 31
　open IPTV, 34
　O-STB, 33
　Pay-TV, 33
　personal IPTV, 34
　P-STB, 32
　S-STB, 33
　TVA, 34
　TV-OTT, 33
　use-cases analysis, 53–55

　video service accessment, 35–36
　V-STB, 33
Open-IPTV services
　CAPEX *vs.* OPEX, 52–53
　cloud design motivations, 50–51
　collaboration, 49, 51
　collaborative architecture for, 47
　future network-service convergence, 48
　life cycle, 47, 48–49
　making domestic cloud, 51
　model analysis, 49
　service actors, 48
　traditional broadcast TV *vs.* web-TV, 49,
　　　50
OpenSocial project, 456
Operational Expenditure (OPEX), 30, 41, 51
　cost analysis, 52
OPNET simulator version 14. 0, 267
Optical Character Recognition (OCR), 452
Orthogonal Frequency Division Multiplexing
　　　(OFDM), 59, 62
Out-of-network caching, 174
Overlay management, 133
Over-the-top (OTT), 191
　CDNs, 191, 192
　networking, 11
　operators, 5
　V-OTT, 33

P

P4P/ALTO initiative, 123–124
Packet Switch (PS), 110
Packet-switched Streaming Service (PSS), 109,
　　　110, 111
Palo Alto Research Center (PARC), 151
Parallax barrier, 93
Parental control services, 230, 249
Passive Optical Networking (PON), 15
Pay-TV, 33
Peer-to-Peer (P2P), 102, 112, 190
　assisted delivery, 221
　CDS, 112–114
　network, 366
Peer-to-Peer Streaming Protocol (PPSP), 113
　tracker protocol, 114
Pending Interest Table (PIT), 151
Per hop behavior (PHB), 477, 486
Perceived Quality of Service (PQoS), 296, 298.
　　　See also Quality of Experience
　　　(QoE)
　Alarm, 334

evaluation, 299
PQoS-aware dynamic cross layer
    adaptation, 301
Perceptual Evaluation of Speech Quality
    (PESQ), 226
Perceptual Speech Quality Measure (PSQM),
    261
Peripheral family access, 491
Peripheral home access, 491
Peripheral public access, 491
Personal Area Network (PAN), 405
Personal Broadcast Service, 393
Personal Computer (PC), 486
Personal Digital Assistant (PDA), 18, 68, 497
Personal Identification Number (PIN), 382
Personal Mobility, 44
Personal user identity management, 378
Personal Video Recording (PVR), 12, 16, 17,
    18, 59, 393
    N-PVR, 104, 228
    services, 4
    while on move, 381
Physical Set-Top-Box (P-STB), 32
Place-shifting TV, 228
Plexus, 157–158
Point-of-presence (PoPs), 219
Policy and Charging Control Function (PCC
    Function), 306
    functionalities, 306–307
    PCEF, 307
    PCRF, 112, 306, 307
Policy and Charging Enforcement Function
    (PCEF), 307, 313
Policy and Charging Rules Function (PCRF),
    112, 307
    decisions, 306
Policy Decision Function (PDF), 315
Portability, 47, 52
    content, 22
    device, 411
    service, 45
Portable TV, 58
Power Line Communication (PLC), 229
PPSP WG, 113, 114
PQoS-aware management system, 299
Presence Information Data Format (PIDF),
    408
Presence services, 394
Principal, 153, 170
Privacy Enhancing Technologies (PETs), 389
Privacy problem, 45
Privacy protection, 389, 416–417, 456

Program Map Table (PMT), 96, 97
Project Canvas. *See* Youview
Promotion of Digital Broadcasting (Dpa), 96
Protocol Independent Multicast (PIM), 37
Provider route forwarding rule, 166, 172
Provider-defined network location identifier
    (PID), 135–136
    providers, 174, 190, 215
Proxy CSCF (P-CSCF), 484
PSQA, 261, 264
    testbeds for, 266–267
    training process of, 268
Public health, 506
Public Switched Telephony Network (PSTN),
    78
Publish Subscribe Internet Technology
    (PURSUIT), 156, 182
Publish-Subscribe Internet Routing Paradigm
    (PSIRP), 154
    burden of resource discovery overheads,
        160
    components, 154
    forwarding, 155
    naming content, 155
    network monitoring and management, 160
    PURSUIT, 156
    rendezvous identifiers, 155
Purchased VoD, 71
Pure media players, 432
Pure-play service providers, 216–217

Q

QoE assessment of IPTV services
    challenges, 240
    EEG analysis, 245, 246
    emotions measurement dimensions,
        244–245
    Emotiv EPOC device, 245, 246
    factors influencing service quality, 247,
        248–249
    human mental states, 244
    IPTV service, 237–238, 239
    measuring human response, 241
    methods for data mining/correlating user
        emotions, 243, 244
    motion-controlled interaction, 246, 247
    QoE dimension, 240
    QoE *vs.* service personalization, 238, 239
    QoS constraints, 238, 239
    redefined approach to, 237
    role of human, 240, 241

QoE assessment of IPTV services (*continued*)
  traffic management uses, 247
  user's attitude, 242
  user's emotion utilization, 244
  user's opinion measurement, 242–243
  users' subjective impressions, 241
QoE Q-Learning-based Adaptive Routing
      (QQAR), 261
  algorithm, 262–263
  architecture and server selection, 263–264
  data routing, 265
QoE/PQoS model, 317. *See also* Video service
      adaptation
  cross-layer adaptation procedure, 320, 321
  multimedia adaptation dependence, 319
Qriocity, 81
16-Quadrature Amplitude Modulation
      (16-QAM), 62
Quadrature Phase-Shift Keying (QPSK), 62
Quadruple-play bundle, 4, 11
Quality of experience (QoE), 5, 20, 224, 226,
      233, 258–259. *See also* Content
      Distribution Network (CDN);
      Internet Protocol Television
      (IPTV); Perceived Quality of
      Service (PQoS)
  of audiovisual services, 232–233
  CDA-QQAR architecture, 268
  common mistakes, 233–234
  components, 233
  e2e quality, 259
  end-user, 125
  evaluation methods, 226–227
  feedback, 264
  full reference methods, 235
  measurements, 236, 261
  monitoring and reporting, 221
  objective QoE assessment models, 235
  QoE-aware architecture, 268, 302
  QoS and, 259–261
  for system design and engineering process,
      298
  3D video, 276, 277
  U2U-QoE, 259
  viewer groups, 225
  VQA algorithms, 234, 235
Quality of Service (QoS), 5, 159, 277
  device capability, 21
  domestic cloud, 51–52
  DV algorithm, 268
  e2e QoS, 259, 377
  feedback, 478

  in IPTV, 20, 22, 107, 238, 248, 391–393
  in managed networks, 46
  in mobile TV services, 70
  Network capacities and, 63
  network technology and, 476
  PQoS, 298
  QoE and, 233, 237, 259–261
  scalability and, 144
  SOMR algorithm, 268
  SPF algorithm, 268
  in unmanaged model, 31, 47
  in WiFi networks, 64
Quantization parameter (QP), 364–366
Quick Response Code (QRCode), 382
Quincunx format, 94

R

Radio access network (RAN), 110
Radio Frequency Identifiers (RFID), 24
Radio Signal Strength Indicator (RSSI), 406
Radio-Frequency (RF), 419
  signal-based localization, 406
Random Neural Network (RNN), 261, 266
Real Time Messaging Protocol (RTMP), 213
Really Simple Syndication (RSS), 388, 465
Real-Time Communication in Web browsers
      (RTCWeb), 115
Real-time media streams, 145
Real-Time Streaming Protocol (RTSP), 64
  in Live-TV, 40
  PSS server, 110
Real-Time Transport Control Protocol
      (RTCP), 478, 480, 486
Real-Time Transport Protocol (RTP), 478,
      480, 486
Received Signal Strength Indicator (RSSI),
      406, 412
Red alarm, 305, 306
Rehabilitation in home. *See also* Information and
      Communications Technology (ICT)
  AT, 491–492
  current care and monitoring in home, 492
  digital television, 496–497
  game consoles, 494
  telemonitoring, 493
  tracking technology, 494–495
  visualizations, 495–496
Relative size cue, 278
  HVS sensitivity for, 283–284
Remote control, 84–85
  of IPTV services, 229, 248

Rendezvous identifiers, 155, 182
Rendezvous Points (RPs), 154, 155
REpresentational State Transfer (REST), 458
Request for Comments (RFC)
Request routing, 199. *See also* Content—
    ingestion/acquisition
    cdni WG, 183
    HTTP redirection, 198
    RFC 3568, 196
Research activities on ICN, 181
    CCN, 151–153
    DONA, 153–154
    NETINF, 156–157
    Plexus, 157–158
    PSIRP, 154–156
    standardization activities, 183–184
Resolution Handlers (RHs), 154
Resource Admission Control Subsystem
    (RACS), 104
    functions, 103
    in NGN-integrated IPTV functional
    architecture, 105
Resource Description Framework (RDF), 394,
    448
    data in LMF, 459
    data publishing, 450
    Sindice, 455
    sKWRL, 160
Resource Description Framework Schema
    (RDFS), 439, 448
Resource-On-Demand (ROD), 44
RESTful resource management, 458
Retinal blur, 278
    eye tolerance, 284
    HVS sensitivity for, 282–283
Return on Investment (ROI), 247
Reusable Intelligent Content Ontology
    (RICO), 453
Rich Presence Extension to PIDF (RPID), 408
Roaming, 44, 45. *See also* Nomadic IPTV
    service
    agreements between operators, 396
Room configuration, 473–474
Route-by-name paradigm, 171
Routing Area Working Group (rtgwg), 184
Royal National Institute of Blind People
    (RNIB), 287

S

Satellite Digital Multimedia Broadcast
    (S-DMB), 59, 60–61

Scalable and Adaptive Internet Solutions
    (SAIL), 182
    SAIL project, 157
Scalable Enhancement Information (SEI), 352
Scalable Video Coding (SVC), 70
    media encoding, 367
    standard-based approach, 363
    3D video encoder, 364
s-certData, 170
Search, discovery, and content
    recommendation services, 229
Search engines, 454–456
Secure transmission medium, 170
security, 20
Security, network, 144
    access under managed network, 46, 47
    in collaboration between networks and
    applications, 135
    content, 20
    ICN, 147, 158
    information object, 170
    IPTV, 21
    and privacy requirements, 392
    P-STB, 32
    T-commerce, 17, 18
    TV-Anytime, 34
Segment switching, 475–476. *See also*
    Telepresence (TP)
Segmental SNR (SNRseg), 261
Semantic concept detection, 451
Semantic enhancement
    via linked data, 459
    LMF Enhancer, 465
Semantic indexing, 459
Semantic mediation approach, 439–441
Semantic web, 448–449
    Falcon, 455
    languages, 439
    in NoTube project, 447
    Sindice, 455
    Social Semantic Web technologies, 465
    Swoogle, 456
    SWSE, 455
Semantic Web Search Engine (SWSE), 455
Semantically Interlinked Online Communities
    (SIOC), 454
Semantically linked media for interactive
    user-centric services, 446
    audio analysis solutions, 452
    CommonTag, 454
    concept detection, 451
    ConnectME project, 447

Semantically linked media for interactive
    user-centric services (*continued*)
  content-based video analysis, 451
  Diaspora, 456
  egtaMETA, 447
  Falcon, 455
  FOAF, 454
  hypervideo, 447, 448
  Interlinked Media and User-Centric
    Approach, 454
  linked data, 448, 449–451
  LOD, 447, 449–450
  media assets, 446
  Mozilla's Popcorn project, 446n
  multimedia annotation models, 453
  NoTube project, 446–447
  open collaboration annotation model, 453
  Open Link Search, 455
  OpenSocial project, 456
  search engines, 454–456
  semantic web, 448–449
  semantics, 448
  Sigma, 455
  Sindice, 455
  SIOC, 454
  solutions to image analysis, 452
  spatial segmentation, 451
  StatusNet, 456
  structural video analysis, 451
  Swoogle, 456
  temporal video decomposition, 451
  user profile and behavior, 456
  Watson, 455
Semantics, 439, 448
Service
  context, 404
  provider domain, 11
  provider subsystem, 416
Service Agent Managers (SAM), 416
Service Control Functions (SCF), 104
Service Discovery and Selection (SD&S), 13,
  105
Service Discovery Function (SDF), 103
Service Engines (SEs), 218
Service Level Agreement (SLA), 483, 486
Service Router (SR), 218
Service Selection Function (SSF), 103
Service triggering (ST), 53
Service-Oriented Architecture (SOA), 459
Serving CSCF (S-CSCF), 484
  in MCMS and NGN IMS interaction, 304
Serving GPRS Support Node (SGSN)

Session continuity services, 228–229, 248
Session Description Protocol (SDP), 303, 477,
  486. *See also* Session Initiation
  Protocol-based systems (SIP-based
  systems)
  media descriptions, 480
  session header, 480
  streams properties and, 479–480
  in videoconferencing, 479
Session Initiation Protocol (SIP), 13, 303, 486
Session Initiation Protocol-based systems
  (SIP-based systems), 472. *See also*
  Telepresence (TP)
  architecture for TP, 478
  media transport, 480–481
  overview, 477
  protocol stack for videoconferencing, 479
  streams properties and SDP, 479–480
Set-Top-Box (STB), 4, 6, 32, 44, 70, 229, 376,
  431. Physical Set-Top-Box (P-STB)
  home network, 15–16
  O-STB, 33
  P-STB, 32
  in scalable MCAST transmission, 37
  S-STB, 33, 65
  V-STB, 33
  Youview, 89
Sh interface, 306
Shared service control, 231–232, 249
Short Message Service (SMS), 66
  reception, 67
  T-communication, 18
  transmission, 66
Short Path First (SPF), 268, 270
Sigma, 455
Signal-to-Noise Ratio (SNR), 261
Simple Knowledge Organization System
  (SKOS), 460
Simulcasting, 352
Sindice, 455
Singapore Broadcast Corporation (SBC), 59
Site switching, 475. *See also* Telepresence (TP)
sKWRL reasoner, 460
Smart TV, 79, 80, 85–86
Social networks, 424
Social TV services, 228, 230
  CallerID on TV, 230–231
  instant messaging on TV, 231
  IPTV presence, 230
  shared service control, 231–232
  videoconference on TV, 231
  voicemail on TV, 231

Societal Challenges, 504–506, 509
　ICT contribution, 506
　NEM sector contribution, 507
　on public policy, 505
Society of Motion Picture and Television
　　Engineers (SMPTE), 95
Softmax Action Selection method, 264
Software Development Kit (SDK), 79, 81
Software Set-Top-Box (S-STB), 33, 34, 43
Source Bitrate (SBR), 322
　for different AN states, 324
Source BitRate (SBR), 324
SPARQL Protocol and RDF Query Language
　　(SPARQL), 449
　LMF SPARQL, 460
Spatial audio, 351
　encoder, 370
　multiple microphones for, 358
　QoE metric for 3D video, 291
Spatial Resolution (SR), 324
Spatial segmentation, 451
Standard Definition (SD), 10
Standard Definition Television
　　(SDTV), 62
Standard Developing Organizations (SDOs),
　　102, 205
　activities on IPTV, 107
　ATIS IIF, 109
　ITU-T IPTV-GSI, 108–109
　Open IPTV Forum, 107–108
Standard Optimal QoS Multi-Path Routing
　　(SOMR), 268
Standardization, 13, 482–483. *See also*
　　Technical fragmentation
　CEA 2014 specifications, 86–87
　contribution network, 95–96
　distribution network, 96–97
　DLNA, 87
　HBBTV, 88
　home network, 97
　HTML 5, 89
　OIPF, 87–88
　Youview, 89
Static caching, 200
StatusNet, 456
Stereoscopic 3DTV, 92
　frame packing, 95
　frame-compatible formats, 93, 94
　lenticular lens sheet, 93
　parallax barrier, 93
　rendering technologies, 92, 93
Stereoscopy, 92

image/video quality metrics, 288–289
store, 151
3D video, 352
Structural comparison (SC), 289
Structural Similarity Index (SSIM), 289, 290
Structural video analysis, 451
Study Groups (SGs), 108
Subscriber Identity Module (SIM), 64, 382, 405
Surrogates, 189, 193
　in CDN architecture, 257
　content discovery, 258
Swoogle, 456

T

Targeted Advertising, 393, 394
T-commerce, 7, 18
T-communication, 17, 18–19
Technical fragmentation, 82
　within connected TV, 85–86
　content protection, 85
　cookie support, 84
　cope with, 86
　decoding capabilities, 84
　fonts, 83
　graphic resolutions, 83
　media player, 84
　performance, 85
　remote control, 84–85
　within same manufacturer, 85
　transport protocol, 85
　web technologies, 83
　website *vs.* widget *vs.* native user interface,
　　82–83
Technology innovations, 219, 221. *See also*
　　Content delivery network (CDN)
Technology vendors, 218
Telecom SudParis, 38
Telecommunication Management Group
　　(TMG), 8
Telecoms & Internet converged Services &
　　Protocols for Advanced Networks
　　(TISPAN), 13
Telemonitoring, 493
Telepresence (TP), 472, 486. *See also* Session
　　Initiation Protocol-based systems
　　(SIP-based systems)
　acoustics and lighting, 477
　conference, 473, 475, 478
　future directions, 485
　IMS and, 483–485
　interoperability, 481

Telepresence (TP) (*continued*)
room configuration, 473–474
segment switching, 475–476
signaling requirements, 473–476
SIP-based systems, 477–481, 478
site switching, 475
standardization, 482–483
technical overview, 472–473
transport requirements, 476
videoconferencing, 482
Television On Demand (TVOD), 435
Television-commerce, 7
Temporal video decomposition, 451
T-entertainment, 17, 19
Terabits per second (Tbps), 218
Term Frequency-Inverse Document Frequency (TF-IDF), 455
Terminal Adaptation Module (TAM), 303
interface, 305
Terms of Reference (ToR), 504
Terrestrial DMB (T-DMB), 59, 61
Terrestrial DVB, 368
Third-Generation Partnership Project (3GPP), 109
aim of, 483
Class 3 handsets, 60
IMS, 102, 303, 472
IMS-based architecture, 111–112
IMS-based distribution, 113
Non-IMS-based architecture, 110–111
Open IPTV Forum, 13
P2P content distribution systems, 113
Sh interface, 306
standards for audiovisual services, 109–111
for support of audiovisual services, 109–110
3GPP TS, 307
3Crowd's multi-CDN platform, 221
3D content preparation and processing, 354–355. *See also* 3D media
content and context-aware 3D media adaptation, 361–363
content-aware depth processing, 358–361
multiview media capturing and postproduction, 355–358
3D immersive multiview media distribution, 366. *See also* 3D multiview coding approach; Hybrid broadband–broadcast approach
bit-torrent use, 366
bottleneck connection rate, 367, 368
chunk generation process, 367
P2P distribution system, 366

3D media. *See also* 3D content preparation and processing; 3D media compression
delivery techniques, 352
formats, 350–351
MVC, 353
MVD, 353
research, 348–350
simulcasting, 352
stereoscopic video coding, 352
TV broadcasting, 353–354
video streaming, 354
3D media adaptation, 361, 362
color-plus-depth format, 362
user perception model, 362–363
3D media compression, 363. *See also* 3D content preparation and processing
hybrid broadband–broadcast approach, 368–371
immersive multiview media distribution, 366–368
multiview coding approach, 363–366
3D media formats, 350
color-plus-depth, 350, 351
DES format, 351
LDV format, 351
MVD format, 351
spatial audio, 351
3D media research
ATTEST project, 349
free-viewpoint vision, 348
immersive 3D media experience, 350
integral imaging, 348
stereoscopic vision, 348
3D-TV services, 349–350
3D multiview coding approach, 363. *See also* 3D immersive multiview media distribution; Hybrid broadband–broadcast approach
Gestalt's theory use, 364
MOS, 365
QP assignment process, 365
QP effect, 364, 365, 366
SVC standard-based approach, 363–364
visual attention model, 364
3D multiview media
camera arrangement models, 355
capturing, 355, 357, 358
linear multicamera set up, 356–357
multiview capture rig, 355
postproduction, 357, 358
3D spatial audio reproduction systems, 351

3D Television. *See also* Stereoscopic 3DTV
  delivery chain, 95
  mobile 3DTV, 98–99
  multiview video formats, 98
  QoE, 97–98
  standardization, 95–97
3D video, 277. *See also* 3D media
  feeling of presence in, 287
  future QoE metrics for, 290–291
  naturalness of, 287
  visual acuity effects of viewers, 287–288
3D Video Coding (3DV), 353
3D video QoE, 277
  affecting factors, 284–285, 287–288
  ambient lighting effects, 285–287
  binocular stereopsis, 278–279
  challenges in, 290
  depth metrics, 289
  depth perception analysis, 277–278
  eye positioning, 280
  future QoE metrics, 290–291
  metrics for visual fatigue prediction,
    289–290
  stereoscopic image/video quality metrics,
    288–289
3D visual experience technique, 288
3D-TV Network of Excellence (3DTV NoE),
  349
Time Division Duplexing (TDD), 63
Time interleaving data, 62
Time of Arrival (TOA), 406
Time of Flight (ToF), 356
Time-shifted TV, 228, 248
Time-To-Live (TTL), 147
T-information, 18
T-learning, 19
TN/AN NQoS mapping, 317, 318–320
Tracker, 113–114
  function, 138
  server, 128
Tracking technology, 494–495
Transactional Video on Demand (TVoD), 388
Transit network operators, 175
  in movie purchase use case, 177, 178
Transit operator. *See* Carriers
Transmission Control Protocol (TCP), 192,
  486
  flow control messages, 40
  four-way handshaking closing, 41
  videoconferencing, 479
Transparent caching, 125, 190
Transport and mobility smart systems, 506

Transport Network Adaptation Module
  (TNAM), 314–315
  in AEM, 310
Transport Network Monitoring Module
  (TNMM), 312
  in AEM, 310
Transport protocol, 85, 110, 354
  MSRP, 481
  RTP, 480
Transport Stream (TS), 60
  ETSI TS, 88, 104, 106, 107, 108, 110, 227
  MPEC-2 TS, 62
  MPEG-2 TS, 371
  MPEG-TS, 41
  3GPP TS, 307
Triple-play, 496
  bundle, 11
  IPTV, 4
  model, 6
Turbo code algorithms, 62
TV converged model, open, 24
TV model, new, 376
TV Over-The-Top (TV-OTT), 33–34
TV-AnyTime (TVA), 34, 395, 431
  metadata, 430–431
2D backward compatibility, 98
2D service compatible. *See* 2D backward
  compatibility
Type–length–value (TLV), 151

U

Ultra Definition (UD), 69
Unicast transmission, 36–37
  traffic load increase, 22
Unified Modeling Language (UML), 414
Uniform Resource Locator (URL), 66, 447
  dynamic, 200
Uniform Resource Names (URNs), 183
Universal Mobile Telecommunications System
  (UMTS), 21
  DL Bearer, 331
  link performance impact, 329
  settings, 322
Universal Plug and Play (UPnP), 419
  remote UI, 87
Unmanaged network, 31
  access under, 46–47
UnManaged service, 34
Unstructured Information Management
  Architecture (UIMA), 450, 461
Urnbis, 183

User Agent Client (UAC), 306
User Agent Profile (UAProf), 394
User Agent Server (UAS), 306
User Authorization Answer (UAA), 43
User Authorization Request (UAR), 43
User behavior, 456
User context, 67, 403
    end-user context, 133
    information, 391
    manager, 416
User equipment (UE), 103, 113, 307, 484
User Generated Content (UGC), 5, 11, 67,
        104, 122, 425
User Interface (UI), 23, 410
    in content provider, 81
    GUI, 229, 458
    issues, 70–71
    and multichannels view, 23
User Mode Linux-Based Networks (UML),
        414
User Network Interfaces (UNI), 108
User perception model, 362–363
User profile, 456
User Profile Server Function (UPSF), 103
User profiling, 384. *See also*
        Content—personalization
    component implementation, 385, 386, 387,
        389
    emotion-centric, 247
    exchanged data, 386
    features, 387, 388–389
    function, 383
    functional architecture, 384–385
    information sources, 384
    IPTV service types, 384
    requirements, 392
User-centric IPTV architecture, 301. *See also*
        QoE/PQoS model
    adaptation engines, 313–316
    DiffServ/MPLS traffic network, 303
    IMS enhancement, 303–307
    IPTV service adaptation steps, 318–319
    MCMS, 307–310
    monitoring engines, 311–313
    NQoS mapping, 318–319
    PQoS-aware management mechanism, 301
    QoE-aware IMS architecture, 302–303
    QoS support and mapping, 316–317
    TN and AN NQoS traffic class mapping,
        320
User-generated content (UGC), 173
User-generated content services, 230

Users' consumption model, 376, 377
User-to-User QoE (U2U-QoE), 259

V

Validations-based testbed, 38
Value-added services, 4, 214
    application acceleration, 214–215
    hypervideo annotation suite, 214, 464
Verizon, 217
Vertical integration, 440
Very-High-Speed DSL (VDSL), 15
Video annotation
    in ConnectMe framework, 463
    via LMF, 461–462
Video encoder, 14
    MPEG, 364
Video encoding, 22
    in DMB, 60
    mobile video encoding settings, 322
Video fast update request. *See* Full Intra
        Request command (FIR command)
Video head-end, 14
Video Home System (VHS), 234
Video on demand (VoD), 95, 191, 226, 301,
        377, 425
Video Quality Assessment (VQA), 234–235
Video Quality Metric (VQM), 261
Video service adaptation, 322. *See also* QoE/
        PQoS model; Multiuser cross-layer
        adaptations
    best-performing encoding sets, 324
    IPLR evolution, 323
    of mobile, 321
    mobile video encoding settings, 322
    UMTS settings, 322
Video services, 33
    IPTV and, 21, 35, 36
    QoE significance, 298
    short, 37
    Near VoD service, 16
Video traffic, 190, 266
Videoconferencing, 482
    interoperability, 481
    protocol stack, 479
    TP, 472, 473, 477
    on TV, 231
Video-on-demand (VOD), 31
    CDN, 198
    considered configurations, 335
    encoders, 95
    for IPTV services, 14

metadata specification, 431
PQoS-aware MCMS system, 301
purchased, 71
scenarios, 39
servers, 4, 14
service, 16, 17
uses, 19
VoD MOS Values, 332
Viera Connect, 81
Vieracast solution. *See* Viera Connect
Virtual Set-Top-Box (V-STB), 33
Vision-based indoor localization, 407
Visited Application Server (VAS), 53
Visited network (VN), 53, 55, 392
Visual attention model, 364, 368
Visual cues, 278
Visual fatigue
    factors affecting, 284–285
    prediction of, 289–290
Visualizations, 495–496
Voice over Internet Protocol (VoIP), 19, 307
Voicemail on TV, 231, 249

W

Warning Alarm, 305
Watson, 455
Wave Field Synthesis (WFS), 351
Web and TV Interest Group, 89, 114
Web Ontology Language (OWL), 439, 448
Web services, 11, 86
Web technologies, 79, 83
    CEA 2014 specifications, 86
    HbbTV specification, 88
    integrating, 206
    semantic, 465
Weber's law, 281

Website acceleration, 214, 218
Website *vs.* widget *vs.* native user interface,
        82–83
WEB-TV, 39
    vs. traditional TV, 49–50
WiFi networks, 64
Wireless access networks, 58
Wireless Application Protocol (WAP), 394
Wireless Local Area Networks (WLANs), 58,
        69, 229
Wireshark tools, 39
Working Group (WG), 205
    ALTO WG, 114
    cdni WG, 183, 205
    IETF ALTO WG, 135
    IETF WG, 113, 183
    PPSP WG, 113, 114
    rtgwg, 184
    W3C Multimedia Annotations WG, 465
World Wide Web (WWW), 226
World Wide Web Consortium (W3C),
        89, 394
    media annotation working group, 453n
    semantic web, 448
    standardization, 83
    Web and TV interest group, 114
Worldwide Interoperability for Microwave
        Access (WiMAX), 21, 303, 320

X

XML HTTP Requests (XHR), 82, 83

Y

Yankee group CDN scorecard, 213
Youview, 89

Milton Keynes UK
Ingram Content Group UK Ltd.
UKHW031124141024
449569UK00006B/458